Nanotechnology Based Microbicides and Immune Stimulators

Suriya Rehman • Ebtesam Al-Suhaimi
Editors

Nanotechnology Based Microbicides and Immune Stimulators

Springer

Editors
Suriya Rehman
Imam Abdulrahman Bin Faisal University
Institute for Research and Medical
Consultation (RMC), Western Campus,
Building A71
Dammam, Saudi Arabia

Ebtesam Al-Suhaimi
Vice presidency for Scientific Research
and Innovation
Imam Abdulrahman Bin Faisal University,
Eastern Campus, Building D1
Dammam, Saudi Arabia

ISBN 978-981-96-0396-1 ISBN 978-981-96-0397-8 (eBook)
https://doi.org/10.1007/978-981-96-0397-8

This Springer imprint is published by the registered company Springer Nature Singapore Pte Ltd.
The registered company address is: 152 Beach Road, #21-01/04 Gateway East, Singapore 189721, Singapore

Preface

The new strategies for treatment are in demand to combat the infectious diseases globally. A number of infectious agents have evolved to resist and skip the immune system that has led to limited benefit from the available therapies, especially with the emergence of COVID-19 pandemic. Researchers are considering endless novel ideas to counter the challenges associated and developing strategies to win the war against the infectious microorganisms. Moreover, the unfolding of co-infections and drug-resistant pathogens has marked up the requirement for advanced diagnostic and therapeutic approaches. These challenges can be dealt with measures that enhance the quality and diversification of an immune system having the wide-spread application. Nanoscience is a multifaceted and important scientific subject that includes multiple disciplines of science and technology at nanoscale and their application has presented much advancement owing to their unique characteristics such as, shape, size, electromagnetic properties, engineerability, and biocompatibility. Nanotechnology and immunotherapy are emerging and revolutionizing areas activating the immune system to overcome the infectious diseases. The integration of nanotechnology in present time medicine has significant edge, specifically in infectious diseases. Several reports indicate the novel prospects of framing immune responses and preventing severe infections via nanoparticles and their role in vaccines and immunotherapy.

This unique book will cover how nanoscale delivery vehicles and biomaterials can be applied to alter and enhance the immune system against various forms of pathogens. Nanoscale carriers are competitive at promoting the immunotherapeutic approaches due to the fact they can be modified to obtain different physical properties, encapsulated agents, and surface ligands. Perhaps, nanoscale diagnostics offer new approaches for sensitive health monitoring that are potable and can guide the use of nanoscale therapies. In addition to treatment, in future it would be achievable to prevent the outbreaks by exploiting nanoscale biomaterials to desired shape, activation and detection of immune system effector function that would lead to practical strategies for the prevention and treatment of infectious diseases. This book also reviews the inbuilt immunoreactive properties of nanomaterials and the advancement in its engineering to modulate the immune system via either immunostimulation or immunosuppression. Nanoparticles offer the best suited option for the diagnostics and therapy owing to their peculiar physiochemical characteristics and controlled, release and delivery systems targeting sick cells and pathogens. During the process, most of the NPs get captured by innate cells resulting in a hindrance; however, this interaction between the NPs and immune cells proves to be an opportunity to regulate the immune response. Furthermore, each chapter layout has an introduction to the current state followed by application of that technology and finishing with its prospects.

The book is contributed by eminent researchers and veterans of international repute working in diverse areas of infectious diseases, immunology, physiology, and nanotechnology. It is the joint efforts of researchers that presents cutting-edge research in their respective areas and dealing with recent developments and predicting the future leads. The book chapters are presented in a way accessible to the teachers, UG/PG students, research scholars, and healthcare workers including pathologists and clinicians worldwide. The book encompasses the broad topics to attract the diverse specialty of students, teachers, and researchers of nanotechnology, immunology, virology, pharmacology, mycology, bacteriology, virology, pathology, biotechnology, and biomedical technology.

The editors are also thankful to their host institute" Institute for Research and Medical Consultation (IRMC), Imam Abdulrahman Bin Faisal University (IAU), Dammam, KSA" for their support to enhance our research and academic rigor. We are extremely grateful to our esteemed contributors for their worthy and timely contributions. Cooperation and support from Springer-Nature throughout the book preparation is highly acknowledged.

Dammam, Saudi Arabia Suriya Rehman
Dammam, Saudi Arabia Ebtesam Al-Suhaimi

Contents

1 **Current Scenario of Antimicrobials Agents and Their Applications** . . 1
Uzma Ali and Syed Mehmood Ali

2 **Immunomodulatory Nanoparticles and Their Advancement in Health** 19
Iram Saba, Kaiser Wani, Khalid Mujassam Batoo, Suriya Rehman, and Saif Hameed

3 **Nanomaterial-Based Vaccines** 37
Insha Nahvi, Mohd Farhan, Suriya Rehman, and Saif Hameed

4 **Application of Nanoparticles as Innate Immune Stimulator** 55
Ifeanyi Elibe Mba, Paul Ekene Chidebelu, and Emeka Innocent Nweze

5 **Activation of Cellular Immunity via Nanotechnology** 75
Samer Mohammed Hassan Alqarni, Raed Albiheyri, and Irfan A. Rather

6 **The Use of Nanomaterials in Activating the Complement System** 85
Vivek P. Chavda, Amol Gholap, Rajashri Bezbaruah, Pankti C. Balar, Damanbhalang Rynjah, Bruce D. Uhal, and Vladimir N. Uversky

7 **Role of Nanoparticle Formulation for the Combination Delivery of Multiple Antigens** 105
Ganesh Lahane, Trupti Ghatage, Jegadheeswari Venkadakrishnan, Twisha Dube, Dishank Arondekar, Rakesh Kumar, Arti Dhar, and Audesh Bhat

8 **Passive Nano Particulate Chemotherapeutic Gene Targeting of Immune Cells** 123
Odate K. Tadros

9 **Nanoscale Diagnostics for Antigen-Specific Cells and Monitoring the Progress of an Immunotherapy** 137
Areej Fatima, Maleeha Azhar, Aroosa Younis Nadeem, Adeeb Shehzad, and Ebtesam A. Al-Suhaimi

10 Prospectus and Concerns of Immunomodulatory Nanotechnologies and Nanoparticles Biocompatibility and Toxicity 165
Shereen A. Arabiyat

11 Innovation in Engineered Nanoparticles for Immune System 191
Sumera Khizar and Abdelhamid Elaissari

12 Limitations and Concerns of Nanotechnology in Obtaining the Desirable Products 217
Umar Manzoor, Saima Masood, Sadaf Nazir, Larub Younis, Adbul Waheed, Syed Ubaid Qurashi, Faheem A Sheikh, and Shafquat Majeed

Correction to: Application of Nanoparticles as Innate Immune Stimulator C1
Ifeanyi Elibe Mba, Paul Ekene Chidebelu, and Emeka Innocent Nweze

Contributors

Raed Albiheyri Department of Biological Sciences, Faculty of Science, King Abdulaziz University, Jeddah, Saudi Arabia

Centre of Excellence in Bionanoscience Research, King Abdulaziz University, Jeddah, Saudi Arabia

Immunology Unit, King Fahad Medical Research Centre, King Abdulaziz University, Jeddah, Saudi Arabia

Syed Mehmood Ali Faculty of Engineering, Department of Biomedical Engineering, Saleem Habib University, Karachi, Pakistan

Uzma Ali Department of Public Health, College of Public Health, Imam Abdulrahman Bin Faisal University, Dammam, Saudi Arabia

Samer Mohammed Hassan Alqarni Department of Biological Sciences, Faculty of Science, King Abdulaziz University, Jeddah, Saudi Arabia

Ebtesam A. Al-Suhaimi Vice presidency for Scientific Research and Innovation, Imam Abdulrahman bin Faisal University (IAU), Dammam, Saudi Arabia

Shereen A. Arabiyat Al-Balqa Applied University, Salt, Jordan

Dishank Arondekar Department of Pharmacy, Birla Institute of Technology and Sciences Pilani, Hyderabad, Telangana, India

Maleeha Azhar Department of Biomedical Sciences, School of Mechanical and Manufacturing Engineering (SMME), National University of Sciences and Technology (NUST), Islamabad, Pakistan

Pankti C. Balar Pharmacy section, L.M. College of Pharmacy, Ahmedabad, Gujarat, India

Khalid Mujassam Batoo King Abdullah Institute for Nanotechnology, King Saud University, Riyadh, Saudi Arabia

Rajashri Bezbaruah Institute of Pharmacy, Assam Medical College and Hospital, Dibrugarh, Assam, India

Audesh Bhat Centre for Molecular Biology, Central University of Jammu, Jammu, Jammu and Kashmir, India

Vivek P. Chavda Department of Pharmaceutics and Pharmaceutical Technology, L.M. College of Pharmacy, Ahmedabad, Gujarat, India

Paul Ekene Chidebelu Department of Microbiology, University of Nigeria, Nsukka, Nigeria

Arti Dhar Department of Pharmacy, Birla Institute of Technology and Sciences Pilani, Hyderabad, Telangana, India

Twisha Dube Department of Pharmacy, Birla Institute of Technology and Sciences Pilani, Hyderabad, Telangana, India

Abdelhamid Elaissari University Claude Bernard Lyon-1, CNRS, ISA-UMR 5280, Villeurbanne, France

Mohd Farhan Department of Basic Sciences, Preparatory Year, King Faisal University, Al Ahsa, Saudi Arabia

Department of Chemistry, College of Science, King Faisal University, Al Ahsa, Saudi Arabia

Areej Fatima Department of Biomedical Sciences, School of Mechanical and Manufacturing Engineering (SMME), National University of Sciences and Technology (NUST), Islamabad, Pakistan

Trupti Ghatage Department of Pharmacy, Birla Institute of Technology and Sciences Pilani, Hyderabad, Telangana, India

Amol Gholap Department of Pharmaceutics, St. John Institute of Pharmacy and Research, Palghar, Maharashtra, India

Saif Hameed Amity Institute of Biotechnology, Amity University Haryana, Gurugram (Manesar), India

Amity Institute of Biotechnology, Amity University Haryana, Gurgaon, Haryana, India

Sumera Khizar University Claude Bernard Lyon-1, CNRS, ISA-UMR 5280, Villeurbanne, France

Rakesh Kumar Department of Biotechnology, Shri Mata Vaishno Devi University, Katra, Jammu and Kashmir, India

Ganesh Lahane Department of Pharmacy, Birla Institute of Technology and Sciences Pilani, Hyderabad, Telangana, India

Shafquat Majeed Multifunctional Nanostructures Laboratory, Department of Nanotechnology, University of Kashmir, Srinagar, India

Umar Manzoor Multifunctional Nanostructures Laboratory, Department of Nanotechnology, University of Kashmir, Srinagar, India

Saima Masood Division of Basic Sciences and Humanities, SKUAST-K, Srinagar, India

Ifeanyi Elibe Mba Department of Microbiology, University of Nigeria, Nsukka, Nigeria

Aroosa Younis Nadeem Department of Biomedical Sciences, School of Mechanical and Manufacturing Engineering (SMME), National University of Sciences and Technology (NUST), Islamabad, Pakistan

Insha Nahvi Department of Life Sciences, College of Science, King Faisal University, Al Ahsa, Saudi Arabia

Department of Basic Sciences, Preparatory Year, King Faisal University, Al Ahsa, Saudi Arabia

Amity Institute of Biotechnology, Amity University Haryana, Gurgaon, Haryana, India

Sadaf Nazir Multifunctional Nanostructures Laboratory, Department of Nanotechnology, University of Kashmir, Srinagar, India

Emeka Innocent Nweze Department of Microbiology, University of Nigeria, Nsukka, Nigeria

Syed Ubaid Qurashi Multifunctional Nanostructures Laboratory, Department of Nanotechnology, University of Kashmir, Srinagar, India

Irfan A. Rather Department of Biological Sciences, Faculty of Science, King Abdulaziz University, Jeddah, Saudi Arabia

Centre of Excellence in Bionanoscience Research, King Abdulaziz University, Jeddah, Saudi Arabia

Suriya Rehman Department of Epidemic Disease Research, Institute of Research and Medical Consultations (IRMC), Imam Abdulrahman Bin Faisal University, Dammam, Saudi Arabia

Damanbhalang Rynjah Girijanada Chowdhury University-Tezpur, Sonitpur, Assam, India

Iram Saba Research and Scientific Center, Sultan Bin Abdulaziz Humanitarian City, Riyadh, Saudi Arabia

Amity Institute of Biotechnology, Amity University Haryana, Gurugram (Manesar), India

Adeeb Shehzad Biodiversity Unit, Research Center, Dhofar University, Salalah, Oman

Faheem A Sheikh Nanostructured and Biomimetic Lab, Department of Nanotechnology, University of Kashmir, Srinagar, India

Odate K. Tadros Department of Allied Health Sciences, Al-Balqa Applied University, Salt, Jordan

Bruce D. Uhal Department of Physiology, College of Human Medicine, Michigan State University, East Lansing, MI, USA

Vladimir N. Uversky Department of Molecular Medicine and Byrd Alzheimer's Research Institute, Morsani College of Medicine, University of South Florida, Tampa, FL, USA

Jegadheeswari Venkadakrishnan Department of Pharmacy, Birla Institute of Technology and Sciences Pilani, Hyderabad, Telangana, India

Adbul Waheed Multifunctional Nanostructures Laboratory, Department of Nanotechnology, University of Kashmir, Srinagar, India

Kaiser Wani Department of Biochemistry, College of Science, King Saud University, Riyadh, Saudi Arabia

University Institute of Biotechnology, Chandigarh University, Mohali, India

Larub Younis Multifunctional Nanostructures Laboratory, Department of Nanotechnology, University of Kashmir, Srinagar, India

About the Editors

Suriya Rehman is currently working as Research Faculty in the Department of Epidemic Diseases Research, Institute for Research and Medical Consultation, Imam Abdulrahman Bin University, Dammam, Saudi Arabia. She has obtained her Ph.D. in Microbial Toxicology from Jamia Hamdard University, New Delhi, India, and Council for Scientific Industrial Research, IIIM Jammu/Srinagar, J&K, India. She is currently a full-time research scientist who also teaches master's courses and supervises multiple theses. She has gained an extensive research experience in the field of infectious diseases, multidrug resistance (MDR), nanotechnology, drug development, and cancer therapeutics. She has published over 100 peer-reviewed papers in international journals of repute, edited two books, and guest editor for few journals. She has six US patents in her name and is a member of several international scientific associations, editorial board of international journals, and a reviewer. She has been awarded as one of the top researchers by Imam Abdulrahman bin Faisal University, Dammam, KSA, and has been adjudged in top 2% world scientist list 2022, 2023, and 2024 by Stanford University.

Ebtesam Abdullah Al-Suhaimi is a Professor of Physiology, Endocrinology and immunology), Imam Abdulrahman bin Faisal University (IAU), Saudi Arabia. She has been the Vice Dean for Scientific Research (2009–2018) and Dean of the Institute for Research and Medical Consultations (2015–2023). She works as advisor for Vice Presidency for Scientific Research and Innovation (2023–Present). She has more than 25 years of research and teaching experience in Endocrinology, Physiology, Immunology, Neuroscience, and Biology. She has authored more than 88 research articles and review articles. Additionally, she has coauthored several book chapters and has edited some books. She is a member of the Endocrine Society, Saudi Biological Society, the Saudi Society of Endocrinology, and other biological fields and has served as an Editorial Member and Reviewer for more than 55 journals and funding agencies. Al-Suhaimi acted as a member in many councils and standing committees at IAU. She has also managed many summer programs for regular and talented students of different levels of education. She is associated with many commercialized projects.

Current Scenario of Antimicrobials Agents and Their Applications

1

Uzma Ali and Syed Mehmood Ali

1.1 Introduction

Antimicrobials, widely recognized as highly efficacious chemotherapeutic agents in the medical domain, have profoundly impacted reducing mortality and controlling infectious diseases, which have long been acknowledged as significant contributors to human morbidity and mortality (Aminov 2010). The introduction of antimicrobial drugs has pointed to a momentous paradigm shift, revolutionizing the management of infectious diseases and shaping the trajectory of humanity. This remarkable antimicrobial chemotherapy advancement has fostered an optimistic expectation that infectious diseases will soon be conquered. However, the emergence and resurgence of infectious diseases, particularly those caused by drug-resistant organisms, have presented an immensely formidable challenge. Consequently, addressing this complex predicament remains a complex conundrum within the clinical practice (Saga and Yamaguchi 2009).

The term "antimicrobials" has expanded beyond its original definition, which exclusively referred to drugs targeting microorganisms. It now encompasses therapeutic agents effective against various microorganisms, including those derived from living organisms and synthetic compounds such as fluoroquinolones and sulfonamides. Two distinct categories exist within antimicrobial agents: bactericidal agents that actively eradicate microorganisms and bacteriostatic agents that impede their growth. An array of antimicrobials has been developed and marketed globally

U. Ali (✉)
Department of Public Health, College of Public Health, Imam Abdulrahman Bin Faisal University, Dammam, Saudi Arabia
e-mail: uasali@iau.edu.sa

S. M. Ali
Faculty of Engineering, Department of Biomedical Engineering, Saleem Habib University, Karachi, Pakistan

S. Rehman, E. Al-Suhaimi (eds.), *Nanotechnology Based Microbicides and Immune Stimulators*, https://doi.org/10.1007/978-981-96-0397-8_1

since the beginning of twentieth century, to combat a broad spectrum of infectious diseases. These drugs function by either eliminating or inhibiting the growth of microbes. Antimicrobial agents are generally classified into three primary groups: disinfectants, antiseptics, and antibiotics. Disinfectants, such as household cleaning products and bleaching agents, are nonselective substances used to prevent the proliferation of microorganisms on inanimate surfaces. In contrast, antiseptics find application in surgical procedures to minimize microbial infections in living tissues. Antibiotics, a specific subset of antimicrobial drugs, are utilized to kill or inhibit the growth and development of microorganisms within the human body. This category includes antibiotics, antivirals, antifungals, and antiparasitic drugs (Aminov 2010).

1.2 Antimicrobials in Pre-antibiotic Era

The research findings challenge the prevailing belief that antibiotic exposure is limited to the modern "antibiotic era" by providing compelling evidence of the antibiotic presence in ancient populations. Bassett et al. (1980) and Nelson et al. (2010) conducted meticulous analyses that successfully identified tetracycline residues in the bones of ancient Sudanese Nubia, back in 350–550 BC. The distribution of tetracycline within these skeletal remains strongly suggests exposure to tetracycline-containing substances through their diet. Furthermore, Cook et al. (1989) conducted a rigorous histological study on samples obtained from the upper bone shafts of late Roman skeletons in Dakhleh, Egypt. This study revealed distinct labeling of fluorocyclines, providing further evidence for the presence of tetracyclines in their diet. The presence of tetracycline in these ancient populations potentially conferred a protective effect, as supported by the observed low incidence of infectious diseases in the Sudanese Nubian population and the absence of bone infections in samples from the Dakhleh region (Armelagos 1969; Cook et al. 1989).

During the early antimicrobial era, traditional and alternative medicines played a significant role in harnessing antimicrobial properties. Traditional Chinese medicine (TCM) stands as an exemplary case, with a history spanning thousands of years. A notable breakthrough during this period was the discovery of Xinghaosu, extracted from Artemisia plants and considered an effective antimalarial drug in the 1970s. Chinese herbalists have long relied on these plants as remedies for diverse ailments. Besides qinghaosu, other herbs employed in TCM demonstrate antimicrobial activity, and uncovering the active compounds within these ancient remedies can expand the range of antibiotics in conventional medicine. Hence, comprehending the impact of traditional and alternative medicine on developing antibiotic resistance becomes crucial for effectively utilizing modern antimicrobials.

1.3 Historical Dynamics of Antimicrobials

During the pre-1940s and 1950s, patients suffering from bacteremia, including *Streptococcus pneumoniae*, faced high mortality rates. The prognosis for these patients was bleak, with tuberculosis mortality rates reaching 50%. However, the introduction of antibiotics brought about a revolutionary change in the prevailing circumstances. Extensive documentation by Austrian and Gold (1964) provides evidence of the rapid emergence of various antibiotic classes in the market during the 1950s and 1960s.

Identifying three pivotal antimicrobials, namely salvarsan, prontosil, and penicillin, is a remarkable achievement and sets a precedent for future drug discovery research. Subsequent researchers have built upon this foundation, developing a diverse range of new antibiotics, many of which have been successfully administered to patients. While the period spanning the 1950s to the 1970s witnessed the discovery of groundbreaking antibiotic classes, the exploration of novel classes has since stagnated. Consequently, the prevailing strategy for combating the emergence and reemergence of antibiotic-resistant pathogens focuses on modifying existing antibiotics, given the scarcity of fresh discoveries. This approach has become the primary method for generating effective drugs against these pathogens (Chopra et al. 2002).

The contributions of Paul Ehrlich and Alexander Fleming are widely recognized as pivotal in the development of modern antibiotics. Ehrlich's "magic bullets" concept aimed to target disease-causing microbes while sparing the host selectively. This concept stemmed from his observation that certain synthetic dyes, like nitrines, could stain specific microbes without affecting others. Ehrlich hypothesized that chemical compounds could be synthesized to target parasites in organisms specifically. In 1904, he initiated a large-scale screening program to discover potential cures for sexually transmitted diseases, particularly syphilis, which were highly prevalent and nearly incurable at the time.

Syphilis, caused by the spirochete *Treponema pallidum*, was commonly treated using inorganic mercury salts, which exhibited limited efficacy and caused significant side effects. Collaborating with chemist Alfred Bertheim and bacteriologist Sahachiro Hata, Ehrlich synthesized various organic arsenic derivatives of the highly toxic drug atoxyl. These compounds underwent testing on rabbits infected with syphilis. In 1909, the team identified the sixth compound of the 606 series, which demonstrated effectiveness in treating syphilis in rabbits. Subsequent limited studies indicated its potential for treating venereal disease in humans. Despite the cumbersome injection process and undesirable side effects, the drug proved to be a groundbreaking discovery. It was marketed as salvarsan and became the primary treatment for syphilis until the introduction of penicillin in the 1940s. A more soluble and less toxic derivative, neosalvarsan, was developed and utilized as an alternative treatment option (Chopra et al. 2002),

Paul Ehrlich's screening methodology has laid the groundwork for drug discovery strategies in the pharmaceutical industry, identifying numerous drugs, including various antimicrobials, that have been successfully applied in clinical practice. This approach played a pivotal role in investigating sulfonamide drugs, with particular emphasis on sulfanilamide (prontosil), which Bayer chemists Josef Klarer and Fritz Mietzsch synthesized. Subsequently, Gerhard Domagk evaluated its antibacterial properties for the treatment of diverse diseases. While prontosil garnered recognition as the first effective drug, its active ingredient, sulfonamide, had already been employed in the dyeing industry for several years and was not eligible for patent protection. Capitalizing on the cost-effectiveness of sulfanilamide production and its non-patentable status, many companies continued to manufacture significant quantities of sulfanilamide derivatives. The prolonged utilization of these early antibiotics has left a profound impact, manifesting as the most prevalent case of drug resistance: sulfonamide resistance is nearly ubiquitous within the class (Chain et al. 2005).

The unexpected event of September 3, 1928, leading to Fleming's discovery of penicillin in 1929, is likely familiar to many. While the antimicrobial properties of molds have been recognized since ancient times, and previous researchers occasionally reported similar observations of *Penicillium*'s antimicrobial activity, Fleming's exceptional perseverance and unwavering belief in this concept truly made a difference. Twelve years after his initial observation, Fleming endeavored to capture the attention of chemists engaged in addressing the challenges associated with purifying and stabilizing the active substances, providing *Penicillium* strains to those who sought them. Although he eventually abandoned the idea in 1940, the same year witnessed the comprehensive papers published by Oxford researchers Howard Flory and Ernest Chain on the purification of penicillin to a level suitable for clinical testing. This protocol ultimately paved the way for penicillin's mass production and widespread distribution in 1945. Fleming's screening method, involving the creation of a barrier on the surface of a medium-sized agar plate to inhibit the growth of pathogenic bacteria, requires significantly fewer resources compared to testing using animal disease models. As a result, numerous scientists in academia and industry have widely adopted this approach for large-scale screening of antibiotic-producing microorganisms. Furthermore, Fleming was among the first to caution about the emergence of penicillin resistance when the treatment duration or dosage is insufficient (Chain et al. 2005).

The introduction of the initial trio of antimicrobials—namely salvarsan, prontosil, and penicillin—represented a groundbreaking milestone that laid the foundation for extended exploration in drug development. The trajectory pursued by subsequent researchers ultimately resulted in the creation of numerous novel antibiotics, some of which exhibited efficacy in patient treatment. The period from the 1950s to the 1970s witnessed a significant epoch in medical history with the discovery of new classes of antibiotics; yet no further categories have been unveiled since then. In response to the diminishing discoveries of new antibiotics, the primary avenues in the pursuit of novel remedies against pathogen resistance to antibacterial agents have revolved around modifying existing antibiotics (Iskandar et al. 2022).

1.4 Goals of Antimicrobial Therapy

The principal goal of antimicrobial therapy is the thorough elimination of bacteria. Antimicrobials that enhance the eradication of bacteria will also optimize the attainment of clinical cures. The direct connection between bacteriologic and clinical success has been evident as studied by Garau (2002). A study that was conducted on

123 patients who were suffering from acute otitis media, 97% clinical success rate was attained and their middle ear fluid culture was found negative after four days of starting the antimicrobial. In the realm of acute exacerbation of chronic bronchitis, a discernible correlation exists between prompt bacterial eradication and favorable clinical outcomes. Achieving bacterial eradication is imperative as it serves as a safeguard against the emergence of antimicrobial resistance (AMR) (Garau 2002).

Dabernat and his colleagues embarked on a significant exploration of the prevalence of penicillin-resistant *S. pneumoniae* strains among young patients suffering from acute otitis media. Their insightful investigation revealed that the proportion of these resilient strains actually increased after treatment and during follow-up, contrary to the initial phase. It is fascinating to observe that the administration of cefixime did not result in a decrease in penicillin-resistant *S. pneumoniae*. Conversely, the utilization of amoxicillin-clavulanate demonstrated a noteworthy reduction in the prevalence of these stubborn strains. The World Health Organization (WHO), recognizing the significance of eradicating bacterial diseases, issued a statement in 2000 emphasizing the critical need to judiciously select appropriate antimicrobial agents when commencing treatment. This strategic approach is meticulously devised to thwart the emergence of antimicrobial resistance and ensure the ongoing effectiveness of treatment protocols. This is not a right philosophy to secure the best agents for last, but rather to use the most effective antimicrobials from the start to eradicate all bacterial pathogens and thus minimize the selection of resistant strains (Dabernat et al. 1998).

1.5 General Principles of Antimicrobial Application

The terms "*antimicrobial,*" "*antibiotic,*" and "*anti-infective*" incorporate an extensive variation of pharmacological agents including antibacterial, antifungal, antiviral, and antiparasitic drugs (Fig. 1.1). The most commonly used ones are the antimicrobials, though other agents follow the same principles as well. The Infectious Diseases Society of America has provided clear evidence-based practice guidelines that guide suitable therapy for a wide range of infectious diseases and syndromes caused by specific microbes, including bacteria, viruses, and parasites. The guidelines provided by the Infectious Diseases Society of America are applicable in the context of characteristics of host, cost, and response to therapy (Leekha et al. 2011).

1.5.1 Antimicrobial Selection and Initiation Criteria

Accuracy in Infectious Disease Diagnosis

There are several criteria for the infectious disease diagnosis such as determining the infection site, host attributes, such as diabetes, age, immune status, etc., and, whenever possible, a microbiological diagnosis. Certain infections are grave and might need prolonged therapy such as sepsis, endocarditis, and meningitis; it is

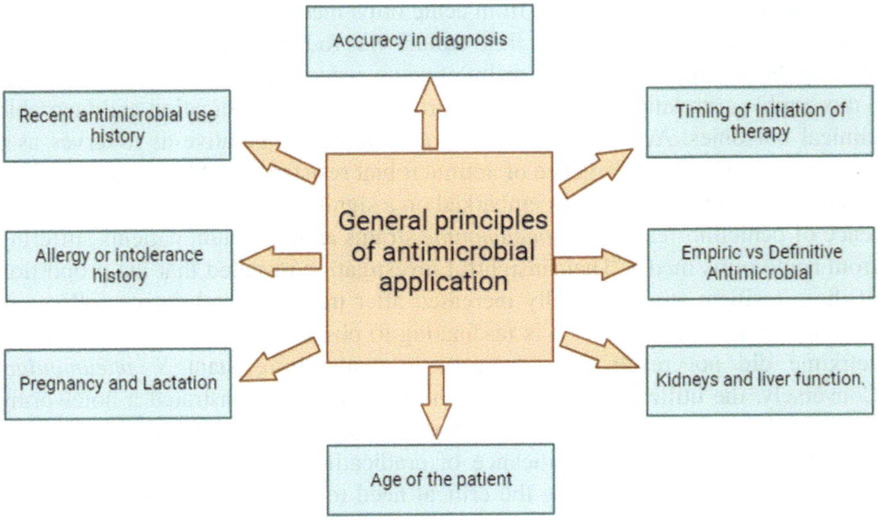

Fig. 1.1 General principles of antimicrobial application

critical to isolate the disease-causing pathogen. Moreover, additional investigations need to be carried out to exclude noninfectious origin of the disease, if the patient is not benefitted for the antimicrobial therapy selected based on clinical presentation. It is obvious that microbiological diagnosis ideally should be dependent on laboratory data such as serology or culture results; however, very often clinical presentation reveals the microbiological etiology. Even in the absence or arrival of positive culture for cellulitis, antimicrobial therapy can be started assuming that causative microorganisms are *Streptococci* or *Staphylococci.* Similarly, community-acquired pneumonia can also be treated empirically before the diagnostic test results arrive (Mandell et al. 2007).

Timing of Initiation of Antimicrobial Therapy

The assessment of clinical severity significantly influences the decision to commence therapy. Immediate initiation of empiric treatment stands as a critical measure in addressing severe clinical conditions, such as sepsis, febrile neutropenia, and bacterial meningitis. This intervention is preferably carried out concurrently with or immediately following the collection of diagnostic specimens. Conversely, a circumspect approach becomes necessary in comparatively stable clinical scenarios, exemplified by subacute bacterial endocarditis and spinal osteomyelitis, characterized by prolonged symptom manifestation. It is judicious to deliberately defer the onset of antimicrobial medication until appropriate specimens have been obtained and forwarded to the microbiology laboratory. Premature administration of antibiotic therapy in these situations risks impeding or delaying the microbiological diagnosis due to its potential to suppress bacterial growth (Surbhi et al. 2011).

Empiric vs Definitive Antimicrobial Therapy
The initial results of the microbiological samples usually become available after 48 to 72 hours; therefore, initial empirical therapy with a broad-spectrum antimicrobial and sometimes with a combination antimicrobial is initiated with the intent to cover multiple possible pathogens, guided by clinical presentation. Delay in empirical therapy or use of inadequate antimicrobial therapy might result in poor outcomes such as prolonged hospital stays, and high morbidity and mortality, especially in critically ill and hospitalized patients for both community- and hospital-acquired infections.

As soon as the microbiology results identify the etiologic pathogen and/or relevant data for antimicrobial susceptibility is received, the antibiotic spectrum should be narrowed to target only the susceptible microorganism. This approach is a critically important element of antibiotic therapy because it helps in the reduction of not only cost and side effects but also toxicity and prevents the antimicrobial resistance emergence in the community (Surbhi et al. 2011).

1.5.2 Host Actors to Be Considered in the Selection of Antimicrobial Agents

Although it is useful for the clinicians to develop familiarity with a few specific antimicrobials, a "one-size-fits-all" approach to antimicrobial selection is not appropriate, and multiple host factors must be taken in to account. There are several guidelines provided by authorities for appropriate dose modification of individual antimicrobials (Gilbert et al. 2010; Saag et al. 2008).

Kidneys and Liver Function
Since the kidneys and liver are the primary organs for drug elimination from human body, the proper functioning of these organs is pivotal during antimicrobial therapy. In cases where the patients have compromised functioning of these organs, antimicrobial dose adjustment is necessary to avoid toxicity and undue drug accumulation. However, antimicrobial doses may need to be raised occasionally to avoid underdosing among young healthy people with rapid renal clearance or those with rapid hepatic metabolism due to enzyme induction caused by simultaneous use of drugs such as rifampin or phenytoin (Gilbert et al. 2010; Saag et al. 2008).

Age
Age of the patients is yet another parameter to determine the antimicrobial therapy. Children and patients who fall in the geriatric age group respond to therapy differently because of variation in body mass index and renal function. In the pediatric age group, the dose is adjusted according to body weight. Patients of both ages handle the drug differently, mainly due to differences in body size and kidney function. Dosages of most children's medications are guided by weight. In elderly patients, serum creatinine level alone does not fully reflect renal function, and

creatinine clearance should be estimated taking into account the age and weight of these patients (Pillai et al. 2010).

Pregnancy and Lactation

Antimicrobial therapy during pregnancy is crucial since it relates to the health of both mother and fetus, so it needs to be given special considerations. Although not enough data is available to estimate the serum level of antimicrobials, during the third trimester specially, there is an increase in the plasma volume and renal blood flow and may result in more quick clearance availability of low serum level of pharmacological agents including the antimicrobials. Despite low availability, higher doses of antimicrobials are not normally recommended in the third trimester of pregnancy. However, in the case of a mother with HIV, the management is done by an increased dose of protease inhibitors as recommended by some experts. As far as the safety of the fetus is concerned, most antimicrobials are either teratogenic or toxic to fetus. Due to limited research on the safety of antimicrobials during pregnancy and their potential effects on the fetus, precautions should be taken before. However, some antimicrobial agents such as penicillins, cephalosporins, and macrolides are generally considered safe for use during pregnancy and are frequently used pregnant women. This has been evident by a multicentric study conducted on over 13,000 pregnant women who were affected by birth defects. There was no association found between birth defect and adverse outcome in these specific antimicrobial agents. On the contrary, a similar study showed an association between various birth defect and agents such as sulfonamides and nitrofurantoin, which were previously not considered teratogenic. Some other antimicrobials, such as tetracyclines and chloramphenicol, should be avoided as they have well established fetal or neonatal adverse effects (Crider et al. 2009).

Allergy or Intolerance History

A history of antimicrobial allergy or intolerance should be routinely taken into consideration in the assessment and treatment of infection.

Recent Antimicrobial Use History

The history of recent infection and antimicrobial therapy, approximately 3 months, must be taken into consideration while selecting another antimicrobial therapy. The reason behind this is that the microorganism that has emerged during the recent infection might be resistant under the selective pressure of a recently used antimicrobial. Therefore, an antimicrobial from a different group must be prescribed (Surbhi et al. 2011).

Oral vs Intravenous Therapy

The route of administration of antimicrobial treatment, whether oral or intravenous, is usually governed by the condition of the patients. Intravenous antimicrobial therapy is usually seen among hospitalized patients who have been admitted amid serious infections such as sepsis or meningitis. The timing for treatment onset must be established based on the requirement of the clinical situation. In crucial instances such as sepsis, febrile neutropenia, and bacterial meningitis, immediate therapeutic measures must be commenced and expedited, both simultaneously and immediately after the collection of diagnostic samples. However, individuals admitted for alternative clinical reasons, such as insufficient bodily fluid, mild to moderate infections, cardiovascular events, and regular gastrointestinal functions, are fitting candidates for oral antibiotics that can be readily assimilated (e.g., managing community-acquired pneumonia with oral fluoroquinolones). Additionally, patients initially undergoing parenteral therapy can safely transition to oral antibiotics after attaining clinical stability.

If oral treatment is employed for invasive infections like pneumonia, pyelonephritis, or hemorrhaging, clinicians are advised to choose agents with exceptional absorption and bioavailability (i.e., the percentage of oral doses that remain unchanged in the blood). Illustrative examples of antibiotics with remarkable bioavailability encompass fluoroquinolones, ligands, trimethoprim-sulfamethoxazole, and metronidazole. However, for more severe infections such as infectious heart disease and central nervous system infections (e.g., meningitis), where elevated drug concentrations are preferable in the bloodstream and cerebral spinal fluids, oral therapy is less dependable and generally not recommended (Baddour et al. 2005).

1.6 Special Situations in Infectious Disease Therapy

1.6.1 Antimicrobial Therapy for Foreign Body–Associated Infections

Contemporary medical treatment has introduced and improved the usage of prosthetic implants and devices, both permanent such as prosthetic joints and artificial heart valves and also temporary including urinary catheters and central venous catheters. These implants, although have surpassed their benefits over disadvantages, are susceptible to develop devise-related infections and result in the formation of biofilms, which are considered "an organized community of bacterial cells bounded in a self-produced polymeric matrix and adherent to an inert or living surface." This results in the alteration of the metabolic state of bacteria growing in the biofilm, making them relatively protected from the effects of antimicrobial therapy (Anderl et al. 2000). If any patient develops infection due to prosthesis or implants, it becomes critical for the physicians to treat them since they might take a longer time to heal, and prolonging antimicrobial therapy may be ineffective with more adverse effects and the emergence of resistant strains of microorganisms, and sometime end up in implant removal as an alternative option for cure. Since implant removal cannot be done easily and often not tolerated by the patients, as an alternative, long-term suppressive antimicrobial therapy is recommended (Lynch and Robertson 2008).

1.6.2 Antimicrobial Agents as Prophylactic or Suppressive Therapy

Antimicrobials are widely used as a prophylactic agent in certain settings where the chances of infection are higher with some specific microorganisms, and an antimicrobial will provide prophylaxis with fewer adverse effects and less likelihood of developing resistance. Some of the scenarios where antimicrobial prophylaxis is appropriate in some instances are discussed in the following:

1.6.3 Presurgical Antimicrobial Prophylaxis

It is a common practice to use antimicrobials as prophylaxis prior to surgeries especially involving the prosthetic implants where chances of infection are high, such as gut surgeries, to minimize the risk of post-surgical site infections. The antimicrobial must give coverage to most likely microorganisms and be found in adequate quantity in the serum at the time of initial incision and during the procedure to minimize the risks of infections. Prophylaxis duration for surgical site infection should not exceed 24 hours in most cases (Bratzler and Houck 2004).

1.6.4 Antimicrobial Prophylaxis in Immunocompromised Patients

The risk of infections is high in patients who have HIV infections, going through chemotherapy or receiving immunosuppressants after organ transplant, since they are immunocompromised. The causative microorganisms are usually predictive, yet associated with high mortality. These infections are caused by predictable organisms at an increased frequency and/or associated (e.g., invasive aspergillosis associated with prolonged neutropenia, *Pneumocystis* pneumonia in the setting of impaired cell-mediated immunity [e.g., AIDS, organ transplant]). Therefore, under these circumstances, extended prophylaxis with antimicrobial therapy is recommended, such as trimethoprim-sulfamethoxazole to prevent *Pneumocystis* pneumonia, unless the restoration of immunity is indicated through immunomarkers (Buskin et al. 1999).

1.6.5 Preventing the Transmission of Communicable Pathogens to Susceptible Contacts

Susceptible hosts are those who can contract the highly infectious disease; so a prophylaxis antimicrobial agent is prescribed to avoid the transmission of the disease. Some of the examples include antiviral agents to limit the transmission of influenza in nursing home residents; ciprofloxacin can be given to close contacts of a patient with meningitis caused by *Neisseria meningitidis*, and macrolides can be prescribed to reduce transmission of pertussis (Wilson et al. 2007).

1.6.6 Traumatic Injuries with a High Probability of Infectious Complications

Certain types of injuries pose a particularly high risk of infection because of disruption of normal barriers and/or delivery of a high inoculum of pathogenic organisms; for example, antibiotic prophylaxis has been shown to be of some benefit and is recommended for certain types of animal bites and after penetrating brain injury. An example of inappropriate antimicrobial "prophylaxis" is prolonged antimicrobial use in those with open but not infected wounds, including surgical wounds. No consensus has yet been reached on the use of antimicrobial prophylaxis in some other settings, such as before invasive procedures in patients with prosthetic joints (Fleisher 1999).

1.7 Judicious Use of Antimicrobial Agents

1.7.1 Cost Considerations in Antibiotic Selection and Antimicrobial Stewardship

There are several factors that determine the cost of antimicrobial agents other than the purchase price of the agent. This may include the administrative cost, duration of stay in the hospital, laboratory tests to monitor the serum level and clinical efficacy. Keeping in mind these costs, cost saving strategies must be considered. One of the strategies that can help to reduce the cost is switching from intravenous antimicrobial therapy to oral therapy, which not only is less expensive but also minimizes the adverse effects and shortens hospital stay. Even though the purchase price of some of the oral formulation might be higher than the parenteral equivalent, cost saving is achieved by reduction in hospital stay. For instance, soft tissue and complicated skin infection caused by methicillin-resistant *Staphylococcus aureus treated with* oral linezolid as compared with intravenous vancomycin (McCollum et al. 2007).

The strategy that focuses on cost saving consideration while selecting and continuing the appropriate antimicrobial therapy in acute care setup is part of a broader activity that is referred to as antimicrobial stewardship. Antimicrobial stewardship programs are aimed at "optimizing antimicrobial selection, dosing, route, and duration of therapy to maximize clinical cure or prevention of infection while limiting the unintended consequences, such as the emergence of resistance, adverse drug events, and cost." These programs are computer based directed by a team of infectious disease physicians and pharmacists. It has several components that are used as guidelines that are as follows: prospective audit and feedback of antimicrobial prescriptions to clinicians, formulary restriction, education, use of clinical order sets and guidelines, de-escalation of therapy, and intravenous-to-oral antimicrobial conversion when appropriate. It is a priority that clinicians should be aware of such programs in their clinical setup and institutions (Dellit et al. 2007).

1.7.2 Antibiotic Resistance Prevention

The widespread emergence of antimicrobial has become one of the biggest health concerns globally. The single most reason for the resistance is attributed to inappropriate and widespread usage of antimicrobials, both in community and hospital setup. Previous exposure to antimicrobial therapy has also been found to be a risk factor in drug resistance as commonly seen in community-acquired respiratory infections caused by drug-resistant *S. pneumoniae*. This is noteworthy, given that acute upper respiratory conditions exert a significant impact on the majority of antibiotic prescriptions in ambulance settings. The prescription of these medications frequently transpires in situations where the utilization of antibiotics is not justified. It is obvious that judicious prescription of antimicrobials could widely diminish the chances of antimicrobial resistance and emergence of resistant strains of microorganisms. For these, few initiatives should be taken (Fig. 1.2) such as the avoidance of antibiotic treatment for community-acquired, mostly viral, upper respiratory tract infections; use of narrow-spectrum antibiotics when possible; and use of antibiotics for the shortest duration that is effective for the treatment of a particular clinical syndrome (Gonzales et al. 2007).

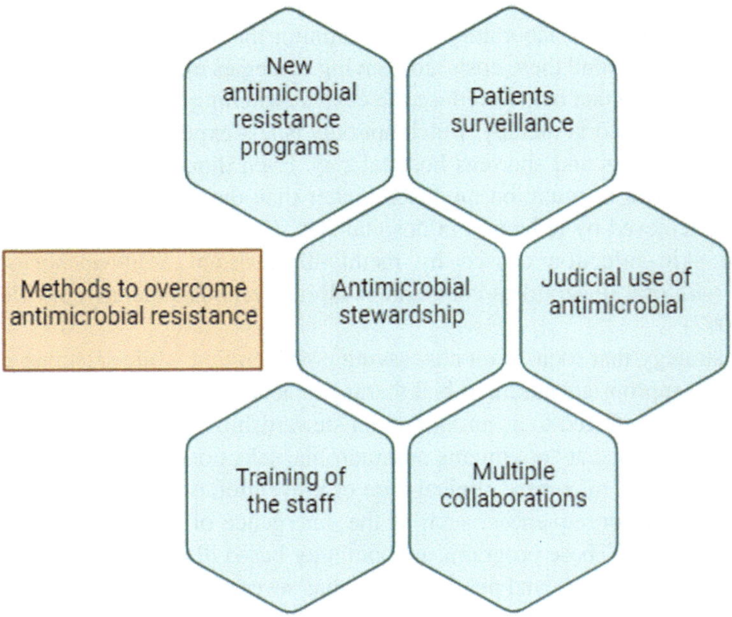

Fig. 1.2 Methods to combat antimicrobial resistance

1.7.3 Common Misuses of Antibiotics

There are certain scenarios where the use of antimicrobials is completely inappropriate and even contraindicated, yet they have been prescribed, resulting in antimicrobial resistance. Few of the scenarios are discussed below (Surbhi et al. 2011).

Prolonged Empiric Antimicrobial Treatment Without Clear Evidence of Infection

One common mistake in the usage of antimicrobials involves the continuation or modification of antibiotic administration, even in the absence of conclusive proof of an established infectious ailment, given that the patient has displayed an evident lack of response to the treatment. Numerous noninfectious, inflammatory, or neoplastic conditions may present symptoms that closely resemble those of infectious diseases. Illustrations include acute Still disease and other disorders affecting connective tissues, which exhibit high fever; fever induced by medication; febrile symptoms accompanying pulmonary embolism; lymphoma; and Wegener granulomatosis, a condition characterized by fever, lung cavity nodules, and recurring sinusitis."

Treatment of a Positive Clinical Culture in the Absence of Disease

Colonization by potentially harmful microorganisms in the absence of concurrent disease symptoms is a prevalent occurrence in specific population segments. Illustrative instances encompass the colonization of the urinary tract in elderly women or individuals with bladder catheters, the colonization of endotracheal tubes in patients undergoing mechanical ventilation, and the colonization of chronic wounds. In such scenarios, it is deemed appropriate to procure cultures solely from these sites if there is a clinical necessity to do so, avoiding the administration of treatment based on positive culture results in the absence of any symptoms or indications of an active infection, such as asymptomatic bacteriuria.

Failure to restrict antimicrobial therapy post-identification of the etiological agent is a common oversight. As previously elucidated, the commencement of treatment often relies on empirical methods, utilizing broad-spectrum agents until the microbiological etiology is determined through culture or other diagnostic tests. Subsequent to acquiring culture and sensitivity data, it is imperative to opt for the most specific antibiotics available for the ongoing therapeutic course. Nonetheless, this transition does not always transpire, particularly when the patient demonstrates improvement during empirical therapy. In such instances, healthcare providers may hesitate to modify the treatment plan despite the positive clinical response (Ibrahim et al. 2000).

Prolonged Prophylactic Therapy

As previously illustrated, the prevention of infection in specific scenarios entails the use of antimicrobial agents, particularly those administered before surgery. Nevertheless, established practices typically advocate for a single preoperative dose of antimicrobial agents, in accordance with established protocols. Prolonging the "prophylaxis" duration only cultivates the potential for the development of antimicrobial resistance. For example, the common practice of extending antimicrobial treatment until the removal of surgical drainage lacks supported, evidence-based proof of its efficacy (Johnson et al. 2008).

The increased utilization of diverse antimicrobial agents across clinical settings poses the risk of hastening the emergence of resistant microorganisms, including multidrug-resistant *Enterococcus faecalis* (*E. faecalis*). One notable example involves the extensive use of fluoroquinolones over the past decade, identified as a key driver of fluoroquinolone-resistant *Clostridioides difficile* strains. These pathogenic strains are now commonly associated with nosocomial infections in healthcare settings. Additionally, changes in treatment policies at individual institutions have resulted in an upsurge in the prescription of levofloxacin for urinary tract infections, contributing to a corresponding increase in fluoroquinolone resistance among outpatient *Escherichia coli* isolates. These findings underscore the necessity for healthcare professionals responsible for managing antimicrobial therapy to adopt a more judicious approach when selecting antibiotics, thereby minimizing the unnecessary exposure of microorganisms to these potent drugs.

1.8 Novel Antimicrobial and Their Application

Antimicrobial resistance is one of the major threats that is faced by the health care system globally and is widely recognized by health-concerned authorities such as the World Health organization and Centers for diseases control and prevention. The global estimated mortality attributed to AMR is massive; i.e., 700,000, and this number is supposed to increase with every passing year with the alarming estimated hike to the ten million deaths per year by 2050. The economic health care burden associated with AMR extends beyond the sufferings of humans and includes the cost of hospitalization and medication. There is an urge for the development of novel antimicrobial agents to combat the resistance threats especially against the six priority pathogens including *Enterococcus faecium, Staphylococcus aureus, Klebsiella pneumoniae, Acinetobacter baumannii, Pseudomonas aeruginosa, and Enterobacte*r species (Chang et al. 2022).

1.8.1 Nanotechnology as a Rescue to Antimicrobial Resistance

Advancement in nanomaterial and technologies has given new horizons and a sustainable solution to antimicrobial resistance due to its unique mode of action on nanoparticles on microbial pathogens, and it is considered as the technology revolution's twenty-first-century weapon. This emerging field of nanotechnology requires immense effort and expenditure for the progression and evolution of novel pharmaceutical drug system. Nanoparticles have multiple benefits over conventional antibiotics, including absorption, durability, controlled release, dispersion, and administration as compared to the massive utility of antimicrobials; "antibiotic nanocarriers" based on liposomal, solid/lipid, terpenoid, polymeric, dendrimeric, and inorganic materials have demonstrated overwhelming outcomes in attaining the sustainable goal of antimicrobial action (Fig. 1.3) (Xuefang et al. 2023).

A suggested strategy to accomplish the antimicrobial goal involves integrating or harmonizing the effects of nanomaterials. This approach aims not only to enhance the therapeutic impact and antibacterial capacity but also to constrain the emergence of resistance. By employing a combination of two or more nanomaterials, a minute structure captures multiple medications within a singular nanostructure or generates amalgamated nanoparticles. The mechanism of action adopted by nanoparticles to fight against antimicrobial-resistant strains are not only novel and unique but also not found in bacterial arsenal of natural defense. The nanostructures are less toxic, economical, and environmentally friendly with enhanced pharmacokinetics which make them ideal for fighting against drug-resistant strains. They are helpful for long lasting, sustainable, and more potent therapeutic window as compared to traditional antibiotics.

Nanoparticles hold distinct chemical and physical properties that make them more adjustable for precise applications. Moreover, they are found to have better chemical activity than the chemical agents used for antimicrobial experiencing resistance and are attributed to their large volumetric surface area. Various methodologies distinguished by diverse attributes, such as morphology, surface coatings, size, and electrical charge, empower researchers to formulate distinctive composite

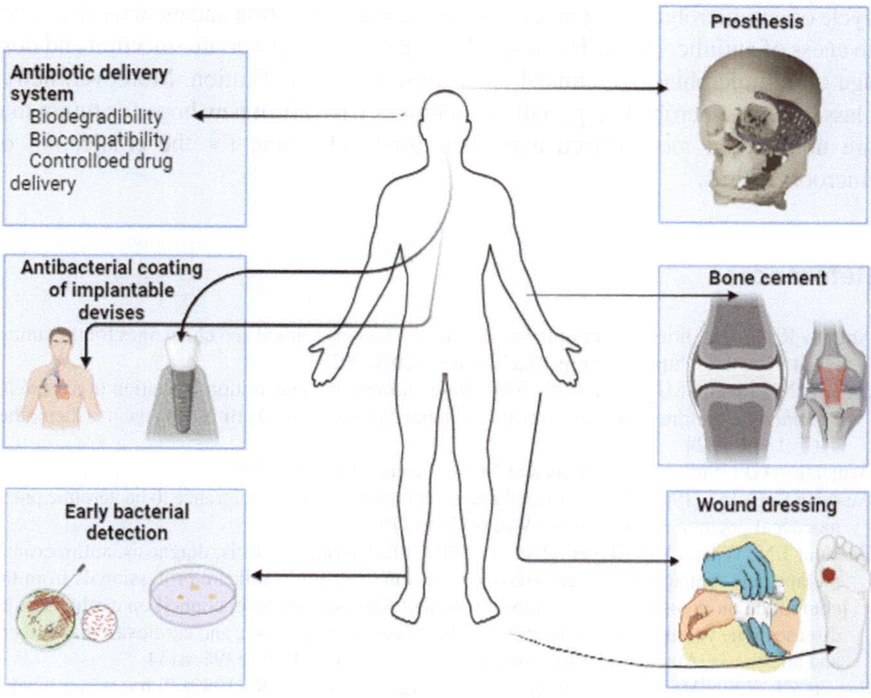

Fig. 1.3 Antimicrobial application of nanoparticles

antimicrobial compounds suitable for diverse antimicrobial applications. The antimicrobial efficacy demonstrated by inorganic nanoparticles (NPs) and carbon-based NPs presents significant potential for future medical, scientific, and industrial exploration and advancement in addressing the challenge of antimicrobial resistance to conventional treatment modalities. Nanoparticles also prove advantageous in the prevention and detection of infections, providing a diverse range of options for biofilm management.

However, prior to the widespread adoption of these nanoparticles in commercial applications, it is imperative to conduct a comprehensive investigation to ascertain their impact on naturally occurring organic tissues and to comprehend their implications for human health and the environment. Rigorous research efforts are essential to ensure the judicious and informed utilization of these innovative materials (Chakraborty et al. 2022).

1.9 Conclusion

To conclude, it is an undeniable fact that antimicrobials have been one of the revolutionary inventions to eradicate and control infectious diseases and to combat with a wide spectrum of microorganisms. However, the emergence of resistant strains of microorganisms has become a global challenge, while efforts in discovering new agents are turning down and getting lessened. There is a need to break the vicious cycle of antimicrobial resistance to maximize the eradication and enhance the effectiveness of antimicrobials. This could be achieved by appropriate selection and dosage of antimicrobials for clinical use and microbial eradication. Moreover, newer classes of antimicrobials especially nanobiotics have given new hope for future that are unique and more effective in their mode of action for the eradication of microorganisms.

References

Aminov R (2010) A brief history of the antibiotic era: lessons learned and challenges for the future. Front Microbiol. https://doi.org/10.3389/fmicb.2010.00134

Anderl JN, Franklin MJ, Stewart PS (2000) Role of antibiotic penetration limitation in *Klebsiella pneumoniae* biofilm resistance to ampicillin and ciprofloxacin. Antimicrob Agents Chemother 44(7):1818–1824

Armelagos GJ (1969) Disease in ancient Nubia. Science 163:225–258

Austrian R, Gold J (1964) Pneumococcal bacteraemia with a special reference to bacteremic pneumococcal pneumonia. Ann Intern Med 60:759–776

Baddour LM, Wilson WR, Bayer AS et al (2005) Infective endocarditis: diagnosis, antimicrobial therapy, and management of complications: a statement for healthcare professionals from the committee on rheumatic fever, endocarditis, and Kawasaki disease, council on cardiovascular disease in the young, and the councils on clinical cardiology, stroke, and cardiovascular surgery and anesthesia, American Heart Association. Circulation 111(23):e394–e434

Bassett EJ, Keith MS, Armelagos GJ, Martin DL, Villanueva AR (1980) Tetracycline-labeled human bone from ancient Sudanese Nubia (A.D. 350). Science 209:1532–1534

Bratzler DW, Houck PM (2004) Antimicrobial prophylaxis for surgery: an advisory statement from the National Surgical Infection Prevention Project. Clin Infect Dis 38(12):1706–1715

Buskin SE, Newcomer LM, Koutsky LA, Hooton TM, Spach DH, Hopkins SG (1999) Effect of trimethoprim-sulfamethoxazole as *pneumocystis carinii* pneumonia prophylaxis on bacterial illness, *pneumocystis carinii* pneumonia, and death in persons with AIDS. J Acquir Immune Defic Syndr Hum Retrovirol 20(2):201–206

Chain E, Florey HW, Gardner AD, Heatley NG, Jennings MA, OrrEwing J, Sanders AG (2005) The classic: penicillin as a chemotherapeutic agent. Clin Orthop Relat Res 439:23–26

Chakraborty N, Jha D, Roy I et al (2022) Nanobiotics against antimicrobial resistance: harnessing the power of nanoscale materials and technologies. J Nanobiotechnol 20:375. https://doi.org/10.1186/s12951-022-01573-9

Chang RYK, Nang SC, Chan HK, Li J (2022) Novel antimicrobial agents for combating antibiotic-resistant bacteria. Adv Drug Deliv Rev 187:114378. https://doi.org/10.1016/j.addr.2022.114378

Chopra I, Hesse L, O'Neill A (2002) Discovery and development of new anti-bacterial drugs. In: Pharmacochemistry library, Trends in Drug Research III, vol 32. Elsevier, pp 213–225

Cook M, Molto E, Anderson C (1989) Fluorochrome labelling in Roman period skeletons from Dakhleh oasis, Egypt. Am J Phys Anthropol 80:137–143

Crider KS, Cleves MA, Reefhuis J, Berry RJ, Hobbs CA, Hu DJ (2009) Antibacterial medication use during pregnancy and risk of birth defects: National Birth Defects Prevention Study. Arch Pediatr Adolesc Med 163(11):978–985

Dabernat H, Megraud GP, Megraud F et al (1998) Effects of cefixime or co-amoxiclav treatment on nasopharyngeal carriage of *Streptococcus pneumoniae* and *Haemophilus influenzae* in children with otitis media. J Antimicrob Chemother 41:253–258. 10

Dellit TH, Owens RC, McGowan JE Jr et al (2007) Infectious diseases society of America and the Society for Healthcare Epidemiology of America guidelines for developing an institutional program to enhance antimicrobial stewardship. Clin Infect Dis 44(2):159–177

Fleisher GR (1999) The management of bite wounds. N Engl J Med 340(2):138–140

Garau J (2002) Treatment of drug-resistant pneumococcal pneumonia. Lancet Infect Dis 2(7):404–415

Gilbert DN, Moellering RC, Jr, Eliopoulos GM, Chambers HF, Saag MS (2010)The Sanford Guide to Antimicrobial Therapy. 40th Sperryville: Antimicrobial Therapy, Inc.

Gonzales R, Malone DC, Maselli JH, Sande MA (2007) Excessive antibiotic use for acute respiratory infections in the United States. Clin Infect Dis 33(6):757–762

Ibrahim EH, Sherman G, Ward S, Fraser VJ, Kollef MH (2000) The influence of inadequate antimicrobial treatment of bloodstream infections on patient outcomes in the ICU setting. Chest 118(1):146–155

Iskandar K, Murugaiyan J, Hammoudi Halat D, Hage SE, Chibabhai V, Adukkadukkam S, Roques C, Molinier L, Salameh P, Van Dongen M (2022) Antibiotic discovery and resistance: the chase and the race. Antibiotics (Basel) 11(2):182. https://doi.org/10.3390/antibiotics11020182

Johnson L, Sabel A, Burman WJ et al (2008) Emergence of fluoroquinolone resistance in outpatient urinary & TDIFSJDIJB DPMJ isolates. Am J Med 1(10):876–884

Leekha S, Terrell CL, Edson RS (2011) General principles of antimicrobial therapy. Mayo Clin Proc 86(2):156–167. https://doi.org/10.4065/mcp.2010.0639. PMID: 21282489; PMCID: PMC3031442

Lynch AS, Robertson GT (2008) Bacterial and fungal biofilm infections. Annu Rev Med 59:415–428

Mandell LA, Wunderink RG, Anzueto A et al (2007) Infectious Diseases Society of America/American Thoracic Society consensus guidelines on the management of community-acquired pneumonia in adults. Clin Infect Dis 44(Suppl 2):S27–S72

McCollum M, Sorensen SV, Liu LZ (2007) A comparison of costs and hospital length of stay associated with intravenous/oral linezolid or intravenous vancomycin treatment of complicated skin and soft-tissue infections caused by suspected or confirmed methicillin-resistant in elderly US patients. MJ5IFS 29(3):469–477

Nelson ML, Dinardo A, Hochberg J, Armelagos GJ (2010) Brief communication: mass spectroscopic characterization of tetracycline in the skeletal remains of an ancient population from Sudanese Nubia 350–550 CE. Am J Phys Anthropol 143:151–154

Pillai SK, Eliopoulos GM, Moellering RC, Section E Jr (2010) Anti-infective therapy: principles of anti-infective therapy. In: Mandell GL, Bennett JE, Dolin R (eds) Mandell, Douglas, and Bennett's principles and practice of infectious diseases, vol 1, 7th edn. Churchill Livingstone/Elsevier, Philadelphia

Saag M, Balu R, Phillips E et al (2008) High sensitivity of human leukocyte antigen-b5701 as a marker for immunologically confirmed abacavir hypersensitivity in white and black patients. Clin Infect Dis 46(7):1111–1118

Saga T, Yamaguchi K (2009) History of antimicrobial agents and resistant bacteria. Japan Med Assoc J 52(2):103–108

Surbhi L, Terrell CL, Edson RS (2011) General principles of antimicrobial therapy. Mayo Clin Proc 86(2):156–167. https://doi.org/10.4065/mcp.2010.0639. PMID: 21282489; PMCID: PMC3031442

Wilson W, Taubert KA, Gewitz M et al (2007) Prevention of infective endocarditis: guidelines from the American Heart Association: a guideline from the American Heart Association rheumatic fever, endocarditis, and Kawasaki disease committee, council on cardiovascular disease in the young, and the council on clinical cardiology, council on cardiovascular surgery and anesthesia, and the quality of care and outcomes research interdisciplinary working group. Circulation 116(15):1736–1754

Xuefang C, Xu J, Ji B, Fang X, Jin K, Qian J (2023) The role of nanotechnology-based approaches for clinical infectious diseases and public health. Front Bioeng Biotechnol 11:1146252. Sec. Nanobiotechnology

Immunomodulatory Nanoparticles and Their Advancement in Health

2

Iram Saba, Kaiser Wani, Khalid Mujassam Batoo, Suriya Rehman, and Saif Hameed

2.1 Introduction

The "immune surveillance" mechanism is crucial in safeguarding our bodies against diverse diseases. It enables our immune system to protect us by constantly monitoring and targeting potential threats to our health. This process is vital in safeguarding our health and against various diseases (Dunn et al. 2002). In theory, immune cells can rapidly recognize viruses, bacteria, and cancer cells as foreign antigens and eradicate them (Agrawal 2019; Alrushaid et al. 2023). Unfortunately, the disconcerting truth is that pathogens have evolved numerous effective strategies which involve hindering "phagocytosis, impeding antigen presentation, or directly destroying immune cells", creating a formidable challenge for the immune system's defense mechanisms to thwart immune clearance (Finlay and McFadden 2006). Ideally, immune cells can recognize and eliminate foreign antigens such as viruses, bacteria,

I. Saba
Research and Scientific Center, Sultan Bin Abdulaziz Humanitarian City, Riyadh, Saudi Arabia

Amity Institute of Biotechnology, Amity University Haryana, Gurugram (Manesar), India

K. Wani
Department of Biochemistry, College of Science, King Saud University, Riyadh, Saudi Arabia

University Institute of Biotechnology, Chandigarh University, Mohali, India

K. M. Batoo
King Abdullah Institute for Nanotechnology, King Saud University, Riyadh, Saudi Arabia

S. Rehman
Department of Epidemic Disease Research, Institute of Research and Medical Consultations (IRMC), Imam Abdulrahman Bin Faisal University, Dammam, Saudi Arabia

S. Hameed (✉)
Amity Institute of Biotechnology, Amity University Haryana, Gurugram, Haryana, India
e-mail: shameed@ggn.amity.edu

S. Rehman, E. Al-Suhaimi (eds.), *Nanotechnology Based Microbicides and Immune Stimulators*, https://doi.org/10.1007/978-981-96-0397-8_2

and cancer cells through targeted killing mechanisms (Schirrmacher 2019). It has been a well-established fact that several pathogens have successfully evaded immune recognition by employing diverse and practical strategies, such as hindering "antigen presentation, impeding phagocytosis, and actively turning off immune cells" (Garg and Agostinis 2017). Immunotherapy treats infections by modulating a person's immune system. Effective prevention of immune-related disorders or diseases demands investigating the broad range of immune cells, cytokines, and enzymes that may affect human immunological conditions (Gordon et al. 2014). However, aberrant immune response intensity, which includes immunosuppression and immuno-stimulation, will end up in the disease (He et al. 2023). Immunosuppression can be caused by an imbalance in any component of the immune system, resulting in a decrease in immune function and a failure to effectively clear pathogens, thus leading to infection or disease (Ponnappan and Ponnappan 2011). Immuno-stimulation could improve the capacity to resist pathogens; however, if it is hypersensitive, it might trigger a severe adverse reaction such as autoimmune disease (Landesman-Milo and Peer 2012).

The idea that nanosystems can influence the activity of the immune system for therapeutic purposes has dramatically altered healthcare over the past 50 years. The concepts of "therapeutic immunomodulation", "immunotherapy", and "immunoengineering" are widely accepted in extensive biomedical applications. Several properties of nanoplatforms have been demonstrated in studies (Khatun et al. 2023; Saba et al. 2022). These properties include:

A. Efficiently transporting antigens and adjuvants to identical antigen-presenting cells (APCs) or intracellular compartments (Saba et al. 2022).
B. Prolonging the lifespan of bioactive cargo molecules by preventing enzymatic degradation during circulation in the bloodstream (Hale and Medina 2022).
C. Increasing tumor tissue accumulation through the size-dependent enhanced permeability and retention (EPR) impact (Wang et al. 2019).
D. Modifying the surface to target particular organs or cells (Chen et al. 2017).
E. Exhibiting "stimuli-sensitive behavior" to ensure safe transport and "intelligent drug release" (Gao et al. 2019).
F. Allowing for higher drug doses since they minimize accumulation in non-target organs and tissues (Musetti and Huang 2018).
G. Using surface antigen and costimulatory molecule binding to create artificial antigen-presenting cells (aAPCs) for robust T cell activation (Kim et al. 2012).
H. Providing various administration routes, such as intranasal delivery or through microneedle patches under the skin (Ahamad et al. 2021).
I. Inherently possessing immune-modulating properties due to their nanoparticle composition (Sırıken et al. 2018).

The immunomodulatory responses to nanomaterials (NMs) can have both beneficial and detrimental consequences, as illustrated in Fig. 2.1. As a result, gaining a complete and exact knowledge of the interactions between nanomaterials (NMs) and the immune system is critical. These insights are crucial for the secure

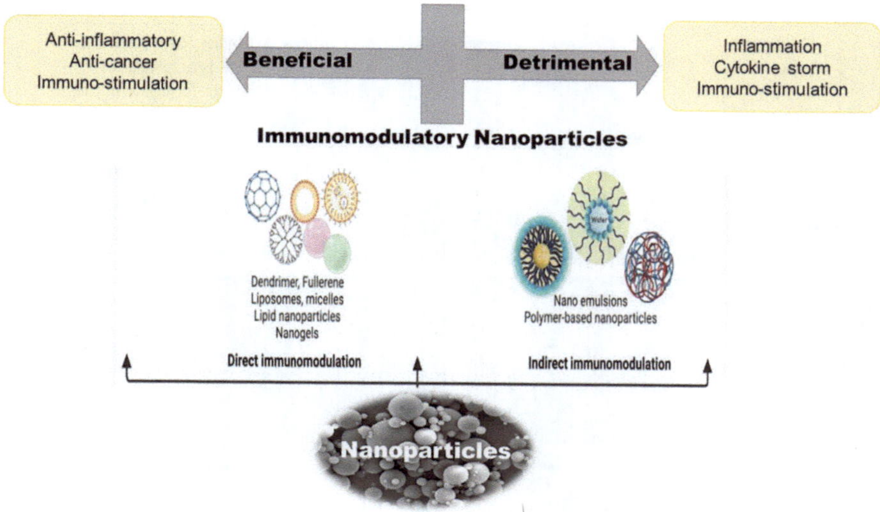

Fig. 2.1 Immunomodulatory responses to nanomaterials (NMs). Immunomodulatory responses to nanoparticles include both direct and indirect immune system interactions. The physical contact of nanomaterials with immune cells influences their activation, cytokine release, and overall immunological responses in direct immunomodulation. Indirect immunomodulation, on the other hand, entails the influence of nanomaterials on the surrounding environment, which results in subsequent effects on immune cells

implementation of immunomodulatory nanomaterials (iNMs) in health applications. In the literature, the authors found gaps and discrepancies in underlying mechanisms through which immunomodulatory nanoparticles exert their effects and the challenges associated with the biocompatibility of immunomodulatory nanoparticles. Here, we aimed to offer a thorough overview of immunomodulatory nanoparticles, covering their categories, general characteristics, size and shape considerations, surface modifications, and immunotherapeutic aspects. This chapter addresses the immunomodulatory effects of diverse NMs and their potential application in health. The article discusses the immune system's involvement in how NMs interact with the immune system to eliminate unhealthy cells actively or passively. It also discusses the latest developments in immune nanomedicines (iNMs). The focus is on understanding the interactions between nanomaterials and the immune system, emphasizing elucidating the mode of action.

2.2 Categories of Immunomodulatory Nanosystems (iMNs)

Immunomodulatory nanosystems can be classified into four distinct classes, as depicted in Fig. 2.2. These systems are designed with biodegradable and immunologically inert components. The first category encompasses polymeric nanostructures, lipids, and carbohydrate-mediated immunomodulatory nanosystems. Polymeric nanoparticles in this category are amazing nano-carriers, acting as

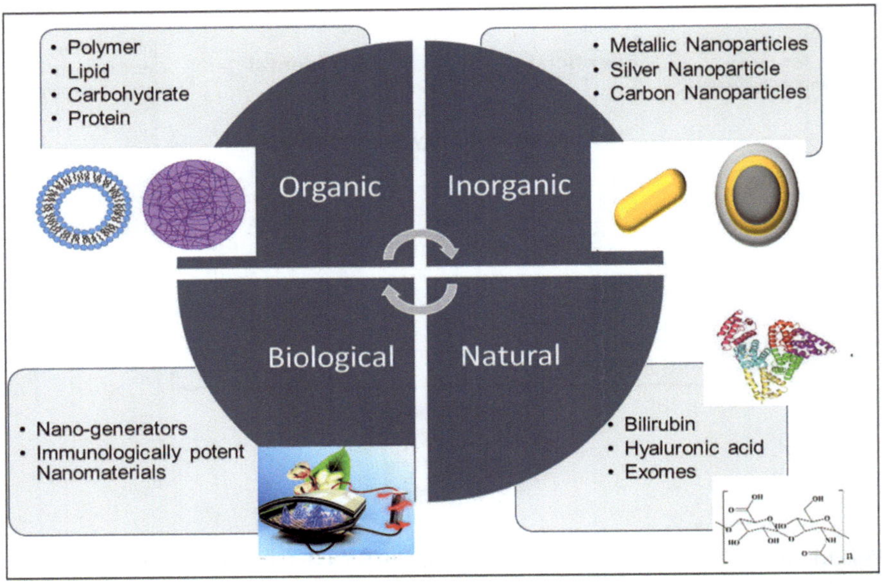

Fig. 2.2 Types of immunomodulatory nanosystems. Immunomodulatory nanosystems fall into many groups, each with benefits in biomedical applications. Organic nanosystems, which are often lipid or polymer-based platforms for drug delivery, offer tunable properties for controlled release and improved biocompatibility. For example, metal or metal oxide nanoparticles have different physicochemical features that enable targeted medication administration and imaging. Biological nanosystems, such as virus-like particles or exosomes, transport drugs and modulate immune responses using natural architecture. Natural nanomaterials from plants or other biological sources are environmentally friendly and low in toxicity

adjuvants and delivery vehicles for various payloads, including antigens and other co-adjuvants. These nanoparticles facilitate the transport of adjuvants and antigens to immune cells while employing a sustained release mechanism that protects against the detrimental effects of harsh environmental conditions, including enzymatic processes. Studies have shown that polymer-based nanosystems interact with antigen-presenting cell (APC) receptors like Dectin1, which sets off a chain reaction involving the cytosolic DNA sensing pathway (cGAS-STING) and the activation of the "NLRP3 (NOD-like receptor family, pyrin domain-containing protein 3) inflammasome" (Kandasamy et al. 2023). The interaction between polymeric nanoparticles and APC surfaces causes dendritic cells (DCs), a "Th1/Th2-mediated immune response", and the release of "type 1 interferons (IFNs), chemokines (CSCL10/IP-10), and cytokines (IL-12, IL-4, IL-10, and TNF-alpha)" (Khatun et al. 2023). Furthermore, it has been observed that polymeric nanoparticles possess the potential to stimulate the immune system by promoting DC maturation, macrophage activation, and activation of IgG/IgA antibodies (Grego et al. 2020; Svindland et al. 2013).

One of the types of IMNs is made from metallic nanoparticles like "silica, copper (Cu), iron oxide (FeO2), gold (Au), silver (Ag), and carbon nanotubes" (Han et al.

2019; Ann Maria et al. 2023; Aldossary et al. 2022). The ability of different functional groups and ligands to bind to metallic NPs is improved. Their natural magnetic and optical properties, as well as the fact that they have a large surface area to volume ratio, make them more useful for therapy. They are good options for making nano-vehicles because the chemicals used to make them are easy to control, and they are easy to attach adjuvants or antigens to. A recent study found that metallic nanoparticles, like silica nanoparticles, can change the immune system by causing the host to release more inflammation cytokines, like IL-1 and IL-18, which can start and spread inflammatory pathways (Luo and Chang 2015; Dhanda et al. 2023). In natural defense, sending a warning signal to Toll-like receptors (TLRs) is important. This process involves recognizing "pathogen-associated molecular patterns (PAMPs) and danger-associated molecular patterns (DAMPs)", as well as identifying virus nucleic acids (Svindland et al. 2013; Gómez Gallego et al. 2017; Escamilla-Tilch et al. 2013).

Metallic nanoparticles, such as "silica, copper (Cu), iron oxide (FeO_2), gold (Au), silver (Ag), and carbon nanotubes", are a second type of nanosystem that can change how the immune system works (Rehman and Jermy 2022). The ability of these metallic nanoparticles to bind to different functional groups and compounds is improved. Their magnetic and optical qualities and the fact that they have a lot of surface area for their size increase their medicinal potential (Dar et al. 2015, 2017). They have a lot of promise as nano-vehicle mixtures because they can be controlled, and it is easy to attach adjuvants or antigens to them. Recent studies have shown that metal nanoparticles, like silica nanoparticles, can change how the immune system works. These nanoparticles can cause the body to release more proinflammatory cytokines, like IL-1 and IL-18, which can start and spread inflammation pathways (Arango Duque and Descoteaux 2014). For the natural defense to work, it is important to signal through "Toll-like receptors (TLRs)".

The third group of immunomodulatory nanosystems includes "bilirubin, rosmarinic acid, hyaluronic acid, cell membranes, and extracellular vesicle-mediated nanoparticles", all of which are naturally occurring organic substances. Together, these organic components and nanoparticles (NPs) mimic the properties and functions of native cells, allowing them to work together in a beneficial way (Vieira et al. 2017). These NPs get past cellular hurdles and slow down the body's ability to get rid of them quickly, ensuring they get to where they should go. In a study by Chen et al., "virus-like nanoparticles filled with model antigens made from the spike protein of avian coronavirus were successful" (Chen et al. 2016). The results showed that these nanoparticles helped immunity regulation by getting antigens to the lymph nodes of the host in an effective way. This led to a strong T cell reaction in the spleen and a strong antibody response.

The fourth group of immunomodulatory nanosystems is made up of hybrid systems that are made by cross-linking or covering different parts. These include "polymer-polymer hybrids, hybrid cell membrane-coated nanoparticles, and protein-polymer hybrid nanoparticles". These combination nanosystems have amazing additive properties and treat diseases well (Ansari et al. 2023). Polymer-hybrid nanosystems have gotten a lot of interest as possible tools for preventing and getting rid of diseases (Khalili et al. 2022). Viral diseases are affected by how they are made, how they move, where they live, and how the body's immune system reacts to them. Hybrid nanosystems can get strong humoral and cellular defensive reactions against viruses (Guo et al. 2021). These nanosystems boost the immune system by making it easier for antigens to move and be released at particular places, making more antigens available in lymph nodes, making dendritic cells mature, and making it easier for antigen-presenting cells (APCs) and T cells to talk to each other. So, immune responses are triggered, which get rid of viral diseases through antigen-specific CD4+ and CD8+ T cell responses and increase of CD40 and CD86 markers (Speiser et al. 2023; Muntjewerff et al. 2020).

2.3 General Characteristics of Immunomodulatory Nanoparticles

Nanoparticles (NPs) have conceivable medical scenarios. The immune system is a crucial defense mechanism, safeguarding entities against foreign contaminants. NPs modulate the function of the immune system, triggering "immunosuppression" or "immunostimulation". These modulating effects may be suitable or disastrous. "Compositions, sizes, and surface chemistry" subsequently impact this immunomodulation. In this context, we explored the correlation between the physicochemical attributes of NPs considered for clinical applications and their behaviors in modulating the immune system.

2.3.1 Size and Shape

Size is probably the most important thing to consider when defining a nano-scaled material, and it is also a key part of protection. Because of this, a lot of research has been done on how the amount of NPs affects the molecular reaction of the host. Tan et al. (2013) examined what size means for how NPs are spread out. As the size of a particle has the opposite effect on the diffusion coefficient, they found that smaller NPs bind faster than bigger ones. But fluid flow causes bigger NPs to experience more pressure, which makes it less likely that they will stick together. The way NPs stick together depends on the particle, the stress rate, and the shape of the vessel. Because their contact area is bigger, NPs with smaller sizes and rod shapes are better at binding (Tan et al. 2013). So, particle size is very important in how these things are absorbed, moved, and kept.

At the nanoscale, size is one of the most important characteristics of a material, and it also has a big effect on how the immune system responds. Because of this, a lot of research has been done to find out how the size of nanoparticles (NP) affects the molecular reaction of the host. In a study by Tan et al. (2013), the amount of NPs was used to examine how they were spread out. As the diffusion coefficient is the opposite of particle size, it was seen that smaller NPs connect more quickly than larger ones. But fluid flow causes bigger NPs to have more drag, which makes it less likely that they will stick together (van Ommen et al. 2012). The binding of NPs is affected by things like the particles' properties, the pressure rate, and the vessel's shape (Liu et al. 2012).

Lundqvist et al. (2008) looked at how exposure to TiO_2 nanoparticles in the lungs affected inflammation in the lungs caused by LPS and the systemic inflammation that followed, focusing on size dependence. Different formulas were tested, such as TiO_2 NPs with 15, 50, and 100 nm sizes, lipopolysaccharide (LPS) alone, and NPs and LPS together. The study's authors said that giving NPs and LPS smaller than 50 nm together led to higher amounts of proinflammatory cytokines and chemokines in the lungs and circulation system than when LPS was given alone. Exposure to nanoparticles alone did not cause the significant production of cytokines like those listed above. Also, the change was more noticeable when the nanoparticles were smaller than when they were bigger. This suggests that nanoparticles and lipopolysaccharides work together to reduce inflammation at the systemic level.

Also, it has been suggested that the increased immune reaction in nanoliposome formulas may be because antigen-presenting cells (APCs) can naturally collect nanoliposomes better than bigger liposomes (Nakanishi et al. 1999). Also, it has been found that smaller SLNs are more active, which makes it easier for macrophages and DCs that live in tumors to take up SLNs. The metabolic characteristics of the drug that has been enclosed may explain why it has a stronger effect on the immune system, as shown in reference (Pan et al. 2009). The cases given above show how important particle size is when it comes to how a host reacts to nanoparticles. When making nanoscale uses and nanomaterials, it is important to think about size from a biological point of view.

In addition to size, a nano-scaled biomaterial's sensitivity is also affected by its shape. A lot of research has been done on the role of NP shape in binding (Alsharif et al. 2020), cell entry (Huang et al. 2013), drug delivery (Liu et al. 2012), and preferred targeting (Adriani et al. 2012). Zhang and his colleagues have made cylindrical and needle-shaped "PLGA-PEG" NPs and researched how they affect the way HepG2 cells work (Zhang et al. 2017). They showed that both kinds of NPs got into cells through endocytosis and stayed in membrane-bound vesicles once they were inside. Only the needle-shaped NPs had sharp ends, and the authors think this made a big difference in how the NPs and lipid bilayer membranes interacted. Cytotoxicity was much higher for needle-shaped NPs than for circular NPs. The lysosome membrane rupture caused the needle-shaped NPs to be harmful to cells. The current study showed that the shape and size of NPs have a big effect on how cells respond physiologically (Zhang et al. 2017).

These studies show that host cells are more likely to take in spheroidal nanoparticles than ones that aren't spherical. This is important because the form of a material changes its toxicity. Nanoparticles with a needle or barrel shape are more dangerous than those with a ball or rod form. Since both internalized and non-internalized NPs are harmful, these results also show that cellular absorption and cytotoxicity have nothing to do with each other. But, future studies need to look at more mathematics, chemicals, and the biochemical reactions of biological types.

2.3.2 Surface Modification

Several scientific studies have shown that changing the surface, including changing the charge or chemistry of the surface, can have a big effect on how nanoparticles and organic fluids, like blood, talk to each other (Gupta and Gupta 2005; Ali et al. 2016). This can change how proteins stick to substances and cause different immune reactions. Also, it has been shown that the level of contact with the immune system is highly related to how slippery the material is (Lin et al. 2015; Andorko and Jewell 2017). If the surfaces of nano-sized substances aren't changed to stop opsonin from sticking to them, macrophages are more likely to eat them. This means that they leave the system very quickly. Opsonization is when opsonins are wrapped around a particle or live thing outside the body. This makes it easier for phagocytic cells to find (Owens III and Peppas 2006). People have thought of different ways to hide NPs from phagocytosis. The most common way is PEGylation, which involves connecting or adsorbing PEG chains to the surface of a particle to make a protection layer that stops opsonin from sticking to the particle so that phagocytic cells can't see it (ten Hagen et al. 2002).

By changing the way plasma proteins stick to the surface of NPs, it is possible to make them much more compatible with blood (Cagliani et al. 2019). The surface charge or hydrophobicity of the NPs has the most effect on how proteins stick to them. For example, when poly-hexadecyl cyanoacrylate nanoparticles (NPs) were heavily covered with polyethylene glycol (PEG), proteins were taken up less (Peracchia et al. 1999). Studies have shown that "as the surface charge of polymer-based NPs goes up, proteins stick to them less" (Lück et al. 1998). However it has also been shown that when the number of charges on the surface increases, more proteins stick to the surface (Gessner et al. 2002). This means that adding different functional groups can change the hydrophobicity of a material by changing one physical property, such as the surface charge. Verma et al. say that unbiased charged ligands can help reduce random interactions between NPs and their biological surroundings (Verma and Stellacci 2010) if placed in the right way on the surfaces of NPs.

Even though neutral functional groups are good, most charged functional groups make it easier for NPs and cells to interact with each other and help avoid interactions with biological components that aren't meant to happen. "Neutral and negatively charged NPs are less likely to stick to negatively charged cell membranes than positively charged NPs" (Verma and Stellacci 2010). This makes them less likely to be taken inside the cell. Also, tests on positively charged NPs show that cationic particles can get through cell walls no matter what kind of particle they are, which could be a factor in their mortality (Verma and Stellacci 2010). NPs with harmless surface coats, like PEG, don't interact much with cells and don't get inside the cells very much. The way the molecules of PEG are arranged on the surface of PEGylated NPs is a key part of how the NPs communicate with cells. Even though the negatively charged cell membrane binds to negatively charged NPs, it is interesting that even negatively charged NPs can get into cells. But nanoparticles with a positive charge are especially good at getting through cell walls and settling in the cytoplasm

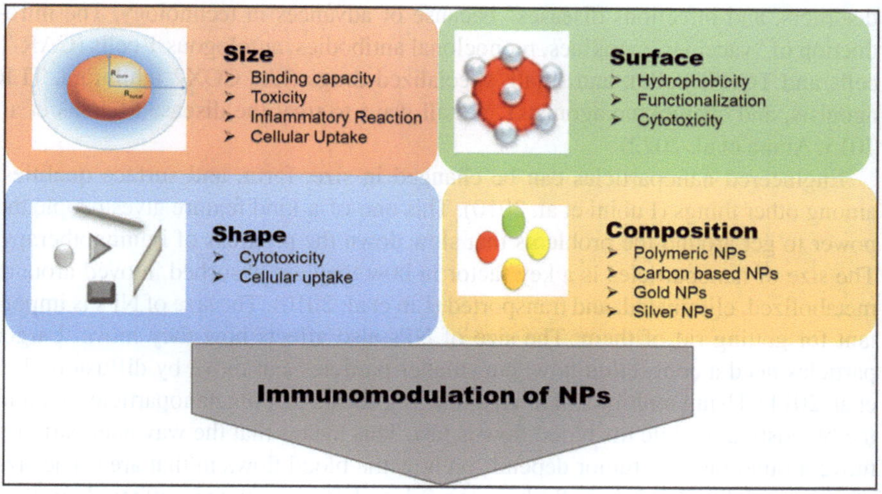

Fig. 2.3 Immunomodulation of nanoparticles. Modifying the size, shape, surface, and composition of immunomodulatory nanoparticles is critical for optimizing their activity. Size changes affect cellular absorption and biodistribution, whereas shape changes refine these interactions. Surface alterations and compositional changes are critical in improving biocompatibility and imparting specialized targeted abilities. These changes enable researchers to create nanomaterials with precise immunomodulatory characteristics, enhancing therapeutic effects

or nucleus (He et al. 2010). Studies on positively charged NPs have shown over and over again that cationic particles can get through cell walls. This may be a reason why they kill cells. Figure 2.3 shows how nanomaterials' size, form, surface, and makeup can be changed to make them more immunomodulatory.

The surface of nanoparticles can be changed on purpose to affect the molecular and biological parts of the blood in a certain way. In future research, you can only change one number at a time so that you can learn about the most important structural factors for protein binding. There are many parts to the connection in question. By looking into the particle's structure-activity relationship, especially in terms of how it interacts with blood cells and protein parts, many problems can be solved, and a new group of immunomodulatory nanoparticles can be made.

2.4 Immunotherapeutic Aspects of Nanoparticles

Immunotherapy allows parts of the immune system to treat diseases, such as cancer, by either stopping or starting immune reactions. There are two main types of immunotherapy: active immunotherapy and silent immunotherapy. In active immunotherapy, the immune response changes, while in passive immunotherapy, immune effector molecules or antigens are given to start an immune reaction (Lesterhuis et al. 2011). Immunotherapy is being looked at more and more as a possible treatment for diseases like "cancer, AIDS, autoimmune disorders, inflammatory disorders, and infectious diseases" because of advances in technology. The introduction of "vaccines, cytokines, monoclonal antibodies, autologous T cells (CAR-T cells and TCR T cells), and small, specialized drugs (like COX2 inhibitors, TLR agonists, and chemokine agonists)" are all ways to treat the disease (Adams et al. 2015; Amna et al. 2022).

Engineered nanoparticles can be changed in size, form, and surface qualities, among other things (Fubini et al. 2010). This one-of-a-kind feature gives people the power to get around the problems that slow down the progress of immunotherapy. The size of nanoparticles is a key factor in how they are absorbed, moved around, metabolized, eliminated, and transported (Lin et al. 2010). The size of NPs is important for getting rid of them. The size of NPs also affects how they move. Larger particles need a convection flow, but smaller particles can move by diffusion (Toy et al. 2011). Using multimodal in vivo imaging shows that big nanoparticles get into the bloodstream while the blood flow is fast. This means that the way nanoparticles move around inside a tumor depends on how the blood flows in that area. The size of nanoparticles (NPs), usually between 0.1 and 10 nanometers, affects how they move through the body (De La Vega and Häfeli 2015). The liver organ has less aggregation of nanoparticles with larger sizes, while the lung system has more aggregation of nanoparticles with smaller sizes. The shape of the nanoparticles has a big effect on what happens to them in the end, including how they move through the body, where they go, and how they get there (Toy et al. 2013; Mie et al. 2014). The absorption and spread of NPs are made easier by how they are set up. It has been shown that nanoparticles can only get into solid tumors with the right shape, size, and aspect ratio. How immunogenic NPs are depends on their size and shape (Kulkarni and Muddapur 2014; Mailander and Landfester 2009).

Mice, "RAW264.7 cells", and "bone marrow-derived dendritic cells (BMDCs)" showed different immune responses when exposed to gold nanoparticles of different sizes and shapes (spherical, rod-shaped, and cubic) (Agarwal et al. 2015). These immune responses "depended on the size and shape of the nanoparticles". Antigens set off the TH1 response pathway on spherical nanoparticles, while the TH2 response pathway was set off by rod-shaped nanoparticles (Kumar et al. 2015). The "size, form, and surface charge of nanoparticles" can change how immunotherapy works. Froman et al. found that cationic and anionic nanoparticles linked to ovalbumin caused lung vaccination, with cationic nanoparticles boosting humoral and mucosal immunity more than anionic nanoparticles (Kreft and Jetz 2007). The biodistribution and immune reaction of nanoparticle ligands are affected by their density. Liu and his colleagues noticed that the quantity of polyethylene glycol changed how lipid-calcium-phosphate nanoparticles were spread out in hepatocytes and Kupffer cells. To study the biological effects of nanoparticle elasticity, they also used polyethylene glycol-based hydrogel nanoparticles with different elasticity moduli (Circulation et al. 2015). When a person has an autoimmune disease, the immune system sends out a reaction that attacks the body's cells (Janeway Jr et al. 2001).

Immunosuppressant drugs are often used to treat diseases caused by the immune system attacking itself. They work by stopping the defense system from doing its job. Earlier research has shown that nanoparticles (NPs) can also stop the immune system from reacting to inflammatory illnesses (Liu et al. 2022). Carbon nanotubes (CNTs) stop T cells and natural killer cells from becoming active (Dumortier 2013). Also, it has been found that encephalitogenic Th17 cells can be stopped by internalizing multi-walled CNTs. Fullerenes, another type of NP made of carbon, can stop human mast cells and basophils from having an allergic reaction (Moraes et al. 2013). Gold nanoparticles have demonstrated the capability to halt inflammation in the memory response of human monocytes previously primed with BCG (Della Camera et al. 2022). Iron oxide NPs have been found to stop BALB/c mice splenocytes from making IL-6 and IL-17 when exposed to ovalbumin. Also, Treg cells, known to stop immunity cells from working and help with autoimmune diseases, have been studied as a way for NPs to control autoimmune diseases. So, immunosuppression based on NPs could be a good way to treat inflammatory diseases (Swartzwelter et al. 2020; Hsiao et al. 2018). The studies listed above give us important information about how nanoparticles cause immunotoxicity. They explain how nanoparticles interact with different parts and cells of the immune system and how they affect both natural and adaptive immunity. These studies have shown that nanoparticles can increase the production of reactive oxygen species, deplete superoxide dismutase, make nitric oxide, and turn on Toll-like receptors, nuclear factor kappa B, and NOD-like receptor family pyrin domain containing 3, all of which are involved in immune system responses. Since the "physical and chemical properties of nanoparticles" affect both natural and adapted reactions, it is possible to change and improve these qualities to make effective immunotherapeutic approaches.

2.5 Promising Clinical Applications of Nanosized-Immunomodulators

In cutting-edge biological research and innovation, the focus has moved to highly specific tailored treatments to control different immunity processes. Nanotechnology has become a potential way to send components that have been changed chemically or biochemically, such as immunomodulators and immunomodulatory molecules (Jantan et al. 2015; Alhazmi et al. 2021). Nanotechnology could improve things by making them last longer, letting them work better over time, and making it easier for them to break down without being harmful (Chakravarty and Vora 2021; Irvine et al. 2015; Saba et al. 2021). For example, giving nano-sized liposomes with immunomodulators like cytokines (such as "IFN-, IFN-g, TNF-, or IL-2") in them lengthens the time it takes for the plasma to absorb the molecules (Verma et al. 2022). This increases the likelihood of immunomodulators persisting and accumulating at tumor sites, capitalizing on the permeability and retention effect. When supplied by nanoengineered particles that use advanced genome editing and molecular modification methods, these specific cytokines and growth factors act as immunity boosters to improve how immune cells work. These particles can also carry nucleic acids like siRNA to fix genes linked to disease in living organisms (Smith et al. 2017).

In a similar study, the physiological qualities of an agent that causes immunogenic cell death (ICD) were better when it was delivered on nanodiscs. Pathological situations with unexpected immunostimulatory signals have been linked to persistent tumors and strong T cell reactions to proteins found in tumors. So, it looks like using nanodiscs with ICD-inducing drugs is a good idea (Kasturi et al. 2017). "Poly(lactic-glycolic acid) (PLGA)" is a biodegradable polymer nanoparticle that the U.S. Food and Drug Administration has cleared for use in safe drug delivery systems (Acharya and Sahoo 2011). Because of its slow and controlled release, low toxicity, and ability to work with the body, PLGA has been chosen to treat diseases and conditions (Jain 2000). Notably, in clinical studies, giving "PLGA-based nanoparticles with TLR 4/7/8 ligands" as immunoadjuvants through the vaginal route improved the defense against SIVsmE660 in some monkeys (Xu et al. 2011).

Using nano-vehicles, especially chitosan-based nano-carriers, to improve the immune response to pneumococcal DNA vaccines through oral delivery is a new and clever way to get them to people. Some groups of vaccinated mice showed better mucosal, systemic, and cellular immune reactions with this method, which suggests it could be used in clinical study methods (Xu et al. 2014). In the last 20 years, it has become clear that using nanoscale carriers with immunomodulators has a lot of promise for clinical uses and biological methods in immunotherapy (Cheng et al. 2020; Rahman et al. 2022). These carriers can boost the immune system and help people with "HIV, TB, type 1 diabetes, multiple sclerosis, and rheumatoid arthritis" deal with their diseases better (Dacoba et al. 2017; Allawadhi et al. 2022; Varadé et al. 2021).

Although immunomodulatory nanoparticles show promise for therapeutic uses, possible hazards and adverse consequences must be studied and investigated. Unintentional immune activation may pose a serious concern since it may result in autoimmune responses or systemic inflammation. Nanoparticles can cause adverse reactions by inducing immunological responses that differ according to size, content, and surface features. Moreover, questions remain about possible chronic impacts and long-term biocompatibility. Furthermore, the toxicity of nanoparticles must be addressed, particularly concerning inorganic compounds that may provide difficulties for clearance and biodegradation. In-depth preclinical and clinical research is necessary to evaluate and reduce these hazards fully.

2.6 Conclusion

In conclusion, studying immunomodulatory nanoparticles provides a new area in biomedical research with far-reaching consequences for healthcare improvement. This chapter has sought a full review of the topic, including classifications, features, size and form concerns, surface alterations, immunotherapeutic elements, and prospective clinical applications. Nanomedicine is an innovative therapy that uses nanotechnology-based delivery systems to get the right amount of drugs to the right places where they work (Ansari et al. 2021). It has a lot of promise to improve treatment methods by improving drug transport, lowering side effects, better molecular imaging, and allowing theranostics. Researchers have found that nano-sized drug delivery methods that interact with antigens are taken up by mucosal surfaces, gut-associated lymphoid tissue, and phagocytic cells, which work well to boost cellular and humoral immune reactions. Many topics need additional research, including mechanistic insights into the interactions between nanoparticles and immune cells, biocompatibility problems, and the critical need for substantial in vivo investigations. The convergence of scientific discovery and clinical application will reshape the immunotherapy landscape in the coming years. The possibility of harnessing the capabilities of nanotechnology for focused, efficient, and patient-centered therapies is enticing.

References

Acharya S, Sahoo SK (2011) PLGA nanoparticles containing various anticancer agents and tumour delivery by EPR effect. Adv Drug Deliv Rev 63:170–183

Adams JL, Smothers J, Srinivasan R, Hoos A (2015) Big opportunities for small molecules in immuno-oncology. Nat Rev Drug Discov 14:603–622

Adriani G, de Tullio MD, Ferrari M et al (2012) The preferential targeting of the diseased microvasculature by disk-like particles. Biomaterials 33:5504–5513

Agarwal R, Jurney P, Raythatha M et al (2015) Effect of shape, size, and aspect ratio on nanoparticle penetration and distribution inside solid tissues using 3D spheroid models. Adv Healthc Mater 4:2269–2280

Agrawal B (2019) Heterologous immunity: role in natural and vaccine-induced resistance to infections. Front Immunol 10:2631

Ahamad N, Kar A, Mehta S et al (2021) Immunomodulatory nanosystems for treating inflammatory diseases. Biomaterials 274:120875

Aldossary HA, Rehman S, Jermy BR et al (2022) Therapeutic intervention for various hospital setting strains of biofilm forming Candida Auris with multiple drug resistance mutations using nanomaterial Ag-Silicalite-1 Zeolite. Pharmaceutics 14:2251

Alhazmi HA, Najmi A, Javed SA et al (2021) Medicinal plants and isolated molecules demonstrating immunomodulation activity as potential alternative therapies for viral diseases including COVID-19. Front Immunol 12:637553

Ali A, Zafar H, Zia M et al (2016) Synthesis, characterization, applications, and challenges of iron oxide nanoparticles. Nanotechnol Sci Appl 9:49–67

Allawadhi P, Singh V, Govindaraj K et al (2022) Biomedical applications of polysaccharide nanoparticles for chronic inflammatory disorders: focus on rheumatoid arthritis, diabetes and organ fibrosis. Carbohydr Polym 281:118923

Alrushaid N, Khan FA, Al-Suhaimi EA, Elaissari A (2023) Nanotechnology in cancer diagnosis and treatment. Pharmaceutics 15:1025

Alsharif N, Eshaghi B, BrM R, Brown KA (2020) Physiologically relevant mechanics of biodegradable polyester nanoparticles. Nano Lett 20:7536–7542

Amna T, Hassan MS, Gharsan FN, Rehman S, Sheikh FA (2022) Nanotechnology in drug delivery systems: ways to boost bioavailability of drugs. In: Nanotechnology for infectious diseases. Springer, pp 223–236

Andorko JI, Jewell CM (2017) Designing biomaterials with immunomodulatory properties for tissue engineering and regenerative medicine. Bioeng Transl Med 2:139–155

Ann Maria CG, Agnihotri AS, Fatima T, Hameed S (2023) Unlocking the antimicrobial, antifungal, and anticancer power of chitosan-stabilized silver nanoparticles. BioNanoScience 13:1933

Ansari MA, Thiruvengadam M, Farooqui Z et al (2021) Nanotechnology, in silico and endocrine-based strategy for delivering paclitaxel and miRNA: prospects for the therapeutic management of breast cancer. Semin Cancer Biol 69:109–128
Ansari MA, Tripathi T, Venkidasamy B et al (2023) Multifunctional nanocarriers for Alzheimer's disease: befriending the barriers. Mol Neurobiol 61:3042–3089
Arango Duque G, Descoteaux A (2014) Macrophage cytokines: involvement in immunity and infectious diseases. Front Immunol 5:491
Cagliani R, Gatto F, Bardi G (2019) Protein adsorption: a feasible method for nanoparticle functionalization? Materials 12:1991
Chakravarty M, Vora A (2021) Nanotechnology-based antiviral therapeutics. Drug Deliv Transl Res 11:748–787
Chen H-W, Huang C-Y, Lin S-Y et al (2016) Synthetic virus-like particles prepared via protein corona formation enable effective vaccination in an avian model of coronavirus infection. Biomaterials 106:111–118
Chen J, Ding J, Wang Y et al (2017) Sequentially responsive shell-stacked nanoparticles for deep penetration into solid tumors. Adv Mater 29:1701170
Cheng K, Kang Q, Zhao X (2020) Biogenic nanoparticles as immunomodulator for tumor treatment. Wiley Interdiscip Rev Nanomed Nanobiotechnol 12:e1646
Circulation T, Anselmo A, Zhang M et al (2015) Elasticity of nanoparticles in Fl fluences. ACS Nano 9:3169–3177
Dacoba TG, Olivera A, Torres D, Crecente-Campo J, Alonso MJ (2017) Modulating the immune system through nanotechnology. Semin Immunol 34:78–102
Dar R, Qazi P, Saba I et al (2015) Cytotoxic potential and molecular characterization of fungal endophytes from selected high value medicinal plants of the Kashmir Valley–India. Drug Res 66:121–125
Dar RA, Saba I, Shahnawaz M, Qazi PH, Khan IA (2017) Antimicrobial potential of fungal endophytes from selected high value medicinal plants of the Kashmir valley–India. J Psychopharmacol 6:307–310
De La Vega JC, Häfeli UO (2015) Utilization of nanoparticles as X-ray contrast agents for diagnostic imaging applications. Contrast Media Mol Imaging 10:81–95
Della Camera G, Liu T, Yang W et al (2022) Induction of innate memory in human monocytes exposed to mixtures of bacterial agents and nanoparticles. Int J Mol Sci 23:14655
Dhanda N, Thakur P, Kumar R et al (2023) Green-synthesis of Ni-Co nano ferrites using Aloe Vera extract: structural, optical, magnetic, and antimicrobial studies. Appl Organomet Chem 37:e7110
Dumortier H (2013) When carbon nanotubes encounter the immune system: desirable and undesirable effects. Adv Drug Deliv Rev 65:2120–2126
Dunn G, Bruce A, Ikeda H, Old LJ, Schreiber RD (2002) Cancer immunoediting: from immunosurveillance to tumor escape. Nat Immunol 3:991–998
Escamilla-Tilch M, Filio-Rodríguez G, García-Rocha R et al (2013) The interplay between pathogen-associated and danger-associated molecular patterns: an inflammatory code in cancer? Immunol Cell Biol 91:601–610
Finlay BB, McFadden G (2006) Anti-immunology: evasion of the host immune system by bacterial and viral pathogens. Cell 124:767–782
Fubini B, Ghiazza M, Fenoglio I (2010) Physico-chemical features of engineered nanoparticles relevant to their toxicity. Nanotoxicology 4:347–363
Gao S, Tang G, Hua D et al (2019) Stimuli-responsive bio-based polymeric systems and their applications. J Mater Chem B 7:709–729
Garg AD, Agostinis P (2017) Cell death and immunity in cancer: from danger signals to mimicry of pathogen defense responses. Immunol Rev 280:126–148
Gessner A, Lieske A, Paulke BR, Müller RH (2002) Influence of surface charge density on protein adsorption on polymeric nanoparticles: analysis by two-dimensional electrophoresis. Eur J Pharm Biopharm 54:165–170

Gómez GDM, Urcuqui-Inchima S, Hernandez LJC (2017) Silica nanoparticles induce NLRP3 inflammasome activation in human primary immune cells. Innate immun 23:697–708
Gordon JR, Ma Y, Churchman L, Gordon SA, Dawicki W (2014) Regulatory dendritic cells for immunotherapy in immunologic diseases. Front Immunol 5:7
Grego EA, Siddoway AC, Uz M et al (2020) Polymeric nanoparticle-based vaccine adjuvants and delivery vehicles. In: Nanoparticles for rational vaccine design, pp 29–76
Guo Z, Kubiatowicz LJ, Fang RH, Zhang L (2021) Nanotoxoids: biomimetic nanoparticle vaccines against infections. Adv Ther 4:2100072
Gupta AK, Gupta M (2005) Synthesis and surface engineering of iron oxide nanoparticles for biomedical applications. Biomaterials 26:3995–4021
Hale MM, Medina SH (2022) Biomaterials-enabled antithrombotics: recent advances and emerging strategies. Mol Pharm 19:4453–4465
Han T, Nag A, Mukhopadhyay SC, Xu Y (2019) Carbon nanotubes and its gas-sensing applications: a review. Sensors Actuators A Phys 291:107–143
He C, Hu Y, Yin L, Tang C, Yin C (2010) Effects of particle size and surface charge on cellular uptake and biodistribution of polymeric nanoparticles. Biomaterials 31:3657–3666
He M, Wang M, Xu T et al (2023) Reactive oxygen species-powered cancer immunotherapy: current status and challenges. J Control Release 356:623–648
Hsiao Y-P, Shen C-C, Huang C-H, Lin Y-C, Jan T-R (2018) Iron oxide nanoparticles attenuate T helper 17 cell responses in vitro and in vivo. Int Immunopharmacol 58:32–39
Huang C, Zhang Y, Yuan H, Gao H, Zhang S (2013) Role of nanoparticle geometry in endocytosis: laying down to stand up. Nano Lett 13:4546–4550
Irvine DJ, Hanson MC, Rakhra K, Tokatlian T (2015) Synthetic nanoparticles for vaccines and immunotherapy. Chem Rev 115:11109–11146
Jain RA (2000) The manufacturing techniques of various drug loaded biodegradable poly (lactide-co-glycolide)(PLGA) devices. Biomaterials 21:2475–2490
Janeway C Jr, Travers P Jr, Walport M (2001) Immunobiology: the immune system in health and disease, Immunological memory. Garland Science, New York
Jantan I, Ahmad W, Bukhari SNA (2015) Plant-derived immunomodulators: an insight on their preclinical evaluation and clinical trials. Front Plant Sci 6:655
Kandasamy G, Karuppasamy Y, Krishnan UM (2023) Emerging trends in nano-driven immunotherapy for treatment of cancer. Vaccine 11:458
Kasturi SP, Kozlowski PA, Nakaya HI et al (2017) Adjuvanting a simian immunodeficiency virus vaccine with Toll-Like receptor ligands encapsulated in nanoparticles induces persistent antibody responses and enhanced protection in TRIM5α restrictive macaques. J Virol 91:e01844–e01816
Khalili L, Dehghan G, Sheibani N, Khataee A (2022) Smart active-targeting of lipid-polymer hybrid nanoparticles for therapeutic applications: recent advances and challenges. Int J Biol Macromol 213:166–194
Khatun S, Putta CL, Hak A, Rengan AK (2023) Immunomodulatory nanosystems: an emerging strategy to combat viral infections. Biomater Biosyst 9:100073
Kim Y-C, Park J-H, Prausnitz MR (2012) Microneedles for drug and vaccine delivery. Adv Drug Deliv Rev 64:1547–1568
Kreft H, Jetz W (2007) Global patterns and determinants of vascular plant diversity. Proc Natl Acad Sci 104:5925–5930
Kulkarni N, Muddapur U (2014) Biosynthesis of metal nanoparticles: a review. J Nanotechnol 1: 510246
Kumar S, Anselmo AC, Banerjee A, Zakrewsky M, Mitragotri S (2015) Shape and size-dependent immune response to antigen-carrying nanoparticles. J Control Release 220:141–148
Landesman-Milo D, Peer D (2012) Altering the immune response with lipid-based nanoparticles. J Control Release 161:600–608
Lesterhuis WJ, Haanen JB, Punt CJ (2011) Cancer immunotherapy–revisited. Nat Rev Drug Discov 10:591–600

Lin D, Tian X, Wu F, Xing B (2010) Fate and transport of engineered nanomaterials in the environment. J Environ Qual 39:1896–1908
Lin X, Zuo YY, Gu N (2015) Shape affects the interactions of nanoparticles with pulmonary surfactant. Sci China Mater 58:28–37
Liu Y, Tan J, Thomas A, Ou-Yang D, Muzykantov VR (2012) The shape of things to come: importance of design in nanotechnology for drug delivery. Ther Deliv 3:181–194
Liu J, Liu Z, Pang Y, Zhou H (2022) The interaction between nanoparticles and immune system: application in the treatment of inflammatory diseases. J Nanobiotechnol 20:1–25
Lück M, Paulke BR, Schröder W, Blunk T, Müller R (1998) Analysis of plasma protein adsorption on polymeric nanoparticles with different surface characteristics. J Biomed Mater Res 39:478–485
Lundqvist M, Stigler J, Elia G, Lynch I, Cedervall T, Dawson KA (2008) Nanoparticle size and surface properties determine the protein corona with possible implications for biological impacts. Proc Natl Acad Sci 105:14265–14270
Luo Y-H, Chang LW, Lin P (2015) Metal-based nanoparticles and the immune system: activation, inflammation, and potential applications. Biomed Res Int 2015:143720
Mailander V, Landfester K (2009) Interaction of nanoparticles with cells. Biomacromolecules 10:2379–2400
Mie R, Samsudin MW, Din LB, Ahmad A, Ibrahim N, Adnan SNA (2014) Synthesis of silver nanoparticles with antibacterial activity using the lichen Parmotrema praesorediosum. Int J Nanomedicine 9:121
Moraes AS, Paula RF, Pradella F et al (2013) The suppressive effect of IL-27 on encephalitogenic Th17 cells induced by multiwalled carbon nanotubes reduces the severity of experimental autoimmune encephalomyelitis. CNS Neurosci Ther 19:682–687
Muntjewerff EM, Meesters LD, Van den Bogaart G (2020) Antigen cross-presentation by macrophages. Front Immunol 11:1276
Musetti S, Huang L (2018) Nanoparticle-mediated remodeling of the tumor microenvironment to enhance immunotherapy. ACS Nano 12:11740–11755
Nakanishi T, Kunisawa J, Hayashi A et al (1999) Positively charged liposome functions as an efficient immunoadjuvant in inducing cell-mediated immune response to soluble proteins. J Control Release 61:233–240
Owens DE III, Peppas NA (2006) Opsonization, biodistribution, and pharmacokinetics of polymeric nanoparticles. Int J Pharm 307:93–102
Pan X, Chen L, Liu S, Yang X, Gao J, Lee R (2009) Antitumor activity of G3139 lipid nanoparticles (LNPs). Mol Pharm 6(1):211–220
Peracchia M, Harnisch S, Pinto-Alphandary H et al (1999) Visualization of in vitro protein-rejecting properties of PEGylated stealth® polycyanoacrylate nanoparticles. Biomaterials 20:1269–1275
Ponnappan S, Ponnappan U (2011) Aging and immune function: molecular mechanisms to interventions. 14(8):1551–1585
Rahman MM, Islam MR, Akash S et al (2022) Recent advancements of nanoparticles application in cancer and neurodegenerative disorders: at a glance. Biomed Pharmacother 153:113305
Rehman S, Jermy R (2022) Methods of treatment using titanium and silver nanoparticles made with an intracellular extract of fomes fomentarious. Google Patents
Saba I, Wani K, Syed A, Rehman S (2021) Role of microbial nanotechnology in bioremediation of heavy metals. In: Microbial nanotechnology: green synthesis and applications, pp 307–319
Saba I, Wani K, Rehman S, Singh V (2022) Platform technologies based on virus-like particles (VLPs) for infectious diseases. In: Nanotechnology for infectious diseases. Springer, pp 541–553
Schirrmacher V (2019) From chemotherapy to biological therapy: a review of novel concepts to reduce the side effects of systemic cancer treatment. Int J Oncol 54:407–419
Sırıken B, Yavuz C, Güler A (2018) Antibacterial activity of Laurus nobilis: a review of literature. Med Sci Discov 5:374–379

Smith TT, Stephan SB, Moffett HF et al (2017) In situ programming of leukaemia-specific T cells using synthetic DNA nanocarriers. Nat Nanotechnol 12:813–820

Speiser DE, Chijioke O, Schaeuble K, Münz C (2023) CD4+ T cells in cancer. Nat Cancer 4:317–329

Svindland SC, Pedersen GK, Pathirana RD et al (2013) A study of Chitosan and c-di-GMP as mucosal adjuvants for intranasal influenza H5N1 vaccine. Influenza Other Respir Viruses 7:1181–1193

Swartzwelter BJ, Barbero F, Verde A et al (2020) Gold nanoparticles modulate BCG-induced innate immune memory in human monocytes by shifting the memory response towards tolerance. Cells 9:284

Tan J, Shah S, Thomas A, Ou-Yang HD, Liu Y (2013) The influence of size, shape and vessel geometry on nanoparticle distribution. Microfluid Nanofluid 14:77–87

ten Hagen TL, Seynhaeve AL, van Tiel ST, Ruiter DJ, Eggermont AM (2002) Pegylated liposomal tumor necrosis factor-α results in reduced toxicity and synergistic antitumor activity after systemic administration in combination with liposomal doxorubicin (Doxil®) in soft tissue sarcoma-bearing rats. Int J Cancer 97:115–120

Toy R, Hayden E, Shoup C, Baskaran H, Karathanasis E (2011) The effects of particle size, density and shape on margination of nanoparticles in microcirculation. Nanotechnology 22:115101

Toy R, Hayden E, Camann A et al (2013) Multimodal in vivo imaging exposes the voyage of nanoparticles in tumor microcirculation. ACS Nano 7:3118–3129

van Ommen JR, Valverde JM, Pfeffer R (2012) Fluidization of nanopowders: a review. J Nanopart Res 14:1–29

Varadé J, Magadán S, González-Fernández Á (2021) Human immunology and immunotherapy: main achievements and challenges. Cell Mol Immunol 18:805–828

Verma A, Stellacci F (2010) Effect of surface properties on nanoparticle–cell interactions. Small 6:12–21

Verma R, Purohit D, Jalwal P, Kaushik D, Pandey P (2022) Advancements in the field of oral, intravenous, and inhaled immunomodulators using nanotechnology. In: Immunomodulators and human health. Springer, pp 187–207

Vieira S, Vial S, Reis RL, Oliveira JM (2017) Nanoparticles for bone tissue engineering. Biotechnol Prog 33:590–611

Wang Q, Zhang P, Li Z et al (2019) Evaluation of polymer nanoformulations in hepatoma therapy by established rodent models. Theranostics 9:1426

Xu J, Dai W, Wang Z, Chen B, Li Z, Fan X (2011) Intranasal vaccination with chitosan-DNA nanoparticles expressing pneumococcal surface antigen a protects mice against nasopharyngeal colonization by Streptococcus pneumoniae. Clin Vaccine Immunol 18:75–81

Xu Z, Wang Y, Zhang L, Huang L (2014) Nanoparticle-delivered transforming growth factor-β siRNA enhances vaccination against advanced melanoma by modifying tumor microenvironment. ACS Nano 8:3636–3645

Zhang B, Sai Lung P, Zhao S, Chu Z, Chrzanowski W, Li Q (2017) Shape dependent cytotoxicity of PLGA-PEG nanoparticles on human cells. Sci Rep 7:7315

Nanomaterial-Based Vaccines

3

Insha Nahvi, Mohd Farhan, Suriya Rehman,
and Saif Hameed

3.1 Introduction

Immunization is considered the most economical way to combat infectious diseases. But nowadays conventional vaccines due to their low immunogenicity mostly provide partial protection only. Nanoparticle-based vaccinations are one of the most interested fields these days in medical community as they allow the space for modification and delivery of cellular environments. Recently, we have seen the major success of COVID-19 mRNA vaccines where nanotechnology was the most important factor in the design of mRNA vaccines for their widespread use in humans.

As of 2022, many infectious diseases are responsible for the death of millions of people worldwide. Taken together, these millions of people are dying due to new as

I. Nahvi (✉)
Department of Life Sciences, College of Science, King Faisal University,
Al Ahsa, Saudi Arabia

Department of Basic Sciences, Preparatory Year, King Faisal University,
Al Ahsa, Saudi Arabia

Amity Institute of Biotechnology, Amity University Haryana, Gurgaon, Haryana, India
e-mail: iahmad@kfu.edu.sa

M. Farhan
Department of Basic Sciences, Preparatory Year, King Faisal University,
Al Ahsa, Saudi Arabia

Department of Chemistry, College of Science, King Faisal University, Al Ahsa, Saudi Arabia

S. Rehman
Department of Epidemic Diseases Research, Institute for Research and Medical Consultations,
(IRMC), Imam Abdulrahman Bin Faisal University, Dammam, Saudi Arabia

S. Hameed (✉)
Amity Institute of Biotechnology, Amity University Haryana, Gurgaon, Haryana, India
e-mail: shameed@ggn.amity.edu

S. Rehman, E. Al-Suhaimi (eds.), *Nanotechnology Based Microbicides and Immune Stimulators*, https://doi.org/10.1007/978-981-96-0397-8_3

well as old infectious diseases with a high impact on healthcare sectors. The pursuit of effective vaccines stands as the cornerstone in combating the spread and devastating consequences of numerous infectious diseases. An ideal vaccine would be safe, stable, and capable of eliciting a potent immune response with minimal doses. Despite the prevalence of vaccines that utilize attenuated or inactivated whole organisms, several alternative vaccine approaches have emerged, highlighting promising immunogenic potential. Despite the challenges, subunit vaccines and third-generation vaccines represent a promising new approach to vaccination. They offer the potential to be safer, more effective, and longer lasting than traditional vaccines. As research in this area continues, subunit vaccines and third-generation vaccines may become more widely available and used to prevent a wider range of diseases. For instance, DNA and RNA vaccines are inexpensive and carry low risks of infection, but they are susceptible to degradation due to delivery changed to target sites. Most recently the global effort to develop a vaccine against COVID-19 took almost 2 years and three clinical trials to approve more than 30 vaccines (WHO 2021; WHO n.d.). In the formulations of several of these vaccinations, organic nanoparticles (NPs) are used to protect and transport the active ingredient (AS). The AS either encodes the viral antigen as a recombinant spike protein or as a nucleic acid (Pati et al. 2018; Roberts et al. 2015).

Adjuvants are frequently needed to boost the immunogenicity of protein-based vaccines, which have already been effectively employed to immunize against a variety of infectious illnesses, including acellular pertussis, tetanus, diphtheria, and pneumococcus. For the past few years, nanoparticles have been used to boost vaccine immunity. Due to their capacity to elicit potent host immunological responses, gold nanoparticles have demonstrated remarkable efficacy in the transport of both viral and bacterial antigens. Flu, HIV, foot and mouth disease, TB, and human immunodeficiency virus (HIV) have all been treated in vivo with gold nanoparticle vaccines. The low cost, great reproducibility, and good safety profiles of metallic and other inorganic nanoparticles make them particularly favorable for vaccine production processes.

A diverse array of nanoparticles possesses inherent physical properties that can trigger an immunological response. Notably, gold, carbon, dendrimers, polymers, and liposome nanoparticles have been shown to induce cytokine and antibody responses. These distinctive properties have paved the way for the utilization of nanoparticles as adjuvants, enhancing the immunogenicity of potential vaccines beyond their role as mere delivery systems. Nanoparticles employed for this purpose, commonly referred to as nano-immuno stimulators or activators (NIS), typically range between 20 and 100 nanometers in size.

Liposomes are spherical vesicles that are composed of a lipid bilayer. They are most frequently used type of nanoparticle (NP) for medical purposes, after PLGA nanoparticles. Liposomes can be used as both vaccination and drug delivery vehicles. When liposomes are placed in water, the hydrophobic tails will align themselves together and the hydrophilic heads will be exposed to the water. This forms a bilayer structure, with the hydrophobic tails in the middle and the hydrophilic heads on the outside. Liposomes can be used to deliver a variety of molecules, including antigens, drugs, and genes. They can also be used to target specific cells.

Liposomes can elicit both cellular and humoral responses. Cellular responses are those that involve the activation of immune cells. Humoral responses are those that involve the production of antibodies. In an in vivo model, it has been demonstrated that the introduction of PEGylated liposomes causes immunoglobulin M (IgM) molecules to react. Liposomes have the ability to deliver vaccines by merging with the target cell membrane in addition to their immunogenic potential. In contrast to other types of nanoparticles, hydrophilic molecules can easily be absorbed into liposomes due to their watery core, while hydrophobic compounds may be conceivably enclosed inside their phospholipid bilayer.

Nanocarriers are small particles that can be used to deliver drugs, genes, and other molecules to cells. They can be made from a variety of materials, including lipids, polymers, and proteins. Nanocarriers have several advantages over traditional delivery methods, including their ability to target specific cells and their ability to protect the delivered molecules from degradation.

When used to deliver antigens, nanocarriers can boost the immune response in a variety of ways. First, nanocarriers can protect the antigens from degradation by enzymes in the body. This allows the antigens to reach the immune cells in their native form, which is more likely to stimulate an immune response. Second, nanocarriers can target the antigens to specific immune cells. This can help to ensure that the antigens are presented to the right cells and that the immune response is directed against the desired target.

One of the important benefits of using nanocarriers to deliver antigens is that they can protect the antigens from early proteolytic degradation. Proteases are enzymes that break down proteins. When antigens are exposed to proteases, they can be broken down before they have a chance to reach the immune cells. This can prevent the immune system from mounting an effective response. Nanocarriers can protect the antigens from proteases by encapsulating them or by coating them with a layer that prevents the proteases from reaching the antigens.

Another benefit of using nanocarriers to deliver antigens is that they can enhance the specificity of antigen delivery to APCs. APCs are cells that present antigens to the immune system. When antigens are delivered to APCs in a non-specific manner, they can stimulate an immune response against a variety of cells, including healthy cells. This can lead to side effects. Nanocarriers can enhance the specificity of antigen delivery to APCs by targeting the antigens to specific receptors on the surface of the APCs. This ensures that the antigens are only presented to the cells that are capable of mounting an effective immune response. Nanoparticles (NPs) are very powerful in the disposition of medications to the target site. These drug carriers also prevent the drug from degradation and halt the damage of the healthier cells. Different NPs like metals, lipids, polymers, emulsions, etc. are being made nowadays. It has always been difficult to create vaccines without using live viruses that can produce a sustained and protective immune response. But it's now solved by biologically derived nanoparticles. Some of the examples of biologically derived

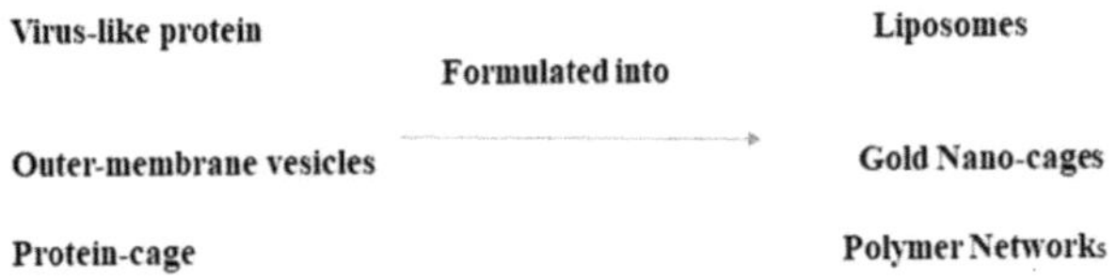

Fig. 3.1 Biologically derived NPs formulated into higher order structures like gold nanocages, liposomes, and polymer networks

NPs are virus-like particles, protein-cage, and outer-membrane vesicles (Fig. 3.1) (Nature Nanotechnology 2020).

Compared to vaccines based on entire pathogens, these biologically generated NPs have a higher safety profile and have been effectively transferred from the benchtop to the clinic thanks to their inherent features, potential for genetic engineering, and high flexibility. Several licensed vaccination platforms are now accessible to the general population. SARS-CoV-2, a virus that first appeared 3 years ago but has already resulted in almost 6.1 million fatalities, is the cause of the recent pandemic. The only means of controlling the pandemic situation is vaccination. Researchers are developing a variety of efficient nanoparticle strategies to tackle SARS-CoV-2. As of the GAVI SARS-CoV-2 vaccine alliance (until July 24, 2021), 17 vaccines have previously been approved for human use, 105 vaccines are currently undergoing various clinical phases, and 184 vaccines are currently in various preclinical stages (GAVI 2022). Quite recently, COVID-19 mRNA nanoparticle-based vaccines were introduced. They use messenger RNA (mRNA) to instruct cells in the body to produce a protein that is unique to SARS-CoV-2. The mRNA is encased in a nanoparticle, which is a tiny particle that helps to protect the mRNA and deliver it to cells. The main advantage of mRNA nanoparticle-based vaccines is that they can be produced quickly and easily. This is because mRNA can be synthesized in the laboratory. As a result, mRNA nanoparticle-based vaccines can be manufactured faster than traditional vaccines, which typically require growing viruses or bacteria in the laboratory.

Additionally, mRNA nanoparticle-based vaccines are very effective at triggering an immune response. This is because mRNA can be directly translated into protein by cells in the body, which means that the body does not need to take any extra steps to produce the SARS-CoV-2 protein. As a result, mRNA nanoparticle-based vaccines can induce a strong immune response with a relatively small dose of mRNA. mRNA nanoparticle-based vaccines are a promising new technology that is fast, effective, and safe. Two COVID-19 mRNA nanoparticle-based vaccines are authorized for use in the United States: the Pfizer-BioNTech vaccine and the Moderna vaccine.

The first vaccines based on mRNA-lipid nanoparticles (LNPs) (developed by PfizerBioNTech and Moderna) are now available for clinical use (GAVI 2022).

3.2 Types of Nanoparticles to Deliver Vaccines

Liposomes, virus-like particles, metallic nanoparticles, polymeric NPs, protein nanocages, inorganic NPs, emulsions, lipid NPs, exosomes, and nanogels are some of the varieties of NPs used to deliver vaccines.

3.2.1 Liposomes

Liposomes as NPs consist of lipids with hydrophilic head and hydrophobic tail (Smith et al. 2015). Additionally, liposomes may delay the deterioration of antigens, improve their uptake by APCs, encourage their release into the cytosol, and even activate the immune system. Charge, size, and lipid composition determine immune response that liposomes can elicit (Wang et al. 2019). Liposomes can mimic the properties of pathogens by coating them with molecules that are found on the surface of pathogens. For example, liposomes can be coated with proteins that are found on the surface of bacteria. This allows the liposomes to be recognized by the immune system as if they were pathogens. The type of immune response that is induced by liposomes depends on the type of lipids that are used to make them. Unsaturated lipids tend to induce Th2 responses, while saturated lipids tend to promote Th1-type immune responses. Th2 responses are associated with the production of antibodies, while Th1-type immune responses are associated with the activation of immune cells. Cationic liposomes deliver subunit vaccines against pathogens, including HIV, influenza, and malaria. They have also been used to deliver subunit vaccines against cancer cells. Cationic liposomes are a promising new approach for delivering subunit vaccines. They offer numerous advantages over traditional delivery methods, including their ability to increase the uptake of the vaccines by APCs and their ability to target the vaccines to specific cells (Wui et al. 2021; Taneichi et al. 2006). It has been noticed that liposome-based vaccines are gentle as compared to traditional vaccines. In a study by Mai Y et al. in 2020, when given liposome-based vaccines, mice responded quickly to antigens like diphtheria toxoid and did not experience the negative side effects of conventional vaccine additives (Joseph et al. 2006).

Liposomes, tiny bubbles made from natural fats can smuggle antigens in many ways. One way is to simply encapsulate the antigens in the liposomes and then inject the liposomes into the body. The liposomes will then be taken up by the immune cells, and the antigens will be released. Another way to deliver antigens using liposomes is to attach the antigens to the surface of the liposomes. This can be done by using a variety of methods, such as covalent bonding or electrostatic interactions. When the liposomes are taken up by the immune cells, the antigens will be displayed on the surface of the cells, which will help to stimulate an immune response.

Liposomes can also be used to deliver adjuvant molecules, which are substances that help to boost the immune response. Adjuvant molecules can be added to the liposomes during their preparation, or they can be added after the liposomes have been prepared. Adjuvant molecules can be used to increase the number of immune cells that are activated by the antigens, and they can also help to increase the production of antibodies.

The combination of multiple antigens and adjuvant molecules in liposomes can lead to a more effective immune response. This is because the different antigens can stimulate different types of immune cells, and the adjuvant molecules can help to boost the immune response. Additionally, the antigens can be displayed on the surface of the liposomes in a highly multivalent manner, which means that they can bind to multiple immune cells at the same time. This can result in effective and quick immune response. One example of adjuvant systems based on liposomes is malaria vaccines which have turned out to be very successful (Mai et al. 2020). Liposomal delivery systems have been used to develop a number of vaccines. For example, liposomes have been used to develop vaccines against hepatitis B, influenza, and human papillomavirus. Liposomes are also being investigated as a potential delivery system for vaccines against HIV and malaria (Van Der Meeren et al. 2018; Ingale et al. 2016). The only virosomal vaccines like hepatitis A, influenza, and flu vaccines are liposome NPs with commercial human use approval (Kheirollahpour et al. 2019).

3.2.2 Virus-Like Particles

Vaccines based on virus-like particles (VLPs) have an empty core that can be used to carry medications or antigens for targeted delivery, and they are synthetic and biodegradable with a size of 80–150 nm. VLPs are made up of proteins that are derived from viruses. The proteins are assembled into particles that have the same shape and structure as the virus. VLPs cannot cause infection (Fig. 3.2). VLPs have been made from many different virus families, they can present multivalent antigens by preserving the antigens' original conformation. VLPs are desirable prospects for continued development as vaccines like hepatitis B vaccine and the HPV vaccine. VLPs were used to design a vaccine against HIV. The VLPs were engineered to express the HIV Env protein. The VLPs were able to activate B cells that express the bnAb VRC01, which is a broadly neutralizing antibody that can block HIV infection. This study suggests that VLPs could be a potential candidate for developing a vaccine against HIV. VLPs are safe, easy to produce, and they can be engineered to express specific antigens. Further research is needed for VLP-based HIV vaccines, but they represent a promising new approach to preventing HIV infection (Zwick et al. 2017). Due to their ease of manufacturing and ability to elicit potent immune responses, VLPs are generating a lot of interest. In order to eliminate the virus' pathogenicity and ensure that it has no negative impact on host cells, the genetic coding for integrase and the virus' RNA are deleted. Bioconjugate VLPs can also be created by pharmacologically binding nonprotein antigens to the membrane of virus (Patel and Swartz 2011). VLPs are a good safety profile and make low-containment manufacture possible. Moreover, when grown in bacterial expression systems, VLPs naturally contain bacterial nucleic acids, which further activate B cells and

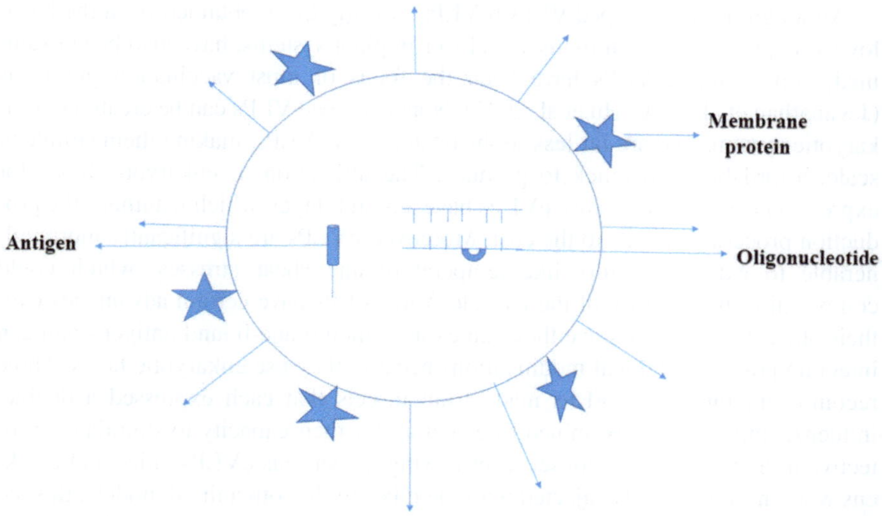

Fig. 3.2 Virus-like particle (VLP)

antigen-presenting cells. Insect viruses, plant viruses, bacteriophages, and other viruses are utilized as platforms for VLP vaccines. The tiny genomes of RNA viruses make them popular VLP platforms because they make it simple to produce novel antigens on their surfaces. One technique for creating VLP-based vaccinations is when VLPs are chemically conjugated to antigen. The study by Basu et al. showed that it is possible to covalently attach ZIKV epitopes to Qβ using a hetero-bifunctional crosslinker. This strategy was able to induce immunity in mice by producing antibodies that neutralized the virus in vitro. However, the mice were not protected from a Zika challenge when evaluated in a mouse model. This suggests that further research is needed to optimize this strategy for use as a vaccine (Yadav et al. 2021).

The use of several VLP-based vaccinations against HBV and HPV has also been authorized. Since the recombinant HBV vaccine is the first VLP-based vaccination to be commercially available, it represents a significant accomplishment (Basu et al. 2018). Recombinant HBsAg and yeast and mammalian cells are used in the vaccine's synthesis of immunogenic VLPs. The selection of the host cell has an impact on both the production process and the VLP structure (Vassileva et al. 2001). Over 15 human papillomavirus varieties are known to cause cervical cancer, and the 16 and 18 HPV types are thought to be responsible for 70% of occurrences worldwide. Because of this, approved VLP-based HPV vaccinations include both HPV-16 and HPV-18 VLPs. In 2006, Gardasil became the first VLP-based HPV vaccination to receive FDA approval (Vassileva et al. 2001). Keeping in view the recent pandemic of COVID-19, the spotlight has turned to innovative vaccine approaches. Among these are virus-like particles (VLPs), offering a promising alternative to conventional vaccines (Roldão et al. 2010).

Although lipid-enveloped VLPs (eVLPs), mostly those obtained from the baculovirus expression system in insect cells or in plant systems, have also been examined, nonenveloped VLPs have been the focus of most vaccination platforms (Jeyanathan et al. 2020; Hu et al. 2022). Nonenveloped VLPs can be created in prokaryotic systems and are far less sophisticated than eVLPs, making them simple to scale, affordable, and quick to produce. The utilization of eukaryotic hosts for expression is required because eVLPs have a lipid bilayer, which lengthens the production process and adds to the cost. Moreover, eVLPs are significantly more vulnerable to external factors like temperature and shear stresses, which could compromise the integrity of the particle. Yet, eVLPs have certain advantages over their other counterparts since they can express membrane-bound antigens and can integrate post-translational modifications because they use eukaryotic hosts. Three recombinant influenza eVLPs made from insects that each expressed a distinct influenza antigen were examined by Hu et al. for their capacity to stimulate a protective immunological response. After mixing the various eVLPs, mice and chickens were intramuscularly injected with the mixture. In both animal models, this led to a potent antibody response against a deadly virus challenge (Jeyanathan et al. 2020).

3.2.3 Polymeric NPs

Traditional vaccines often rely on weakened or inactivated viruses, which can raise safety concerns. Enter the innovative world of polymeric materials, a promising alternative offering a protein-free approach to vaccination. Unlike complex protein production, polymeric materials are easy to synthesize and fabricate, making them cost-effective and scalable. These platforms generally don't trigger unwanted immune responses against the material itself, ensuring a safer experience. Already approved for human use in drugs, PLGA is a versatile and well-studied polymer offering controlled release and biodegradability. PGA is a natural polymer produced by bacteria that boasts excellent biocompatibility and targeting capabilities for specific cells. Polymeric nanoparticles (NPs) are tiny spheres crafted from these materials, holding immense potential for vaccine delivery:

- Biodegradability: They break down naturally within the body, minimizing safety concerns.
- Stability: They can store and protect vaccine antigens for long periods, allowing for easier storage and transportation.
- Safety: Their gentle nature reduces potential side effects.
- Simplicity: They're relatively easy to modify and customize for specific needs.
- Biocompatibility: They play nice with our biological systems, minimizing adverse reactions.
- Predictability: Their behavior can be fine-tuned and controlled for optimal performance.

Also, polymeric nanomaterials have adjuvant effects. They can boost the immune response to the vaccine, requiring lower doses and potentially broader protection. Their delayed release can gradually release the vaccine over time, mimicking natural infection and potentially leading to longer-lasting immunity (Dai et al. 2018). Hyaluronic acid, chitosan, and alginate are the three most commonly used natural nanoparticles; research is currently focused on conjugating animal venom toxins to these NPs (Mehrabi et al. 2018). These natural NPs can improve cellular interactions, targeted therapy, and drug delivery efficiency. Synthetic polymeric NPs, on the other hand, degrade slowly and can be used to achieve goals such as encapsulating antigens for delivery or maintaining antigen release. These polymeric NPs can be made using a range of polymers to create vesicles that can either contain or display antigens. The process of double emulsion-solvent evaporation is frequently used to make these NPs (Lou et al. 2019). A primary emulsion can be created by dissolving the polymer in an organic solvent, adding the antigen, and then swirling the mixture. By adding an emulsifying agent, it is transformed into a water-in-oil-in-water type emulsion, precipitating the polymer around the antigen. Another widely utilized technique for creating synthetic polymer nanoparticle-based vaccination systems is nanoprecipitation (Lou et al. 2019).

Another application of polymers as nanostructured vaccines are PEM (polyelectrolyte multilayers). The electrostatic attraction between the layers drives the assembly process, and the number of layers deposited can control the thickness of the film. Several studies have demonstrated the potential of PEMs for vaccine delivery. For example, one study showed that PEMs co-delivered with a hepatitis B virus antigen induced a strong immune response in mice. Another study showed that PEM technology has crafted a DNA vaccine against the flu's spike protein and mice receiving this vaccine were well-equipped to fend off the real virus challenge (Lou et al. 2019).

PEMs are a promising new technology for vaccine delivery with several potential advantages. Further research is needed to optimize PEMs for vaccine delivery and to evaluate their efficacy in humans.

One group of researchers described a microneedle array that stimulated a powerful CD8+ T cell response and antibody responses in mice after treatment that were ten times greater than those elicited by injection or electroporation (Bilati et al. 2005). Polyethyleneimine (PEI), which is widely utilized in gene therapy applications, is another significant polymer in polymeric NPs (DeMuth et al. 2013). Depending on the intended uses, cytokines and ligands may also be used in these processes to modify vaccinations. PEI-based vaccines are currently being developed to treat a variety of cancers and infections, with promising results (Shen et al. 2017).

3.2.4 Protein Nanocages

Protein nanocages are symmetrical, macromolecular enclosures that are formed by the "self-assembly" of numerous protein subunits, which spontaneously arrange themselves into specific patterns. These traits frequently enable them to imitate native diseases because they have high-density repetitive patches on their surfaces that B-cell receptors may be able to identify and crosslink. Protein subunit vaccines may also be thought of as nanocages, which are analogous to VLPs. Ferritin is a ubiquitous protein that is found in all living cells. It is responsible for storing iron, and it does so by forming a hollow, spherical cage with a diameter of about 12 nanometers. The cage is made up of 24 subunits, and each subunit is composed of two polypeptide chains. Ferritin-based nanocages have a number of advantages over traditional vaccine delivery methods. First, ferritin-based nanocages are very stable and can be easily stored and transported. Second, they are non-immunogenic, which means that they do not trigger an immune response when they are injected into the body. Third, ferritin-based nanocages can be easily loaded with antigens, which are the proteins that the body learns to recognize as foreign during vaccination. Ferritin-based nanocages have been shown to be effective in a number of animal studies. For example, they have been shown to be effective in protecting mice from influenza, hepatitis B, and HIV (Wei et al. 2020). Some of the potential benefits of using ferritin-based nanocages for vaccine development are: (a) Improved safety: Ferritin-based nanocages are non-immunogenic, which means that they do not trigger an immune response when they are injected into the body. This can help to minimize the challenge of side effects, (b) Increased efficacy: Ferritin-based nanocages can be easily loaded with antigens, which are the proteins that the body learns to recognize as foreign during vaccination, (c) Improved delivery: Ferritin-based nanocages are small and can easily be transported to the target cells in the body. This can help to improve the effectiveness of the vaccine.

Overall, ferritin-based nanocages are a promising new platform for vaccine development. They have a number of advantages over traditional vaccine delivery methods. Being versatile makes engineered ferritin an important vaccine platform for displaying antigens from various pathogens like influenza, rotavirus, hepatitis B, and SARS-CoV-2 (Wang et al. 2020; Wuertz et al. 2021).

Several ferritin-based vaccines are undergoing clinical trials nowadays. Investigations on animals have shown that ferritin nanocages demonstrated impressive correlates of protection.

3.2.5 Inorganic Metal and Nonmetal Nanoparticles

Inorganic nanoparticles are used in drug delivery, TB, and HIV vaccine development. Vaccines become more stable when they are carried by gold, silver, or silica NPs delivery system. Unveiling the size-shape dance: Inorganic NPs act as key players in optimized vaccine delivery (Yassine et al. 2015). With their ability to manipulate size, shape, and surface characteristics, gold nanoparticles have also generated interest as vaccines. For instance, using the sexual stage antigen Pfs25 linked to gold Nps, Kumar and colleagues generated vaccinations that inhibit the transmission of malaria and this study suggested that gold nanoparticles may develop effective vaccines against malaria (Niikura et al. 2013). In addition, gold nanoparticles that have been coated with viral glycans maybe helpful in delivering vaccines to dendritic cells. By targeting dendritic cells with vaccines, it is possible to elicit a more robust immune response. This makes it easier for T cells to present antigens to them, resulting in a stronger cellular antiviral immune response (Kumar et al. 2015). Also, one more study indicated that polystyrene beads with immobilized antigen altered the ratio of type 1/type 2 cytokines. According to Maysinger et al., microglial cells' inflammatory response depends on their morphology (Climent et al. 2017). In a study, researchers investigated the effect of different nanoparticle shapes on the immune response. They found that nanourchins, which are a type of nanoparticle with a spiky surface, exposure led to a surge in interleukin-1 (IL-1) production, potentially amplifying inflammatory processes in the body. Spherical and rod-shaped nanoparticles did not have the same effect. It is believed that the spiky surface of the nanourchins is responsible for their ability to increase IL-1 production. The spikes may damage cells and tissues, which can trigger an immune response. This study suggests that the shape of nanoparticles can have a considerable effect on the immune response. Nanoparticles with a spiky surface, such as nanourchins, may be more effective at stimulating the immune system than nanoparticles with other shapes. This could be useful for developing new vaccines and treatments for diseases (Hutter et al. 2010).

Carbon nanoparticles are a type of nanoparticle that can be made from a variety of carbon-based materials, such as graphite, carbon nanotubes, and fullerenes. They have a number of advantages for vaccine development, including being small and easily transported to the target cells in the body, easily modified to carry different antigens, and can be used to target particular cells or tissues. Carbon nanoparticles could be used to deliver a vaccine against hepatitis B.

Carbon nanoparticles can be made into different shapes, including nanotubes and mesoporous spheres. The different shapes of carbon nanoparticles can also be used to target different cells or tissues in the body. For example, nanotubes can be used to target cancer cells, while mesoporous spheres can be used to target cells in the liver (Bachmann and Jennings 2010). The use of carbon and silica nanoparticles for vaccine delivery is a promising new approach to vaccine development. These nanoparticles are created by deliberately changing their structural properties to change how they interact with cells. Calcium phosphate nanoparticles are an additional kind of inorganic nanoparticles. They can be made between 50 and 100 nm in size and are non-toxic (Wang et al. 2011).

Other than above-mentioned varieties of nanoparticles, various other nanomaterials like emulsions, lipid nanoparticles, exosomes, etc. are also in use for the formulation of vaccines. MF59TM, an approved oil-in-water emulsion, is frequently used in the manufacture of vaccines (Ingle et al. 2018; He et al. 2002). For use in influenza vaccinations, this vaccine adjuvant has undergone thorough testing and is very efficient. Lipid-based nanoassemblies have been utilized to carry medications as well as DNA and RNA, and lipid-based nanoparticles represent the least risk for in vivo applications. Lipid nanoparticles are now the most promising medication delivery system for oral administration (Kheirollahpour et al. 2019).

3.3 Mechanisms Enhancing Vaccination Responses with Nanomaterials

The engineering of nanomaterials to important immune system cells and tissues is a vital component in the building of vaccines against HIV, malaria, and TB. Conventional drug delivery applications typically focus on delivering a drug to a specific cell type. For example, a drug that is used to treat cancer might be targeted to cancer cells. This can be done by attaching a ligand to the drug that binds to a receptor on the cancer cell. To trigger a strong immune response, a successful immunogenic nanomaterial vaccine requires effective engagement with a diverse array of cell populations (Das and Chaudhury 2011). This is because the immune system is a complex system that involves many different cells like antigen-presenting cells (APCs), macrophages, T cells, and B cells, working together. In order to be immunogenic, a nanomaterial vaccine needs to interact with all of these cells. The nanoparticles can also be designed to release antigens slowly, which can help to keep the immune system activated for a longer period of time (Das and Chaudhury 2011).

Antigens can be significantly more persistent in the bloodstream, in lymphoid tissues, or inside APCs by being encapsulated or conjugated within nanomaterials. It has long been understood that increasing an antigen's persistence can increase its immunogenicity, and several nanotechnological methods have been investigated to accomplish this. Demento et al. used a protein called ovalbumin and tiny plastic particles named PLGA to show how slowly releasing antigens can affect the immune system (Fries et al. 2021). Improved immunogenicity was made possible by the prolonged antigen release, and the PLGA nanoparticles' rate of degradation could be controlled to further prolong the ovalbumin release.

The immune system has two main ways of fighting off infection. Humoral immunity, a key arm of our defense system, relies on antibodies, specialized proteins that bind to and neutralize antigens, the foreign molecules triggering immune responses. Cell-mediated immunity is mediated by T cells, which help to kill cells that are infected with viruses or bacteria. In order to develop a protective immune response, the body needs to produce antibodies that are specific to the antigens of the infectious agent. This process is called antibody affinity maturation. Antibody affinity maturation is a process of natural selection in which antibodies that are more effective at binding to the antigens are more likely to survive and reproduce. The development of high-affinity antibodies is a slow process that can take several weeks or even months. However, the process can be accelerated by using nanoparticles as vaccines. Nanoparticles can help to increase the number of antigens that are presented to the immune system, which can help to speed up the process of antibody affinity maturation. Nanoparticles can help to improve antigen presentation by APCs (Demento et al. 2012). Nanoparticles can be coated with antigens, which can help to increase the number of antigens that are presented to APCs. This can help to improve T cell activation and the development of a strong cellular immune response.

The development of depots at injection sites can also help to boost the defensive mechanism of vaccines. Depots are areas of the body where the vaccine is

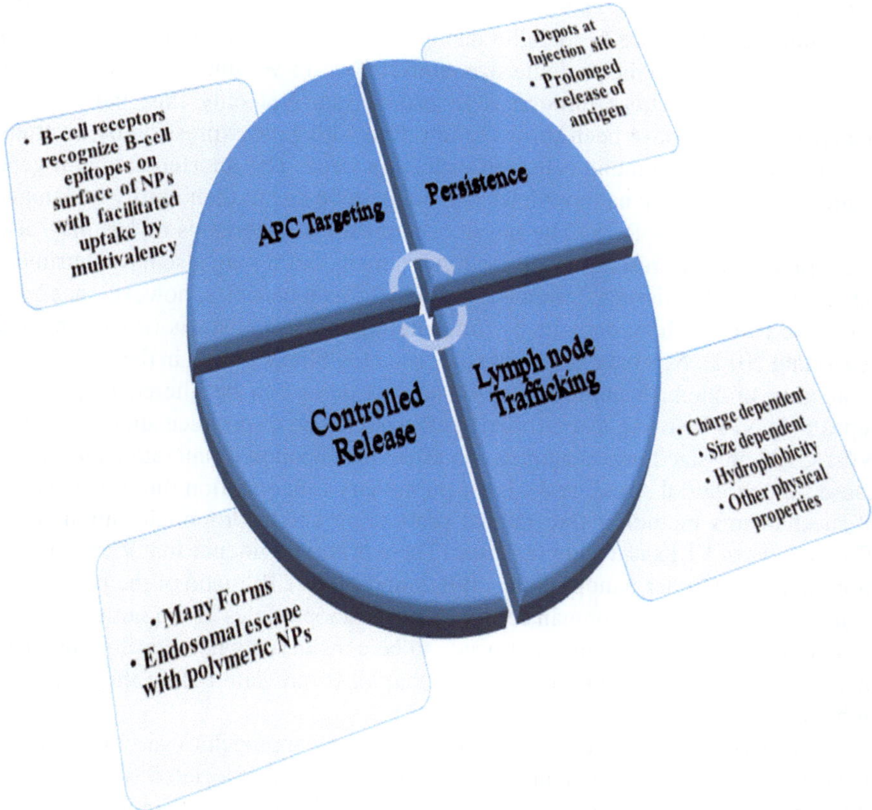

Fig. 3.3 Mechanisms to improve nanomaterial vaccine delivery

concentrated. This can help to increase the amount of time that the vaccine is available to the immune system, which can help to boost the immune mechanism. The development of depots can be achieved by using nanoparticles as vaccines. The combination of nanoparticles and depots can help to improve the immune response to vaccines. This can help to develop more effective vaccines for infectious agents (Fig. 3.3).

3.4 Risks/Side Effects and Nanocytotoxicity

Although nanoparticles work uniquely with particular organs or cells, there are some issues associated with them. The most important aspect influencing a nanoparticle's cytotoxicity is the type of cell it is in. One of the issues is that they can easily pass lipid bilayers to enter the human body and may interact with sensitive organs (Reddy et al. 2007). Metallic nanoparticles have an impact on innate immunity as well. Metallic nanoparticles have been shown to be cytotoxic and genotoxic to a variety of cells, including human cells, animal cells, and bacteria. The multifaceted nature of metallic nanoparticle cytotoxicity and genotoxicity is believed to stem from their minute size, high surface-to-volume ratio, and potential to elicit oxidative stress through ROS production. Not only this, but metallic nanoparticles also have been shown to interfere with gene expression by a variety of cells, including immune cells and cancerous cells. The interference with gene expression by metallic nanoparticles is thought to be due to their ability to damage cells and to generate ROS. The physicochemical characteristics of NP may also encourage antigen-presenting cells to bind to them, but it may also have detrimental effects on other tissues. Intranasal delivery of nanoparticles, however, has been shown to result in lung damage by inducing mitochondrial stress (Pourmand and Abdollahi 2012). NPs may group together and clog blood vessels in the host, which is another problem. Manufacturing of nanoparticles can be altered to generate repeatable NPs by using a controlled synthesis process to produce almost uniform NPs. < 100 nm-sized nanomaterials can affect the function of lateral/distal organs, cause mitochondrial stress, and trigger pulmonary inflammation through a variety of mechanisms including free radical production and hydrophobic interactions. The toxicity of VLPs is a complex issue. There is some evidence that VLPs may be more toxic than other nanoparticles. It is found that VLPs made of the hepatitis B virus were more toxic to human cells than gold nanoparticles of the same size. The toxicity of VLPs is a concern that needs to be considered when developing VLP-based vaccines. It is important to ensure that VLPs are safe before they are used in humans.

Nanoparticle-based vaccines are a promising new approach to vaccine development, but there are some potential risks and side effects associated with their use other than toxicity. Some of these risks are immune reactions where nanoparticles can sometimes trigger an immune response, even if they are not specifically designed to do so. This can lead to side effects such as inflammation, redness, and pain at the injection site. In some cases, it can also lead to more serious side effects, such as allergic reactions. Also, in biodistribution where nanoparticles can sometimes accumulate in unintended parts of the body, such as the liver or spleen. This can lead to side effects, such as liver damage or spleen enlargement.

3.5 Conclusion

NP-based vaccines represent a groundbreaking advancement in vaccine technology. These novel vaccines offer several advantages over conventional vaccines, including the ability to safeguard vaccine antigens from degradation, enhance antigen uptake and processing by antigen-presenting cells (APCs), and possess adjuvant capabilities. Additionally, nanoparticle-based vaccines can be administered via mucosal or other alternative routes of immunization. These vaccines have demonstrated the ability to induce effective mucosal and cellular cross-protective immunity, providing broader protection against a wider range of pathogens. Notably, nanoparticle-based vaccines have exhibited efficacy in animal models of various diseases, including cancer, viral infections, and bacterial infections. Various nanoscale platforms, such as liposomes, virus-like particles, and polymeric nanoparticles, are important in the vaccine market. However, more research is needed to develop NPs-based vaccines that are safe and effective in humans. Nanoparticle-based vaccines emerge as a revolutionary paradigm in vaccine development, holding the promise of enhancing vaccine efficacy and safety. While further research is warranted to translate this technology into safe and effective human vaccines, the potential benefits of nanoparticle-based vaccines are substantial.

References

Bachmann MF, Jennings GT (2010) Vaccine delivery: a matter of size, geometry, kinetics and molecular patterns. Nat Rev Immunol 10:787–796

Basu R, Zhai L, Contreras A, Tumban E (2018) Immunization with phage virus-like particles displaying Zika virus potential B-cell epitopes neutralizes Zika virus infection of monkey kidney cells. Vaccine 36:1256–1264. https://doi.org/10.1016/j.vaccine.2018.01.056

Bilati U, Allémann E, Doelker E (2005) Development of a nanoprecipitation method intended for the entrapment of hydrophilic drugs into nanoparticles. Eur J Pharm Sci 24:67–75

Climent N et al (2017) Loading dendritic cells with gold nanoparticles (GNPs) bearing HIV-peptides and mannosides enhance HIV-specifc T cell responses. Nanomedicine 14:339–351

Dai S, Zhang T, Zhang Y, Wang H, Deng F (2018) Zika virus baculovirus-expressed virus-like particles induce neutralizing antibodies in mice. Virol Sin 33:213–226. https://doi.org/10.1007/s12250-018-0030-5

Das S, Chaudhury A (2011) Recent advances in lipid nanoparticle formulations with solid matrix for oral drug delivery. AAPS PharmSciTech 12:62–76

Demento SL et al (2012) Role of sustained antigen release from nanoparticle vaccines in shaping the T cell memory phenotype. Biomaterials 33:4957–4964

DeMuth PC et al (2013) Polymer multilayer tattooing for enhanced DNA vaccination. Nat Mater 12:367–375

Fries CN, Curvino EJ, Chen J-L, Permar SR, Fouda GG, Collier JH (2021) Advances in nanomaterial vaccine strategies to address infectious diseases impacting global health. Nat Nanotechnol 16:385–398

GAVI (2022) The COVID-19 vaccine race; VaccinesWork, 2022. https://www.gavi.org/vaccineswork/covid-19-vaccine-race. Accessed 1 Feb 2022

He Q, Mitchell A, Morcol T, Bell SJD (2002) Calcium phosphate nanoparticles induce mucosal immunity and protection against herpes simplex virus type 2. Clin Vaccine Immunol 9:1021–1024

Hu J, Zhang Q, Peng P, Li R, Li J, Wang X et al (2022) Baculovirus-derived influenza virus-like particle confers complete protection against lethal H7N9 avian influenza virus challenge in chickens and mice. Vet Microbiol 264:109306. https://doi.org/10.1016/j.vetmic.2021.109306

Hutter E, Boridy S, Labrecque S, Lalancette-Hébert M, Kriz J, Winnik FM, Maysinger D (2010) Microglial response to gold nanoparticles. ACS Nano 4:2595–2606

Ingale J et al (2016) High-density array of well-ordered HIV-1 spikes on synthetic liposomal nanoparticles efciently activate B cells. Cell Rep 15:1986–1999

Ingle SG, Pai RV, Monpara JD, Vavia PR (2018) Liposils: an effective strategy for stabilizing paclitaxel loaded liposomes by surface coating with silica. Eur J Pharm Sci 122:51–63

Jeyanathan M, Afkhami S, Smaill F, Miller MS, Lichty BD, Xing Z (2020) Immunological considerations for COVID-19 vaccine strategies. Nat Rev Immunol 20:615–632

Joseph A, Itskovitz-Cooper N, Samira S, Flasterstein O, Eliyahu H, Simberg D et al (2006) A new intranasal influenza vaccine based on a novel polycationic lipiddceramide carbamoyl-spermine (CCS): I. Immunogenicity and efficacy studies in mice. Vaccine 24(18):3990

Kheirollahpour M, Mehrabi M, Dounighi NM, Mohammadi M, Masoudi A (2019) Nanoparticles and vaccine development. Pharm Nanotechnol 8:6–21

Kumar R et al (2015) Nanovaccines for malaria using plasmodium falciparum antigen Pfs25 attached gold nanoparticles. Vaccine 33:5064–5071

Lou B, De Beuckelaer A, Boonstra E, Li D, De Geest BG, De Koker S, Mastrobattista E, Hennink WE (2019) Modular core-shell polymeric nanoparticles mimicking viral structures for vaccination. J Control Release 293:48–62

Mai Y, Guo J, Zhao Y, Ma S, Hou Y, Yang J (2020) Intranasal delivery of cationic liposome-protamine complex MRNA vaccine elicits effective anti-tumor immunity. Cell Immunol 354:104143

Mehrabi M, Dounighi NM, Rezayat SM, Doroud D, Amani A, Khoobi M, Ajdary S (2018) Novel approach to improve vaccine immunogenicity: mannosylated chitosan nanoparticles loaded with recombinant hepatitis B antigen as a targeted vaccine delivery system. J Drug Deliv Sci Technol 44:19–26

Nature Nanotechnology (2020) Nanomedicine and the COVID-19 vaccines. Nat Nanotechnol 15(12):963

Niikura K, Matsunaga T, Suzuki T, Kobayashi S, Yamaguchi H, Orba Y, Kawaguchi A, Hasegawa H, Kajino K, Ninomiya T (2013) Gold nanoparticles as a vaccine platform: influence of size and shape on immunological responses in vitro and in vivo. ACS Nano 7:3926–3938

Patel KG, Swartz JR (2011) Surface functionalization of virus-like particles by direct conjugation using Azide-alkyne click chemistry. Bioconjug Chem 22:376–387

Pati R, Shevtsov M, Sonawane A (2018) Nanoparticle vaccines against infectious diseases. Front Immunol 9. https://doi.org/10.3389/fimmu.2018.02224

Pourmand A, Abdollahi M (2012) Current opinion on nanotoxicology. DARU J Pharm Sci 20:95

Reddy ST et al (2007) Exploiting lymphatic transport and complement activation in nanoparticle vaccines. Nat Biotechnol 25:1159–1164

Roberts RA, Eitas TK, Byrne JD, Johnson BM, Short PJ, McKinnon KP et al (2015) Towards programming immune tolerance through geometric manipulation of phosphatidylserine. Biomaterials 72:1e10

Roldão A, Mellado MCM, Castilho LR, Carrondo MJ, Alves PM (2010) Virus-like particles in vaccine development. Expert Rev Vaccines 9:1149–1176

Shen C, Li J, Zhang Y, Li Y, Shen G, Zhu J, Tao J (2017) Polyethylenimine-based micro/nanoparticles as vaccine adjuvants. Int J Nanomedicine 12:5443

Smith JD, Morton LD, Ulery BD (2015) Nanoparticles as synthetic vaccines. Curr Opin Biotechnol 34:217e2

Taneichi M, Ishida H, Kajino K, Ogasawara K, Tanaka Y, Kasai M et al (2006) Antigen chemically coupled to the surface of liposomes are cross-presented to CD8þ T cells and induce potent antitumor immunity. J Immunol 177(4):2324

Van Der Meeren O et al (2018) Phase 2b controlled trial of M72/AS01 EVaccine to prevent tuberculosis. N Engl J Med 379:1621–1634

Vassileva A, Chugh DA, Swaminathan S, Khanna N (2001) Effect of copy number on the expression levels of hepatitis B surface antigen in the methylotrophic yeast Pichia Pastoris. Protein Expr Purif 21:71–80

Wang T, Zou M, Jiang H, Ji Z, Gao P, Cheng G (2011) Synthesis of a novel kind of carbon nanoparticle with large mesopores and macropores and its application as an oral vaccine adjuvant. Eur J Pharm Sci 44:653–659

Wang N, Chen M, Wang T (2019) Liposomes used as a vaccine adjuvant-delivery system: from basics to clinical immunization. J Control Release 303:130–150

Wang W, Zhou X, Bian Y, Wang S, Chai Q, Guo Z et al (2020) Dual-targeting nanoparticle vaccine elicits a therapeutic antibody response against chronic hepatitis B. Nat Nanotechnol 15:406–416. https://doi.org/10.1038/s41565-020-0648-y

Wei J, Li Z, Yang Y, Ma G, Su Z, Zhang S (2020) An Apoferritin hemagglutinin conjugate vaccine with encapsulated nucleoprotein antigen peptide from influenza virus confers enhanced cross protection. Bioconjug Chem 31:1948–1959. https://doi.org/10.1021/acs.bioconjchem.0c00308
WHO (2021) COVID-19 Vaccine Tracker and Landscape. Accessed 4 Aug 2021
WHO (n.d.). https://www.who.int/publications/m/item/draf-landscape-of-covid-19-candidate-vaccines
Wuertz KM, Barkei EK, Chen W-H, Martinez EJ, Lakhal-Naouar I, Jagodzinski LL et al (2021) A SARS-CoV-2 spike ferritin nanoparticle vaccine protects hamsters against alpha and beta virus variant challenge. NPJ Vaccin 6. https://doi.org/10.1038/s41541-021-00392-7
Wui SR, Ko A, Ryu JI, Sim E, Lim SJ, Park SA, Kim KS, Kim H, Youn H, Lee NG (2021) The effect of a Tlr4 agonist/cationic liposome adjuvant on varicella-zoster virus glycoprotein e vaccine efficacy: antigen presentation, uptake and delivery to lymph nodes. Pharmaceutics 13:390
Yadav R, Zhai L, Kunda NK, Muttil P, Tumban E (2021) Mixed bacteriophage Ms2-L2 Vlps elicit long-lasting protective antibodies against Hpv Pseudovirus 51. Viruses 13:1113. https://doi.org/10.3390/v13061113
Yassine HM, Boyington JC, McTamney PM, Wei C-J, Kanekiyo M, Kong W-P et al (2015) Hemagglutinin-stem nanoparticles generate heterosubtypic influenza protection. Nat Med 21:1065–1070. https://doi.org/10.1038/nm.3927
Zwick MB et al (2017) Dense array of spikes on HIV-1 Virion particles. J Virol 91:1–19

Application of Nanoparticles as Innate Immune Stimulator

4

Ifeanyi Elibe Mba, Paul Ekene Chidebelu, and Emeka Innocent Nweze

4.1 Introduction

Immunosurveillance by the cells of the innate immune system helps to check the presence of nonself molecules and mount an appropriate response to the stimuli. Although the absence of memory limits the innate immune reaction, emerging evidence suggests that innate immune memory is inducible in which the invading immunogens, like the molecular patterns associated with microbes, vaccination, or damage-associated molecular patterns (DAMPs), tend to reprogram the immune cells for long-lasting response when subsequently provoked in their inactive states (Muñoz-Wolf and Lavelle 2021). The innate immune reaction is prominent in understanding diseases (Rehman et al. 2021). The innate immunity has been effectively explored in the ongoing advances in diagnosis and treatment of different ailments. It is also being used as biomarkers in some detection techniques to predict the progression or attenuation of diseases during treatments (Hecht et al. 2024).

Advances in nanomedicine have also included the development of efficient nanocarrier drug delivery systems, which use nanoparticles (NPs), as active pharmaceutical ingredients for drug and vaccine delivery and biodistribution through a continuous flow synthesis (Huanbutta et al. 2024). NPs provide a suitable medium for the efficient delivery of drugs, especially those with low polarity and other high molecular weight therapeutic biomolecules such as nucleic acids or their gene products, while also serving as vehicles for simultaneous compounds with combinatorial efficacy at the target sites (Silva et al. 2017). Therapeutic molecules with multiple targets are also deliverable using the nanocarriers.

The original version of the chapter has been revised. A correction to this chapter can be found at https://doi.org/10.1007/978-981-96-0397-8_13

I. E. Mba · P. E. Chidebelu · E. I. Nweze (✉)
Department of Microbiology, University of Nigeria, Nsukka, Nigeria
e-mail: emeka.nweze@unn.edu.ng

S. Rehman, E. Al-Suhaimi (eds.), *Nanotechnology Based Microbicides and Immune Stimulators*, https://doi.org/10.1007/978-981-96-0397-8_4

Nanoparticles are efficient immunostimulators, and as nanostimulators, they are potent inducers of ferroptosis, a downstream product of lipid peroxidation, in a glutathione-mediated redox-signal (Deng et al. 2024; Zheng et al. 2021). In addition to ferroptosis, it was further shown that a hybrid of two metal ions nanostimulators could cause the reactivation of innate immunity within a TNBC microenvironment, suggesting their potential duality in immunotherapy applications (Deng et al. 2024). Following their introduction, the NPs are usually coated with the host-derived protein substances around the particles, forming a large mass of spherical protein corona, which has surface adsorbed DNA, RNA, sugar, or lipid moieties, and as such, can elicit several distinct types of innate immune response facilitated by expression of their varied PRRs (Pondman et al. 2023; Cronin et al. 2020). Nanoparticles can interact with innate immune system, giving rise to inflammatory and anti-inflammatory responses. Most of the interactions between nanoparticles and the immune systems depend on their characteristics, and nanomaterials can also be engineered to interact with some immune cells in certain ways.

Therefore, in this chapter, we explore the different advances in the modes of interactions between nanoparticles and immune cells. We tentatively discussed the interaction between the innate immune system and nanoparticles, with the aim of understanding the recognition pattern, mechanisms of interaction, and how this can be explored for treating different diseases.

4.2 A Brief Overview of Nanoparticles

Nanoparticles (NPs) have become one of the new frontiers in research development. The discovery and application of these fine, very minute particles of 1 to 100 nm in size has proved a real milestone in biology and medicine. These particles are usually derived from the infusion of two or more different or closely related materials. The ultimate goal is to have very fine particle products with enhanced biochemical and physicochemical activities. Transition metal elements constitute major components of nanoparticles. Recently, nanometals synthesized from highly electrovalent transition metals with high electrocatalytic sites have been utilized in reactions for the evolution of oxygen (Yang et al. 2024). The same observation also suggested that oxidation or fluoridation can induce catalytic sites of high entropy in nonmetals. These modifications have generated nanoparticles of varied physicochemical properties, leading to their categorization into different structural and functional types. Their grouping varies with the nature as in either hard or soft NPs; design, origin in which they can be natural or artificial or engineered; dimension; synthesis; or toxicity of the particles (Mobeen et al. 2022). Other important classifications are based on their material compositions, including polymeric, semiconductor, carbon-based, lipid-based, metal, and ceramic nanoparticles (Khan et al. 2019). Consequently, NPs have gained relevance in almost all fields of science, viz.: nanomedicine, nanoagriculture, nanoengineering, and nanopharmacy, among others.

Two major procedures have been adopted for the synthesis of NPs: (1) the destructive top-down protocol, using grinding and other material decomposition techniques, and (2) the building-up or bottom-up method, which is a reverse approach given that the NPs are generated from relatively simpler substances, through the processes of sedimentation and reduction (Khan et al. 2019) (Fig. 4.1). Nanoparticles synthesized from metals predominate those from other sources and have gained wider application. Synthesis of NPs involves a range of methods: chemical, photoinduction, and biochemical (Sarmin et al. 2024). Plant materials (phytochemical compounds) with higher antioxidant properties, such as phenols, influence the reduction of metal ions. Other phytochemicals are also important in improving the metal-chelating ability and stabilization of certain metal nanoparticles (Sarmin et al. 2024). For phytoinduction, many modifications have been adopted, including white LED light-based photocatalysis. Organic materials are also essential components of nanoparticles. Reports have shown the use of biological molecules in the synthesis of NPs. In one of the instances, bacterial cellulose nanofiber has been successfully complexed to metal nanoparticles with potential application in food packaging of low bacterial contamination (Zhou et al. 2024).

Commonly used metals for the synthesis of NPs are often found naturally occurring in plants through bioaccumulation, which has also accounted for the innovation in NP exploration techniques. Thus, using plant materials to synthesize nanoparticles is consistent with their natural ability to complex with metals in their natural environments (Ma et al. 2024). The tremendous rise in nanotechnology innovations has revolutionized NP-based research in the past few decades.

Recent applications of NPs include high-performance protein microarrays, the color printing industry, and biochemical sensors. Their utility in disease treatment and diagnosis has been much highlighted in biomedical fields. Nanomedicine has held sway as an emerging cancer therapy and treatment of other clinical conditions,

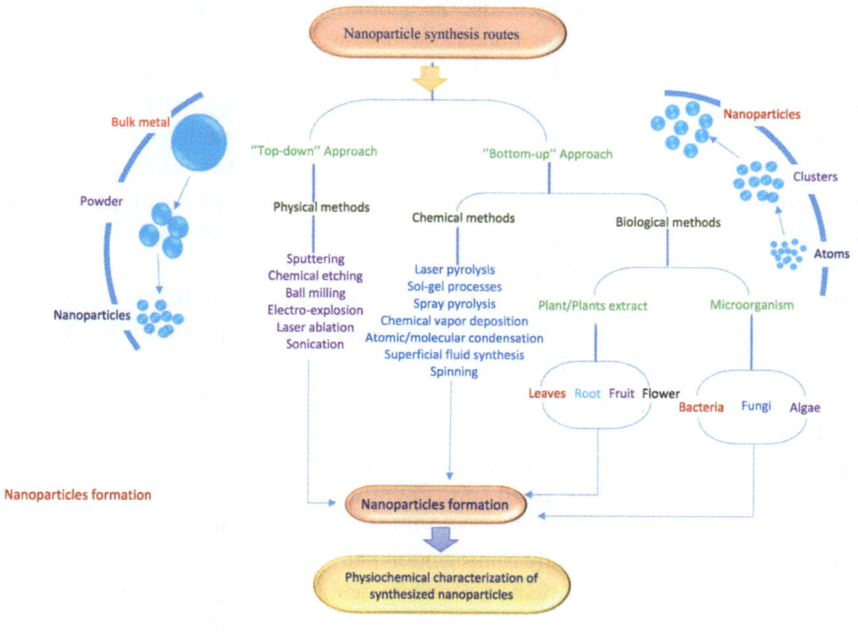

Fig. 4.1 Mechanism of nanoparticle synthesis

including inflammatory and immune diseases and infectious diseases such as COVID-19 (Daramy et al. 2023). Advances in nanomedicine have also included the development of efficient nanocarrier drug delivery systems, which use NPs as active pharmaceutical ingredients for drug and vaccine delivery and biodistribution, through a continuous flow synthesis (Huanbutta et al. 2024). In solar light and electronic device technologies, metal NPs have been incorporated to improve their photoluminescence and conductivity (Bastiamini et al. 2024).

4.3 A Brief Overview of Innate Immune System

The immune system is a composition of defensive barriers, including cells, tissues, or organs, and chemicals with which they interact to confer protection against pathogens and other foreign bodies like toxins and tumors (Marshall et al. 2018). Depending on the specificity of the immune reaction, immunity can be innate (non-specific) or adaptive (specific), and notwithstanding their mechanistic differences, there is a net convergence in their immune functions. Innate immunity is the first array of host defense mechanisms in the eukaryotes and can discriminate between self and non-self-molecules (McMahon et al. 2023). Innate immunity is the only type of immune system found in invertebrates, while in higher plants and animals, it combines efficiently with adaptive immunity (Kim et al. 2024). Initiation of the innate immune response is rapid and faster than the specific immunity. However, it is devoid of immune memory such that a similar cascade of events is repeated on subsequent with the same invading foreign body (Marshal et al. 2018).

Certain tissues and organs constitute the anatomical barrier in innate immunity, including the human skin and the mucous membrane. Physiological barriers are the physicochemical components that mediate immune functions. These comprise biochemical substances like cytokines and chemokines and physical parameters of temperature and pH, which are important factors in innate immunity. Others include an array of cells adapted for immune surveillance and clearance/elimination. Thus, the dendritic cells (DCs), natural killer cells (NK), and macrophages represent the major cells that pose phagocytic, endocytic, and inflammatory non-specific immune barriers to foreign bodies (Marshall et al. 2018; Kim et al. 2024).

Immunosurveillance by the cells of the innate immune system helps to check the presence of nonself molecules and mount an appropriate response to the stimuli. Although the absence of memory limits the innate immune reaction, emerging evidence suggests that innate immune memory is inducible in which the invading immunogens, like the molecular patterns associated with microbes, vaccination, or damage-associated molecular patterns (DAMPs), tend to reprogram the immune cells for long-lasting response when subsequently provoked in their inactive states (Muñoz-Wolf and Lavelle 2021). The central mechanism of innate immunity training revolves around the epigenetic transcription pathway (Leent et al. 2022), such as observed in an inflammation response during which an epigenetically modified innate immune cell demonstrated stronger immune reaction with enhanced inflammatory functions (Muñoz-Wolf and Lavelle 2021; Sonnenberg et al. 2024).

The early phase in the innate defense response to invading pathogens is the release of bioactive cytokines, which attract and mobilize the immune cells to the infection foci, while the appropriate local signals for inflammation are activated in the event of cell injury (Mattiola and Diefenbach 2023). These cells participate in the innate immunity following their stimulation by the innate chemical mediators, especially the interferons type 1 (IFNs), and their subsequent response aims to induce a hostile environment by upregulation of intracellular molecular mechanisms against a pathogen or in an injured site. It could be strictly noted that NK cells are restricted to the innate immune mechanism, whereas the other cell types function as a link between the two types of immunity, wherein some function as antigen-presenting cells (APCs) in neoplasma (Wang et al. 2021).

The antiviral function of innate immunity is widely documented. During viral infections, they play salient roles in a tightly regulated mechanism as the first immune barrier of the host cell (Savan and Gale 2023). Their activation is largely dependent on the signal elicited by the invading virus pathogen-associated molecular patterns (PAMPS), which are transmitted downstream following their binding to the cell receptors for appropriate response (Savan and Gale 2023). Studies have demonstrated the differential antiviral activation and response of innate immune cells. In acute viral infections such as the flavivirus and influenza virus infections, NK antiviral activity is predominantly mediated through the CD56 cell populations and its antibody-dependent cytotoxicity (ADCC) respectively (Björkstöm et al. 2022). Similarly, it was identified that innate immunity positively modulates the antiviral effects of $CD8^+$ cells in HIV infections. In the recent SARS-CoV-2 outbreak, innate immunity is central to the survival of the infected population. It was reported that the infection evokes a network of the host cell's innate immune defense mechanisms, which also worked in concert with adaptive immunity, to inhibit viral pathogenesis (Wang et al. 2023). IFN1 activity against SARS-CoV-2 infection was related to immune programming, which interferes with the viral RNA metabolism in a process mediated by the interferon-secreting genes (ISGs).

The innate immune reaction is prominent in cancer biology and cancer therapy. The ongoing advances in tumor diagnosis and treatment have effectively explored innate immunity. During tumorigenesis, both arms of the immune system synergize to reverse the progression or even eradicate the tumor cells. The innate immune cells are being used as biomarkers in some detection techniques to predict the progression or attenuation of tumors during treatments (Hecht et al. 2024). Evidently, immunotherapy and, lately, immunochemotherapy, or immune-chemo-radiotherapy, are promising alternatives to radiotherapy that have proven ineffective in solid tumors. The immunosurveillance roles of NK cells are unique in that their recognition of tumors is independent of the expression of the tumor cell surface marker, the human leukocyte antigen -I (HLA-I). They also exhibit antitumor activity by secretion of chemotactic factors that deliver the immune cells to the vicinity of the tumor microenvironment (McMahon et al. 2023; Kim et al. 2024). Importantly, NK cells, in their active states, can be tumoricidal, in a direct cytolytic mechanism involving phagocytosis or through their secreted granules such as the perforins and granzymes (Chen et al. 2021a, b).

Costimulation and inhibition interaction observed between innate and adaptive immunity are being investigated as potential antitumor treatment approaches (Cao and Kagan 2023). Also, the interaction of some microbiota members that migrated into the tumor microenvironment with the innate immune cells appears synergistic in antitumor immunity (Mattiola and Diefenbach 2023). However, some innate immune reaction is pro-tumorigenesis. For instance, the recruitment of myeloid-derived suppressive cells (m-MDSCs) highlights one of the limitations of ionizing radiations used in cancer treatment, as the m-MDSCs favor the immune checkpoints expression (Hecht et al. 2024).

4.4 The Effects of Nanoparticles on the Innate Immune System

Nanoparticles interact with different immune cells (Fig. 4.2). Generally, the fate of the inorganic NPs inside the body is determined by their interactions with the immune cells. The innate immune response to nano stimulation can, therefore, promote host immunity, including gradual inactivation and subsequent clearance of the NPs from the body system, rendering their therapeutic purposes impracticable or causing deleterious effects such as system toxicity and autoimmunity (Pondman et al. 2023; Mohammapdour and Ghandehari 2022). The release of free metal ions may be the main bactericidal mechanism of Ag NPs (Swartzwelter et al. 2020). However, the findings by Mohammapdour and Ghandehari (2022) indicated that the release of Ag^+, Au^+, Cd^+, Zn^{+2}, and Fe^{+2} metallic ions NPs can lead to toxicity, increased free radicals, or incorporation of the nanometal-ions into the body to be utilized as part of the essential micronutrients. According to a previous report, the risk of immunotoxicity is relatively high following in vivo administration of iron oxide NPs, with fatal consequences impacting cell motility, disruption of the cytoskeleton, and cell death, raising more safety concerns on their continued therapeutic applications (Safane and Girard 2023). In contrast to the release of ions observed in most metal NPs, the zinc nanomaterial, ZnOPs, exerts its toxicity through reactivation, in which its exposure to enterocytes was associated with high-level expression of the antioxidant gene due to the induction of reactive oxygen species (Petrarca et al. 2015).

Like the endogenous particles and molecules of microbial origins, exogenous particles are immunogenic and offer a platform for retooling innate immunity for a memory-like response (Muñoz-Wolf and Lavelle 2021; Leent et al. 2022). NPs have been utilized in trained innate immunity as found in the reversal of altered phenotype of macrophages, M_2 from tumor-promoting cell to its original antigen-presenting activity.

This feature is being utilized in the proposed induction of innate immunity with the immunomodulatory, papaya mosaic virus-coated, virus-like NPs (PapMV), that specifically enter macrophages to unload their ssRNA contents, which metabolic products function in the activation of TLR7&8 (Carignan et al. 2018). Infusion of the recombinant PapMV and a nucleic acid (ssRNA) results in an innate

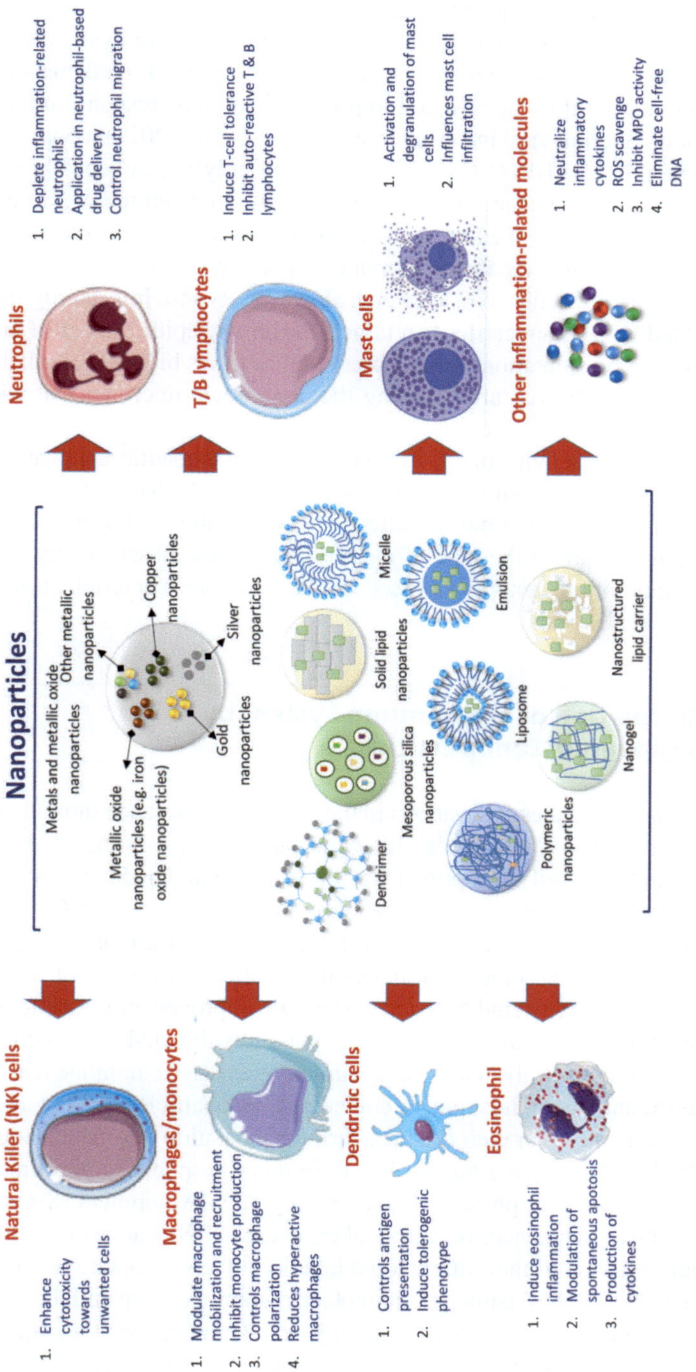

Fig. 4.2 Effect of nanoparticles on different immune cells

immunogenic NP product with in vitro inhibitory activity against pulmonary bacterial or viral infections (Mathieu et al. 2013).

One of the nanoscientific approaches to trained innate immunity in vertebrates is the substitution of the endogenous inducer particles with a synthetic homolog, which can be delivered to the host cell to prime the immune response prior to second stimulation. During viral infection in fish, Samms et al. (2022) reported that the double-stranded RNA (dsRNA), synthetic homolog-polyI:C, can equally stimulate antiviral, inflammatory, and anti-cancer effects of the innate immunity. On the other hand, the therapeutic uses of exogenous immunomodulators can be enhanced using nanocarriers. Commonly, the innate immune response elicited on administration of TLR7 immunotherapy tends to be elevated. However, its antiviral activity improves when delivered with a nanocarrier that targets the airway epithelial cells (Kan et al. 2021). In diet supplementation, it was observed that zinc's higher bioavailability in animal tissue was achieved after loading the amino acid-micronutrient with NPs (Wang et al. 2022).

The antibacterial activity of NPs is one of their therapeutic applications. The high reactivity of some metal nanoparticles (MENPs) is the basis of their bacteriostatic and bactericidal or antibiofilm effects. However, their roles in bacterial biofilm formation depend on the type of the NP since it was observed that while Ag NPs have antibiofilm activity, others could promote bacterial proliferation (Swartzwelter et al. 2020).

4.5 Modulation of the Immune System by Engineered Nanoparticles

Nanotechnology facilitates innovations in the field of nanoscience through a practical modulation of the nanoparticles for wider applications. Some of these challenges that have reduced the optimal use of NPs in nanomedicine include their contribution to cell toxicity and immune suppression (Saafane and Girard 2023). Nanoengineering has attracted a growing interest in nanomedicine, partly necessitated by the need to overcome these inherent limitations in the use of the existing NPs. Therefore, the engineered NPs aim to generate improved, more stable products with enhanced functions. Engineered NPs are basically directed to a specific target, relieving the host cells of the undesirable burden of excessive immune reactions.

With the expanding applications of engineered nanoparticles (ENPs), passive or active acquisition of NPs or their products through inhalation or subcutaneous injection into the body elicits the human innate immune response as the first line of defense, even at the early phase of human embryonic development (Swartzwelter et al. 2020). The consequences of continual exposure to ENPs are usually detrimental, especially because of their immunomodulatory effects on innate immune functions. ENPs like microbial pathogens and other immunogenic materials are limited by undesirable stimulation of innate immunity despite the milestones in developing non-immunologically toxic nucleic acid-based therapies (Chandler et al. 2020).

Most of the NPs are genetically engineered as immune checkpoint blockade interference tools whereby Ding et al. (2023) noted that the encapsulation of siAdari-LNP@mPD1 engineered NPs, with the PD1 protein facilitating efficient delivery, such that the encapsulated nanomaterial can exert its antitumor effects, thereby serving as a potential tumor resistance therapy (Ding et al. 2023). ENPs can also induce or interfere with innate memory induction, and while the former scenario follows a similar pattern as in other foreign particles, the latter has been demonstrated when BCG priming of the human monocyte is done with AU NPs (Swartzwelter et al. 2020).

Excluding the event where they are successfully delivered to their targets, available evidence suggests the fate of ENPs on exposure to immune cells determines their innate immunomodulatory properties. The therapeutic utility of ENPs often relies on innate immune suppression (Malackowski and Hassel 2020). Reports have indicated that by route to their targets inside the body, ENPs can be readily cleared from circulation or survive the innate defense to escape into deeper non-target sites, such as following a compromised integrity of the alveolar epithelial cell tight junction (Cronin et al. 2020). Consequently, the rationale for engineering a nanotherapeutic particle has realistically involved incorporating particles coding for innate immune expression silencers, and in this way, eliminates immune barriers to their delivery targets.

4.6 Interaction Between Nanoparticles and Neutrophils

Neutrophils represent a distinct category of cellular components within the immunological system. They are pivotal in inflammation. They also exhibit limited activity and play a crucial function in combating pathogens (Mantovani et al. 2011). Neutrophils constitute about 50–70% of white blood cells (Liew an Kubes 2019), arising from myeloid progenitor cells. Before entering circulation as mature cells, they undergo three sequential stages in the bone marrow (Lawrence et al. 2018). Subsequently, mature neutrophils are released and form reservoirs (Summers et al. 2010).

In the context of inflammation, circulating neutrophils are recruited to inflammatory sites through adhesion molecules. Following the established steps of recruitment, they accumulate in the inflammatory region. Once activated and recruited, neutrophils employ various mechanisms, including phagocytosis and the release of proteins to eliminate pathogens (Manfredi et al. 2018). Neutrophils are widely acknowledged as the frontline defenders of the innate immune system. They are responsible for scavenging invaders and modulating inflammation. Furthermore, their involvement in initiating and advancing adaptive immune reactions, imparting a more intricate function, has been reported (Liew and Kubes 2019). Given their critical roles in disease, different nanomaterials target the neutrophils. Thus, the activation of neutrophils by nanoparticles has been established (Skallberg et al. 2019). These nanomaterials target the neutrophils and manipulate the inflammatory microenvironment. The mechanisms behind these effects encompass modulating the migration of neutrophils. Nanoparticles also deplete the neutrophils with inflammatory roles. Neutrophil-mediated phagocytosis and ROS generation have been shown to be prevented by AgNps (Huang et al. 2021). Similar findings were previously reported by Liz et al. (2015). Overall, the interaction between neutrophils and nanoparticles has been investigated in several studies (Kang et al. 2021; Chen et al. 2023a, b; Saafane et al. 2022; Cruz et al. 2022; Vanharen et al. 2023).

4.7 Interaction Between Nanoparticles and Macrophages

Macrophages serve as vital components within the innate immune system. It is widely distributed across various tissues, exhibiting remarkable heterogeneity and adaptability. Their distinctive roles encompass diverse aspects of an organism (Wynn et al. 2013; Locati et al. 2020). Macrophages are known to exhibit diversity based on their activation and functional characteristics. Some are classified as being involved in inflammation, while others are involved in wound healing (Funes et al. 2018; Shapouri-Moghaddam et al. 2018; Murray 2017). The polarization of classically and alternatively activated macrophages plays a pivotal role in modulating the balance between pro-inflammatory and anti-inflammatory reactions. Although the mechanisms behind these activities are not completely known, the activation mechanism is acknowledged to be vital in various physiological processes and pathologies (Locati et al. 2020; Shapouri-Moghaddam et al. 2018).

Due to the pivotal roles played by macrophages in various pathogenic processes, numerous nanoparticle formulations have been created to engage with these immune components (Gaspar et al. 2019). For an efficient microphage target and to accomplish specific roles concurrently, nanoparticles must be adorned with diverse ligands on their surface. NPs also mobilize monocytes while enhancing permeability to microvesicles. Moreover, the inherent features of macrophages can be harnessed by nanoparticles to improve targeting effectiveness and also avoid detection by immune cells. Thus, nanoparticles can be modified and engineered for the regulation of macrophage recruitment during inflammatory processes. Chemokines and angiotensin receptors play a vital role in mobilizing and recruiting macrophages. Thus, nanomaterials targeting these molecules exhibit anti-inflammatory effects. Several investigations have also reported Nanoparticles influencing macrophage-associated inflammation (Chen et al. 2020; He et al. 2021; Kraynak et al. 2022; Muhammad et al. 2022; Zhou et al. 2023).

4.8 Other Effects of Nanoparticles on the Innate Immune System

Although nanoparticles interact with macrophages, neutrophils, and other immune proteins, an alternative approach exists by which nanoparticles can impact immune responses (Fig. 4.2). These include interactions with mast cells, eosinophils, and dendritic cells. There is also the modulation of natural killer (NK) cell activity, among other pathways. For example, lipid nanoparticles (Alam et al. 2023), polymer-coated nanoparticles (Shapoval et al. 2021), silver nanoparticles (Aldossari et al. 2015), and iron oxide nanoparticles (Sabareeswaran et al. 2016) have all been shown to influence mast cell activation, degranulation, and mast cell infiltration. Also, the effect of gold nanoparticles (Mehanna et al. 2022; Vanharen et al. 2023), titanium oxide nanoparticles (Nie et al. 2021; Murphy-Marion and Girard 2018), copper oxide nanoparticles (Holan et al. 2019), and silver nanoparticles (Vallières et al. 2016) on eosinophil have also been established. Moreover, dendritic cells, crucial for antigen presentation, bridge innate and adaptive immunity (Waisman et al. 2017; Gardner et al. 2020). These dendritic cells have been harnessed for the development of various nanoparticle platforms. This is owing to their antigen-presenting ability in addition to their immune system modulatory ability.

Dendritic cells are important in immune tolerance, presenting a promising strategy for targeting dendritic cells to alleviate inflammation. A study by Yeste et al. (2016) illustrates how these dendritic cells could interact with nanoparticles to activate other immune cells and trigger inflammation. Another study has shown how nanoparticles could prevent dendritic cell activation with effect on the cells (Deng et al. 2017). A cationic liposome with the potential to impact the efficiency of antigen presentation by dendritic cells is also available. This nanomaterial can elevate lysosomal pH in dendritic cells. It reduces antigens' degradation and enhances T cell responses (Gao et al. 2017).

Natural killer cells are categorized as innate lymphoid cells. They play crucial roles in the innate immunological response. They possess the ability to eliminate abnormal cells through various mechanisms (Gao et al. 2017; Mikelez-Alonso et al. 2021). Consequently, these cells can be harnessed in nanoparticle tactics for targeted cell elimination. A bispecific Au NPs (BiAb-AuNPs) have demonstrated the enhancement of NK cell cytotoxic immune responses against HIV-infected T cells. This formulation involves the conjugation of anti-HIVgp120 and IgG antibodies onto gold nanoparticles (Mikelez-Alonso et al. 2021). Furthermore, a different approach involving a nanoparticle-based cell tracking method is also available. These nanomaterials are directly labeled onto natural killer cells. Upon NK cell death, elevated ROS levels induce an NIR-II signal. This linkage of fluorescent signals with NK cell viability shows the potential of these nanotools in the application of nanoparticle strategies for NK cell-related treatments, particularly in addressing inflammatory diseases (Liao et al. 2021). Additional mechanisms, such as the adjustment of neural-immune cell crosstalk, prevention of hypoxia-induced apoptosis, and several other mechanisms, have also been investigated recently (Huang et al. 2020; Zachs et al. 2019). However, their efficacy in treating diseases awaits further confirmation through in vivo studies.

The behavior of nanoparticle as hapten was previously proposed by Zolink et al. in an attempt to explain how antibodies are produced by nanoparticles (Fig. 4.3). According to the model, the nanoparticle conjugation with a protein (e.g., HAS) triggers endocytosis. Upon nanoparticle endocytosis, the B cells present the protein through HLA to helper T cells. The presentation is also via interaction with T cell

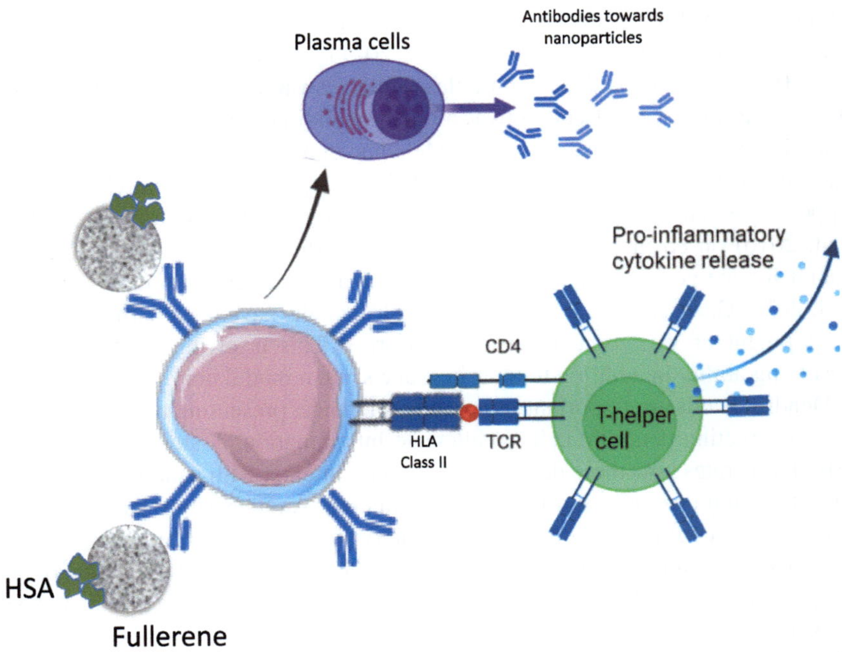

Fig. 4.3 The hypothetical model of nanoparticle antigenicity proposed by Zolink et al.

receptors. Upon activation of the T cells, there is the production of cytokines. This also triggers the plasma cells to make antibodies specific to the nanoparticles.

4.9 Discussion and Perspectives

Nanotechnology's rapid advancement presents a novel avenue for influencing immune responses, particularly in addressing various diseases. As mentioned earlier, nanoparticle platforms exhibit the capability to engage with nearly every element within the immune system, offering a ground-breaking tool for addressing immune-related ailments. Traditionally, inflammatory disease management depends on the use of anti-inflammatory drugs (Smolen et al. 2016; Rendon and Schakel 2019). However, these treatments primarily aim at palliation and often come with significant side effects. By modulating various properties of nanoparticles, including shape, size, biodegradability, and surface charge, and incorporating other molecules like antibodies, cell membranes, enzymes, etc., NPs can possess diverse therapeutic characteristics and limitless potential for mitigating inflammation.

This chapter provides an overview of the interactions between nanoparticles and innate immune system components. The findings suggest that, armed with knowledge about the pathogenic mechanisms of specific diseases, efficient nanoparticle formulations that target the implicated immune system are crucial. This targeted approach enables accurate interference with the inflammatory environment.

Moreover, it can pave the way for more effective and tailored interventions. Nonetheless, the utilization of nanoparticles for immune modulation confronts several hurdles. Firstly, modulatory nanoparticles inhibit immune proteins from curbing excessive immunological responses. While this approach aims to suppress inflammation, it poses the risk of inducing immune dysfunction or immunodeficiency. It also increases the susceptibility to opportunistic infections and malignancies. Secondly, unintended reactions may occur despite efforts to design NPs that specifically interact with different cells. Thirdly, a substantial gap exists in linking in vivo studies to the actual reality of what happened in the human body (van der Walk et al. 2015).

Nanomaterials are used as drug delivery agents. It can shield the drug from rapid clearance by immune cells. However, the use of these nanomaterials as a drug delivery agent is not without side effects. Bridging the gap between pre-clinical results and clinical outcomes is vital for fully harnessing the potential of nanoparticles. Furthermore, physical characteristics, such as shape, size, and surface features, can influence the effects of NPs (Mba and Nweze 2021). Thus, modifying the properties of nanoparticles is a vital approach to improving their efficacy and pharmacokinetics. The surface feature of nanoparticles specifically has been shown to influence immune interaction. Nanoparticles with hydrocarbon chains have been shown to facilitate the production of cytokines (TNF-α, IL-6). Moreover, those with an acid group usually have more ability to elevate complement activation (Gonzalez-Garcia et al. 2022).

Furthermore, despite the continuous emergence of innovative NP approaches, the availability of clinically approved nanotherapeutics remains limited owing to numerous unknown factors. NPs offer a range of customizable characteristics, such as diverse size and shape. They also vary in their charge and surface modifications. For instance, nanoparticle charge holds significance in processes like endosomal escape. Moreover, surface marker modifications can target specific cell types (Mitchell et al. 2021). Any approach that could help promote the safety and specificity of nanomaterials and aid in selecting suitable nanomaterials tailored to a patient's need is indispensable.

Furthermore, the formulation of nanomaterials at the industrial level must adhere to good manufacturing practice standards to ensure consistency in products (Chen et al. 2021a, b). Navigating the transition of NPs from laboratory experimentation to practical application in treating different diseases has consistently presented challenges. Typically, NP formulations already in the market for use or currently in clinical trials predominantly feature liposomes and polymeric particles. Several of these nanomaterials are also under clinical trials for treating several ailments (Jia et al. 2023; Chen et al. 2023a, b). Importantly, care must be given to nanoparticles used as carrier agents. The safety, stability, and seamless connectivity with carriers' therapeutic molecules are vital. Enhancing carriers' ability to extend half-life and facilitating controlled release of functional components is very vital.

Recently, several innovations in nanoparticle design covering the area of innate immune modulation have been introduced. One noteworthy development is the emergence of nanotools. This nanomaterial leverages the inherent capabilities of sourced cells for their environmental influence. Cell membranes, sourced from various cell types, including leukocytes and erythrocytes, can be used to coat NPs. The diverse range of membranes and different core compositions result in a unique platform with versatile applications, including drug delivery, imaging, detoxification, and several other roles. Other strategies involving the transformation of nanoparticles for immune modulation and treatment of different diseases are also available (Fang et al. 2022; Mihai et al. 2018; Navarro Chica et al. 2020; Kwiatkowski et al. 2020; Nakkala et al. 2021; Li et al. 2024). However, it is crucial to approach all these novel formulations with caution, considering their structural complexity and the fact that certain important properties have not been thoroughly explored. Transitioning from laboratory settings to clinical applications also requires more comprehensive studies and evaluations.

Finally, more recent studies are showing how nanoparticles could be harnessed to improve therapeutic purposes based on neutrophils and other immune cells. In fact, the manipulation of neutrophils for use as a drug delivery system is emerging as a current area of research focus (Wu et al. 2022; Liu et al. 2022). Overall, nanoparticles interacting with immune cells could lead to nanoparticle clearance facilitated by macrophages. Conversely, it could lead to systemic inflammation and also immunotoxicity. However, these interactions can be exploited for vaccine formulation and other immunotherapeutic purposes (Pondman et al. 2023). In a study showing how nanoparticles loaded with drugs can be engulfed by neutrophils, Luo et al. (2023) demonstrated how nanoparticles can leverage the potential and their interaction with neutrophils for efficient drug delivery. Several other research on the interaction between nanomaterial and immune cells and how they can be leveraged therapeutically are available (Wang et al. 2021; Zhang et al. 2018; Chen et al. 2024).

4.10 Conclusion

Despite promising pre-clinical outcomes, our comprehension of NP interactions with the immune system remains limited, necessitating thorough clinical trials and prolonged safety assessments. Reliable techniques for monitoring NP therapeutic effects in the body also warrant development. While immunomodulatory NPs exhibit transformative characteristics for treating diseases, their practical integration into clinical settings requires extensive further progress. However, it is evident that the distinctive structures, adaptability, and stability of nanoparticles (NPs) offer significant potential for targeted accumulation in inflamed tissues and treatment of different diseases.

Examining the diverse interactions between NPs and the innate immune system provides insights into how NPs should be crafted to regulate these interactions and ensuing responses. Significantly, the surface coating, more than the core structure, defines much of the bioactivity of NPs. These agents can be tailored for safety, inertness, or enhanced therapeutic benefits by manipulating the nanoparticle interface. The NP core's composition influences its activity. On the other hand, the surface coating predominantly dictates its bioactivity. Moreover, the aggregation of nanoparticles with other nanomaterials and also with other macromolecules has several effects on the cells. Thus, to tackle the interactions between NPs and the immune system, the initial focus must be on addressing the instability associated with nanoparticle surfaces.

An approach for preventing spontaneous aggregation upon introduction into biological systems is very vital. This is because aggregation can lead to a reduction in efficacy. Traditionally, polymeric coatings have been widely used for this purpose despite being associated with high costs and involving intricate multi-step synthesis approaches. An alternative and efficient solution involves enhancing nanoparticle solubility during the formulation phase. This is vital to ensure stability, eliminate the chances of opsonization, and create more chances for direct immune modulation. Nanomaterials can also be designed to target certain immune cells and prevent complement inactivation and other inflammatory activations.

Overall, manipulating the surface characteristics of nanomaterials is a key strategy to fine-tune the physicochemical attributes of nanomaterials, ultimately influencing their interactions with the immune system. However, these area demands more investigation. Thus, it is crucial to approach all these novel formulations with caution, considering their structural complexity and the fact that some of their inherent properties influencing immune cells remain to be completely explored. Transitioning from laboratory settings to clinical applications requires more comprehensive studies and evaluations. Finally, more in-depth studies are needed on the interaction between innate immune system and nanoparticles. This is important to completely understand the recognition pattern, potential cytotoxicity, interaction mechanisms, and how this can be explored for therapeutic purposes.

References

Alam SB, Wang F, Qian H, Kulka M (2023) Apolipoprotein C3 facilitates internalization of cationic lipid nanoparticles into bone marrow-derived mouse mast cells. Sci Rep 13(1):431

Aldossari AA, Shannahan JH, Podila R, Brown JM (2015) Influence of physicochemical properties of silver nanoparticles on mast cell activation and degranulation. Toxicol In Vitro 29(1):195–203

Bastianini F, Hidalgo AI, Hook DZ, Smith JA, Cumming D, Dumber A (2024) Using ag nanoparticles in the electron transport layer of perovskite solar cells to improve efficiency. Sol Energy 268:112318

Björkstöm NK, Strunz B, Ljunggren HG (2022) Natural killer cells in antiviral immunity. Nat Rev Immunol 22:112–123

Cao LL, Kagan JC (2023) Targeting innate immune pathways for cancer immunotherapy. Immunity 56(10):2206–2217

Carignan D, Sabine Herblot S, Laliberté-Gagné M-È, Bolduc M, Duval M, Savard P, Leclerc D (2018) Activation of innate immunity in primary human cells using a plant virus derived nanoparticle TLR7/8 agonist. Nanomedicine 14(7):2317–2327

Chandler M, Johnson MB, Panigaj M, Afonin KA (2020) Innate immune responses triggered by nucleic acids inspire the design of immunomodulatory nucleic acid nanoparticles (NANPs). Curr Opin Biotechnol 63:8–15

Chen Y, Guan M, Ren R, Gao C, Cheng H, Li Y, Gao B, Wei Y, Fu J, Sun J, Xiong W (2020) Improved immunoregulation of ultra-low-dose silver nanoparticle-loaded TiO_2 nanotubes via M2 macrophage polarization by regulating GLUT1 and autophagy. Int J Nanomedicine 15:2011–2026

Chen W, Schilperoort M, Cao Y, Shi J, Tabas I, Tao W (2021a) Macrophage-targeted nanomedicine for the diagnosis and treatment of atherosclerosis. Nat Rev Cardiol 19:228. https://doi.org/10.1038/s41569-021-00629-x

Chen K-G, Liu A, Jiang CT, Zhao D-K, Ye Q-N, Liao Y-Q, Xu C-F, Shen S, Wang J (2021b) Dual-functional super bispecific nano-antibodies derived from monoclonal antibodies potentiate the antitumor effect of innate immune cells. Nano Today 39:101209

Chen X, Xu J, Ji B, Fang X, Jin K, Qian J (2023a) The role of nanotechnology-based approaches for clinical infectious diseases and public health. Front Bioeng Biotechnol 11:1146252

Chen K, Zhang Z, Fang Z, Zhang J, Liu Q, Dong W, Liu Y, Wang Y, Wang J (2023b) Aged-signal-eliciting nanoparticles stimulated macrophage-mediated programmed removal of inflammatory neutrophils. ACS Nano 17(14):13903–13916

Chen J, Ren T, Xie L, Hu H, Li X, Maitusong M, Zhou X, Hu W, Xu D, Qian Y, Cheng S, Yu K, Wang JA, Liu X (2024) Enhancing aortic valve drug delivery with PAR2-targeting magnetic nano-cargoes for calcification alleviation. Nat Commun 15(1):557

Cronin JG, Jones N, Thornton CA, Gareth JS, Jenkins GJS, Doak SH, Clift MJD (2020) Chem Res Toxicol 33(5):1061–1073

Cruz MA, Bohinc D, Andraska EA, Alvikas J, Raghunathan S, Masters NA, van Kleef ND, Bane KL, Hart K, Medrow K, Sun M, Liu H, Haldeman S, Banerjee A, Lessieur EM, Hageman K, Gandhi A, de la Fuente M, Nieman MT, Kern TS et al (2022) Nanomedicine platform for targeting activated neutrophils and neutrophil-platelet complexes using an α_1-antitrypsin-derived peptide motif. Nat Nanotechnol 17(9):1004–1014

Daramy K, Punnabhum P, Hussain M, Minelli C, Pei Y, Rattray NJ, Perrei Y, Rattray Z (2023) Nanoparticle isolation from biological media for protein corona analysis: the impact of incubation and recovery protocols on nanoparticle properties. J Pharm Sci 2023

Deng Z, Rong Y, Teng Y, Mu J, Zhuang X, Tseng M et al (2017) Broccoli-derived nanoparticle inhibits mouse colitis by activating dendritic cell AMP-activated protein kinase. Mol Ther 25:1641–1654

Deng X, Liu T, Zhu Y, Chen J, Song Z, Shi Z, Chen H (2024) Ca & Mn dual-ion hybrid nanostimulator boosting antitumor immunity via ferroptosis and innate immunity awakening. Bioactive Mater 33:483–496

Ding L, Zhang X, Yu P, Peng F, Sun Y, Wu Y, Luo Z, Li H, Zeng Y, Wu M, Liu X (2023) Genetically engineered nanovesicles mobilize synergistic antitumor immunity by ADAR1 silence and PDL1 blockade. Mol Ther 31(8):2489–2506

Fang L, Liu J, Liu Z, Zhou H (2022) Immune modulating nanoparticles for the treatment of ocular diseases. J Nanobiotechnol 20(1):496

Funes SC, Rios M, Escobar-Vera J, Kalergis AM (2018) Implications of macrophage polarization in autoimmunity. Immunology 154:186–195

Gao J, Ochyl LJ, Yang E, Moon JJ (2017) Cationic liposomes promote antigen cross-presentation in dendritic cells by alkalizing the lysosomal pH and limiting the degradation of antigens. Int J Nanomedicine 12:1251–1264

Gardner A, de Mingo PÁ, Ruffell B (2020) Dendritic cells and their role in immunotherapy. Front Immunol 11:924

Gaspar N, Zambito G, Löwik C, Mezzanotte L (2019) Active nano-targeting of macrophages. Curr Pharm Des 25:1951–1961

González-García LE, MacGregor MN, Visalakshan RM, Lazarian A, Cavallaro AA, Morsbach S, Mierczynska-Vasilev A, Mailänder V, Landfester K, Vasilev K (2022) Nanoparticles surface chemistry influence on protein corona composition and inflammatory responses. Nanomaterials 12(4):682

He Y, de Araújo Júnior RF, Cruz LJ, Eich C (2021) Functionalized nanoparticles targeting tumor-associated macrophages as cancer therapy. Pharmaceutics 13(10):1670

Hecht M, Frey B, Gaipl US, Tianyu X, Eckstein M, Donaubauer A-J, Klautke G, Illmer T, Fleischmann M, Laban S, Hautmann MG, Tamaskovics B, Brunner TB, Becker I, Zhou J-G, Hartmann A, Fietkau R, Iro H, Döllinger M, Gostian A-O, Kist AM (2024) Machine learning-assisted immunophenotyping of peripheral blood identifies innate immune cells as best predictor of response to induction chemo-immunotherapy in head and neck squamous cell carcinoma – knowledge obtained from the CheckRad-CD8 trial. Neoplasia 49:100953

Holan V, Javorkova E, Vrbova K, Vecera Z, Mikuska P, Coufalik P, Kulich P, Skoupy R, Machala M, Zajicova A, Rossner P (2019) A murine model of the effects of inhaled CuO nanoparticles on cells of innate and adaptive immunity – a kinetic study of a continuous three-month exposure. Nanotoxicology 13(7):952–963

Huanbutta K, Suwanpitak K, Weeranoppanant N, Sriamornsak P, Garg K, Sharma S, Singh I, Sangnim T (2024) Continuous flow synthesis: a promising platform for the future of nanoparticle-based drug delivery. J Drug Deliv Sci Technol 105265:1773–2247

Huang H, Feng W, Chen Y, Shi J (2020) Inorganic nanoparticles in clinical trials and translations. Nano Today 35:100972

Huang M, Ye K, Hu T, Liu K, You M, Wang L, Qin H (2021) Silver nanoparticles attenuate the antimicrobial activity of the innate immune system by inhibiting neutrophil-mediated phagocytosis and reactive oxygen species production. Int J Nanomedicine 16:1345–1360

Jia Y, Jiang Y, He Y, Zhang W, Zou J, Magar KT, Boucetta H, Teng C, He W (2023) Approved nanomedicine against diseases. Pharmaceutics 15(3):774

Kang H, Seo J, Yang EJ, Choi IH (2021) Silver nanoparticles induce neutrophil extracellular traps via activation of PAD and neutrophil elastase. Biomol Ther 11(2):317

Khan I, Saeed K, Khan I (2019) Nanoparticles: properties, applications and toxicities. Arab J Chem 12(7):908–931

Kim H-D, Yeh C-Y, Chang Y-C, Cheorl-Ho Kim C-H (2024) Dawn era for revisited cancer therapy by innate immune system and immune checkpoint inhibitors. Biochim Biophys Acta Mol basis Dis 1870:167019

Kraynak CA, Huang W, Bender EC, Wang JL, Hanafy MS, Cui Z, Suggs LJ (2022) Apoptotic body-inspired nanoparticles target macrophages at sites of inflammation to support an anti-inflammatory phenotype shift. Int J Pharm 618:121634

Kwiatkowski AJ, Stewart JM, Cho JJ, Avram D, Keselowsky BG (2020) Nano and microparticle emerging strategies for treatment of autoimmune diseases: multiple sclerosis and type 1 diabetes. Adv Healthc Mater 9(11):e2000164

Lawrence SM, Corriden R, Nizet V (2018) The ontogeny of a neutrophil: mechanisms of granulopoiesis and homeostasis. Microbiol Mol Biol Rev 82. https://doi.org/10.1128/MMBR.00057-17

Leent MMT, Priem B, Schrijver DP, de Dreu A, Hofstraat SRJ, Zwolsman R, Beldman TJ, Netea MG, Mulder WJM (2022) Regulating trained immunity with nanomedicine. Nat Rev Mater 7:465–481

Li Q, Lin L, Zhang C, Zhang H, Ma Y, Qian H, Chen XL, Wang X (2024) The progression of inorganic nanoparticles and natural products for inflammatory bowel disease. J Nanobiotechnol 22(1):17

Liao N, Su L, Zheng Y, Zhao B, Wu M, Zhang D et al (2021) In vivo tracking of cell viability for adoptive natural killer cell-based immunotherapy by ratiometric NIR-II fluorescence imaging. Angew Chem Int Ed Engl 60:20888–20896

Liew PX, Kubes P (2019) The neutrophil's role during health and disease. Physiol Rev 99:1223–1248

Liu J, Liu Z, Pang Y et al (2022) The interaction between nanoparticles and immune system: application in the treatment of inflammatory diseases. J Nanobiotechnol 20:127. https://doi.org/10.1186/s12951-022-01343-7

Liz R, Simard JC, Leonardi LB, Girard D (2015) Silver nanoparticles rapidly induce atypical human neutrophil cell death by a process involving inflammatory caspases and reactive oxygen species and induce neutrophil extracellular traps release upon cell adhesion. Int Immunopharmacol 28(1):616–625

Locati M, Curtale G, Mantovani A (2020) Diversity, mechanisms, and significance of macrophage plasticity. Annu Rev Pathol 15:123–147

Luo Z, Lu Y, Shi Y, Jiang M, Shan X, Li X, Zhang J, Qin B, Liu X, Guo X, Huang J, Liu Y, Wang S, Li Q, Luo L, You J (2023) Neutrophil hitchhiking for drug delivery to the bone marrow. Nat Nanotechnol 18(6):647–656

Ma S, Cao J, Liang H (2024) A study of au-bearing-nanoparticles-enriched plants from the concealed gold deposits and their prospecting significance. Ore Geol Rev 105910:105910

Malchowski T, Hassel A (2020) Engineering nanoparticles to overcome immunological barriers for enhanced drug delivery. Eng Regen 1:35–50

Manfredi AA, Ramirez GA, Rovere-Querini P, Maugeri N (2018) The neutrophil's choice: phagocytose vs make neutrophil extracellular traps. Front Immunol 9:288

Mantovani A, Cassatella MA, Costantini C, Jaillon S (2011) Neutrophils in the activation and regulation of innate and adaptive immunity. Nat Rev Immunol 11:519–531

Marshall JS, Warrington R, Watson W, Kim HL (2018) An introduction to immunology and immunopathology. Allergy Asthma Clin Immunol 14(Suppl 2):49. https://doi.org/10.1186/s13223-018-0278-1

Mathieu C, Rioux G, Dumas MC, Leclerc D (2013) Induction of innate immunity in lungs with virus-like nanoparticles leads to protection against influenza and Streptococcus pneumoniae challenge. Nanomedicine 9(7):839–848

Mattiola I, Diefenbach A (2023) Regulation of innate immune system function by the microbiome: consequences for tumor immunity and cancer immunotherapy. Semin Immunol 66:101724

Mba IE, Nweze EI (2021) Nanoparticles as therapeutic options for treating multidrug-resistant bacteria:research progress, challenges, and prospects. World J Microbiol Biotechnol 37(6):108

McMahon RA, D'Souza C, Neeson PJ, Siva S (2023) Innate immunity: looking beyond T-cells in radiation and immunotherapy combinations. Neoplasia 46:100940

Mehanna ET, Kamel BSA, Abo-Elmatty DM, Elnabtity SM, Mahmoud MB, Abdelhafeez MM, Sabry S, Abdoon A (2022) Effect of gold nanoparticles shape and dose on immunological, hematological, inflammatory, and antioxidants parameters in male rabbit. Veterinary world 15(1):65–75

Mihai MM, Preda M, Lungu I, Gestal MC, Popa MI, Holban AM (2018) Nanocoatings for chronic wound repair-modulation of microbial colonization and biofilm formation. Int J Mol Sci 19(4):1179

Mikelez-Alonso I, Magadán S, González-Fernández Á, Borrego F (2021) Natural killer (NK) cell-based immunotherapies and the many faces of NK cell memory: a look into how nanoparticles enhance NK cell activity. Adv Drug Deliv Rev 176:113860

Mitchell MJ, Billingsley MM, Haley RM, Wechsler ME, Peppas NA, Langer R (2021) Engineering precision nanoparticles for drug delivery. Nat Rev Drug Discov 20:101–124

Mobeen H, Safdar M, Fatima A, Afzal S, Zaman H, Mehdi Z (2022) Emerging applications of nanotechnology in context to immunology: a comprehensive review. Front Bioeng Biotechnol 10:1024871

Mohammapdour R, Ghandehari H (2022) Mechanisms of immune response to inorganic nanoparticles and their degradation products. Adv Drug Deliv Rev 180:114022

Muhammad W, Zhu J, Zhai Z, Xie J, Zhou J, Feng X, Feng B, Pan Q, Li S, Venkatesan R, Li P, Cao H, Gao C (2022) ROS-responsive polymer nanoparticles with enhanced loading of dexamethasone effectively modulate the lung injury microenvironment. Acta Biomater 148:258–270

Muñoz-Wolf N, Lavelle EC (2021) Promotion of trained innate immunity by nanoparticles. Semin Immunol 56:101542

Murphy-Marion M, Girard D (2018) Titanium dioxide nanoparticles induce human eosinophil adhesion onto endothelial EA.hy926 cells via activation of phosphoinositide 3-kinase/Akt cell signalling pathway. Immunobiology 223(2):162–170

Murray PJ (2017) Macrophage polarization. Annu Rev Physiol 79:541–566

Nakkala JR, Li Z, Ahmad W, Wang K, Gao C (2021) Immunomodulatory biomaterials and their application in therapies for chronic inflammation-related diseases. Acta Biomater 123:1–30

Navarro Chica CE, de Haan BJ, Faas MM, Smink AM, Sierra L, de Vos P, López BL (2020) Design and characterization of squalene-Gusperimus nanoparticles for modulation of innate immunity. Int J Pharm 590:119893

Nie P, Wang M, Zhao Y, Liu S, Chen L, Xu H (2021) Protective effect of *lactobacillus rhamnosus* GG on TiO_2 nanoparticles-induced oxidative stress damage in the liver of young rats. Nanomaterials (Basel, Switzerland) 11(3):803

Petrarca C, Clemente E, Amato V, Pedata P, Sabbioni E, Bernardini G, Lavicoli I, Cortese S, Niu O, Otsuki T, Paganelli R, Gioacchino M (2015) Engineered metal based nanoparticles and innate immunity. Clin Mol Allergy 13:13

Pondman K, Le Gac S, Kishore U (2023) Nanoparticle-induced immune response: health risk versus treatment opportunity? Immunobiology 228:2

Rehman S, Ravinayagam V, Nahvi I, Aldossary H, Al-Shammari M, Amiri MSA et al (2021) Immunity, sex hormones, and environmental factors as determinants of COVID-19 disparity in women. Front Immunol 12:680845

Rendon A, Schakel K (2019) Psoriasis pathogenesis and treatment. Int J Mol Sci 20. https://doi.org/10.3390/ijms20061475

Saafane A, Girard D (2023) Interaction between iron oxide nanoparticles (IONs) and primary human immune cells: An up-to-date review of the literature. Toxicol In Vitro 91:105635

Saafane A, Durocher I, Vanharen M, Girard D (2022) Impact of ultra-small silver nanoparticles of 2 nm ($AgNP_2$) on neutrophil biology: $AgNP_2$ alter the actin cytoskeleton and induce karyorrhexis by a mitogen-activated protein kinase-dependent mechanism in vitro and transitorily attract neutrophils in vivo. Chem Biol Interact 365:110096

Sabareeswaran A, Ansar EB, Harikrishna Varma PR, Mohanan PV, Kumary TV (2016) Effect of surface-modified superparamagnetic iron oxide nanoparticles (SPIONS) on mast cell infiltration: an acute in vivo study. Nanomedicine 12(6):1523–1533

Samms KA, Alkie TN, Jenik K, de Jong J, Klinger KM, DeWitte-Orr SJ (2022) Oral delivery of a dsRNA-phytoglycogen nanoparticle complex enhances both local and systemic innate immune responses in rainbow trout. Fish Shellfish Immunol 121:215–222

Sarmin M, Gurung S, Sarkar S, Das S, Hoda M (2024) White LED light-based photocatalysis enhances the green synthesis of silver nanoparticles by *Trigonella foenum-graecum* seed's aqueous extract. Nano-Struct Nano-Objects 37:101091

Savan R, Gale M, Jr (2023) Innate immunity and interferon in SARS-CoV-2 infection outcome. Immunity 56(7):1443–1450

Shapouri-Moghaddam A, Mohammadian S, Vazini H, Taghadosi M, Esmaeili SA, Mardani F et al (2018) Macrophage plasticity, polarization, and function in health and disease. J Cell Physiol 233:6425–6440

Shapoval O, Sulimenko V, Klebanovych A, Rabyk M, Shapoval P, Kaman O, Rydvalová E, Filipová M, Dráberová E, Dráber P, Horák D (2021) Multimodal fluorescently labeled polymer-coated GdF_3 nanoparticles inhibit degranulation in mast cells. Nanoscale 13(45):19023–19037

Silva AL, Peres C, Conniot J, Matos AI, Moura L, Carreira B, Sainz V, Scomparin A, Satchi-Fainaro R, Préat V, Florindo HF (2017) Nanoparticle impact on innate immune cell pattern-recognition receptors and inflammasomes activation. Semin Immunol 34:3–24

Skallberg A, Bunnfors K, Brommesson C, Uvdal K (2019) Neutrophils activated by nanoparticles and formation of neutrophil extracellular traps: work function mapping and element specific imaging. Anal Chem 91(21):13514–13520

Smolen JS, Aletaha D, McInnes IB (2016) Rheumatoid arthritis. Lancet 388:2023–2038

Sonnenberg J, Schwarz D, Lagache SMM, Schnack L, Körner H, Leffers M, Hardege H, Liu Y, Reinecke H, Findeisen HM, Sohrabi Y (2024) Protocol for the induction of innate immune memory in human smooth muscle cells and endothelial cells in vitro. STAR Protocols 5(1):102776

Summers C, Rankin SM, Condliffe AM, Singh N, Peters AM, Chilvers ER (2010) Neutrophil kinetics in health and disease. Trends Immunol 31:318–324

Swartzwelter BJ, Fux AC, Johnson L, Swart E, Hofer S, Hofstatter N, Geppert M, Italiani P, Borsachi D, Dusch A, Himly M (2020) The impact of nanoparticles on innate immune activation by live bacteria. Int J Mol Sci 21(24):9695

Vallières F, Simard JC, Noël C, Murphy-Marion M, Lavastre V, Girard D (2016) Activation of human AML14.3D10 eosinophils by nanoparticles: modulatory activity on apoptosis and cytokine production. J Immunotoxicol 13(6):817–826

van der Valk FM, van Wijk DF, Lobatto ME, Verberne HJ, Storm G, Willems MC et al (2015) Prednisolone-containing liposomes accumulate in human atherosclerotic macrophages upon intravenous administration. Nanomedicine 11:1039–1046

Vanharen M, Mahbeer T, Léveillé A, Méthot A, Samountry P, Girard D (2023) Impact of gold nanoparticles (AuNPs) in human neutrophils in vitro and in leukocytes attraction in vivo: a sex-based analysis. Environ Toxicol Pharmacol 104:104319

Waisman A, Lukas D, Clausen BE, Yogev N (2017) Dendritic cells as gatekeepers of tolerance. Semin Immunopathol 39:153–163

Wang J, Mei T, Liu Y, Zhang Y, Zhang Z, Hu Y, Wang Y, Wu M, Yang C, Zhong X, Chen B, Cui Z, Le W, Liu Z (2021) Dual-targeted and MRI-guided photothermal therapy *via* iron-based nanoparticles-incorporated neutrophils. Biomater Sci 9(11):3968–3978

Wang J, Liang K, Chen L, Su X, Liao D, Yu J and He J (2023) Unveiling the stealthy tactics: mycoplasma's immune evasion strategies. Front Cell Infect Microbiol 13:1247182

Wu J, Ma T, Zhu M, Huang T, Zhang B, Gao J, Lin N (2022) Nanotechnology reinforced neutrophil-based therapeutic strategies for inflammatory diseases therapy. Nano Today 46:101577

Wynn TA, Chawla A, Pollard JW (2013) Macrophage biology in development, homeostasis and disease. Nature 496:445–455

Yang P, An Y, Feng C, Liu Y, Liu S, Gao L, Zhou Y, Li X, Li P, Zeng F (2024) Heterogeneous high-entropy catalyst nanoparticles for oxygen evolution reaction: impact of oxygen and fluorine introduction. Int J Hydrog Energy 51(Part B):1218–1228

Yeste A, Takenaka MC, Mascanfroni ID, Nadeau M, Kenison JE, Patel B et al (2016) Tolerogenic nanoparticles inhibit T cell-mediated autoimmunity through SOCS2. Sci Signal 9:ra61

Zachs DP, Offutt SJ, Graham RS, Kim Y, Mueller J, Auger JL et al (2019) Noninvasive ultrasound stimulation of the spleen to treat inflammatory arthritis. Nat Commun 10:951

Zhang Q, Dehaini D, Zhang Y, Zhou J, Chen X, Zhang L, Fang RH, Gao W, Zhang L (2018) Neutrophil membrane-coated nanoparticles inhibit synovial inflammation and alleviate joint damage in inflammatory arthritis. Nat Nanotechnol 13(12):1182–1190

Zheng H, Jiang J, Xu S, Liu W, Xie Q, Cai X, Zhang J, Liu S, Li R (2021) Nanoparticle-induced ferroptosis: detection methods, mechanisms and applications. Nanoscale 13:2266–2285

Zhou K, Yang C, Shi K, Liu Y, Hu D, He X, Yang Y, Chu B, Peng J, Zhou Z, Qian Z (2023) Activated macrophage membrane-coated nanoparticles relieve osteoarthritis-induced synovitis and joint damage. Biomaterials 295:122036

Zhou S, Peng H, Zhao A, Li T, Yang X, Lin D (2024) Synthesis of bacterial cellulose nanofibers/Ag nanoparticles: structure, characterization and antibacterial activity. Int J Biol Macromol 259(Part 2):129392

Activation of Cellular Immunity via Nanotechnology

5

Samer Mohammed Hassan Alqarni, Raed Albiheyri, and Irfan A. Rather

5.1 Introduction

Cellular immunity activation is the term describing the process of stimulating immune cells such as NK cells and T cells which helps in improving the ability of pathogen detection and eradication from the body or abnormal cells like cancer cells. The skill is to get a grasp of how nanotechnology can lead to cellular immunity being activated through nanotechnology. Nanotechnology employs a material scale that is in the range of 1–1000 nanometers which resembles components of immune system (immune system components and antigens). By the unique features of the nanoparticles such as size, shape, charge, and surface area, the production of immune responses and vaccines may be promoted. Via targeted drug delivery or the innate immunity enhancements, nanoparticles can effectively elicit and improve the

S. M. H. Alqarni
Department of Biological Sciences, Faculty of Science, King Abdulaziz University, Jeddah, Saudi Arabia

R. Albiheyri
Department of Biological Sciences, Faculty of Science, King Abdulaziz University, Jeddah, Saudi Arabia

Centre of Excellence in Bionanoscience Research, King Abdulaziz University, Jeddah, Saudi Arabia

Immunology Unit, King Fahad Medical Research Centre, King Abdulaziz University, Jeddah, Saudi Arabia

I. A. Rather (✉)
Department of Biological Sciences, Faculty of Science, King Abdulaziz University, Jeddah, Saudi Arabia

Centre of Excellence in Bionanoscience Research, King Abdulaziz University, Jeddah, Saudi Arabia
e-mail: ammm@kau.edu.sa

S. Rehman, E. Al-Suhaimi (eds.), *Nanotechnology Based Microbicides and Immune Stimulators*, https://doi.org/10.1007/978-981-96-0397-8_5

immune response (Smith et al. 2013). The importance of nanotechnology in the above context cannot be ignored as it boosts cellular immunity mechanisms against cancer. To solve the problem of failure in treatment through poor response rates and inefficient drug delivery in cancer, we use nano-immunotherapy which is a mixture of immunotherapy and nanotechnology. Nano-immunotherapy is aimed to enhance effect and lower toxicity in cancer treatment using nanoparticles in order to direct processes in cancer-immunity loop (Zuo et al. 2021). Besides, nanobiotechnology for immunology and vaccination gives way to other new possibilities for discovery of adjuvants, drugs for the immune system and vaccinations. When contrasted with micro-structures, nanoparticles proved to be superior in delivering immune agents and increasing immune responses (Keikha et al. 2021). Novel therapies may be achieved by using nanotechnology that interacts with the immune system in a general way. Even though nano-vaccines show good progress in the treatment of diseases which conventional vaccines are not suitable for, we can design nanocarriers to target some of the immune cells. Indeed, the future of immune system treatment strategies is rooted in the relationship between nanotechnology and immunology (Dacoba et al. 2017).

5.2 Nanoparticles and Cellular Immunity

In the realm of contemporary medicine, nanoparticles have emerged as versatile tools, extending beyond their initial roles as mere transporters of antigens. They now serve as critical modulators of immune responses and potent enhancers of cellular immunity. This multifaceted role of nanoparticles is underpinned by their unique ability to interface with the body's immune system in intricate ways. As research progresses, it becomes increasingly evident that the primary function of these nanoparticles transcends the simple conveyance of antigens. They are instrumental in elevating the efficacy of vaccine delivery, optimizing the distribution and presentation of antigens to elicit robust immune responses. Furthermore, nanoparticles are ingeniously designed to interact with various immune cells, thereby fine-tuning the immune response. This is particularly vital in cancer therapy, where they play a transformative role in stimulating immune cells to recognize and attack cancerous cells more effectively.

Moreover, the significance of nanoparticles in cellular immunity is highlighted by their ability to modify the immune system's behavior toward malignant cells. By leveraging their surface properties and customizable sizes, nanoparticles can be engineered to enhance the recognition and destruction of cancer cells, thus bolstering the body's natural defense mechanisms. They act as bridges, connecting the innate and adaptive arms of the immune system, thereby orchestrating a more synchronized and potent attack against tumors. The nanoparticles can also be designed to release their therapeutic payloads in a controlled manner, ensuring sustained stimulation of the immune system without overwhelming it. This controlled release is particularly crucial in maintaining the delicate balance of the immune system, preventing overactivation that could lead to harmful side effects.

Therefore, the integration of nanoparticles into the domain of cellular immunity marks a significant leap forward in the pursuit of more effective immunotherapies. Their multifunctionality as transporters, modulators, and enhancers opens up new avenues in vaccine development and cancer treatment, promising a future where immune-based therapies are more precise, potent, and personalized. As the understanding of these tiny yet powerful entities deepens, it paves the way for innovative approaches to harness the full potential of the immune system in combating a range of diseases.

5.2.1 Nanoparticles Training Immune Cells to Attack Cancer

Scientists have come up with nanoparticles to enhance the ability of immune cells called macrophages to detect cancers by inducing trained immunity. This mechanism was inspired by a BCG vaccination, which triggers trained immunity, that is applied in cancer treatment by researchers using the development of nanoparticles. These nanoparticles, which are branded as "nanobiologics," are known to target the innate immune cells in the bone marrow and they cause fast immune responses against cancer. They are constructed with BCG parts, lipids, and proteins (Duan et al. 2019). The latest studies have drawn a link between nanoparticles loaded with immunomodulatory agents and the ability to stimulate immune cells while altering the tumor microenvironment, which consequently intensifies antitumor immunity. These nanoparticle-based cancer immunotherapies are tailored to kill cancer cells by targeting key innate immunity pathways such as neutrophils, dendritic cells, macrophages, myeloid-derived suppressor cells, and natural killer cells (Li et al. 2022). In a study conducted prior to clinical trials, a nanotechnology platform was used for immunotherapy sensitization by attaching immune-activating structures on the tumor. This style of work shows the ability of nanoparticles to improve the efficiency of irregular treatment of cancer cells by making them sensitive to immune attacks (Lu et al. 2022).

5.2.2 Nanoparticles and Immune System Interactions

Immunologically speaking, nanoparticles carry a diversity of physicochemical characteristics which allows them to interact with the immune system in a positive or negative manner. The nanoparticles synthesized and engineered may interfere with the normal functioning of the immune system, via either boosting the immune response or suppression of it in case of cancers and so on. The immune system can be manipulated for therapeutic and prophylactic purposes using nanoparticles. Understanding the effects of nanoparticle properties on immune responses is crucial for the successful development of immunotherapies and vaccines (Zolnik et al. 2010). The complicated relationship between nanoparticles and the immune system is addressed by the recent study, for which the implications for nanotoxicology and nanomedicine can be twofold (Aljabali et al. 2023; Ernst et al. 2021). Moreover, nanoparticles can interact with immune cells to cause molecular responses that may lead to various toxic effects, alcoholism, quick death caused by autoimmune disorders, and cancer development (Kononenko et al. 2015). While nanoparticles can function as either haptens or adjuvants, researchers have noted that these particles can demonstrate immunosuppressive effects as well. Physical and chemical characteristics of nanoparticles may display a different performance in biological environment, and this can result in misinterpretation or distortion in the data collection, thus altering the reaction of immune system to these nanomaterials (Kononenko et al. 2015). Modulatory potential of nanoparticles for biological immunosubversion has also become an object of study in the most recent research. Specifically, nanoparticles could be engineered to interact directly with target organs and immunologic cells, which can open a door to a myriad of new therapeutic options (Fadeel 2019).

5.2.3 Application in Inflammatory Diseases

Nanomaterial adjuvants act immunomodulatory in the treatment of inflammatory diseases. Nanoparticles play a vital role in treating different types of immunologic disorders by conjugating different materials and altering their physicochemical properties They can not only reach the immune cells but also make the immune responses change. Compared to the short-term positive results obtained, more research on the mechanism of action and the long-term hazards and consequences is needed to ascertain their safety and health effects (Liu et al. 2022).

Another strategy to overcome these issues is nanomedicine with nanoparticles as the carrier of anti-inflammatory drugs into the body and replace the current therapies limitations such as low bioavailability, high dosages, and severe side effects. The goal of these nanocarriers is to promote the intake of anti-inflammatory agents by a mechanism of control of uptake, distribution, metabolism, and elimination in the body (Brusini et al. 2020). Besides that the nanoparticle-based drug carrier systems have been designed for the selective drug delivery to inflamed gut tissue by the oral route in the situation of inflammatory bowel disease. Through nanoparticle-based selective drug delivery to focused inflammatory areas, researchers seek to enhance the efficacy, and minimize the side effects, of the usual anti-inflammatory drugs (Moharana et al. 2022). With nanoparticles being able to improve pharmacokinetics and pharmacodynamics and their ability to selectively target, they are being investigated as potential therapeutics for chronic inflammatory diseases such as autoimmunity, gastritis, nephritis, hepatitis, ulcerative colitis, Alzheimer's disease, atherosclerosis, and allergic reactions. Nanoparticles have an improved absorption rate and dispersion of therapeutic substances. So, we will overcome the drawbacks of conventional anti-inflammatory drugs with this nanoparticle's stability (Jogpal et al. 2022).

5.2.4 Nanobiotechnology in Immunology and Vaccination

The application of nanotechnology in immunology reveals new possibilities for creating more specific and effective drugs, adjuvants, and vaccines. In contrast to traditional methods, nanoparticles are a more effective way for the transfer of immunological chemicals, which can appeal to the activation of CTL cells against tumors or infected cells (Keikha et al. 2021). Nanotechnology emerges as a promising way to enhance vaccination efficacy via better delivery of immunological molecules to target sites when compared to classical techniques. A range of nanoparticles, including virus-like particles (VLPs) and squalene-based oil-in-water emulsions like MF59, have been studied for vaccine development with some even in use for decades and others still in the early phases of their development (Keikha et al. 2021). In this way, nanoparticles can amplify innate immune responses or act as adjuvants to deliver antigens accurately to the immune system, ultimately leading to a higher volume and quality of immune responses produced against infections. Researchers need to be able to accurately trigger and modify immune responses at the nanoscale to make nanomaterials work. This will lead to the development of vaccines, adjuvants, and immunomodulatory drugs for diseases that are infectious and non-infectious in nature (Smith et al. 2013).

5.3 Future of Nanotechnology in Cellular Immunity

5.3.1 Immune System Targeting

Nanoparticles as optimum carriers can be developed to selectively target immune cells or molecules which permits treatment precision and attention to the detriment of general systemic problems. It is possible that the finding may trigger off the invention of immune therapies which can be useful in the treatment of diseases such as cancer and autoimmune disorders (Liu et al. 2022). Furthermore, the versatility of nanoparticles allows for the incorporation of various functional elements that can enhance the effectiveness of immune therapies. For instance, nanoparticles can be engineered to carry specific markers or ligands that bind exclusively to receptors present on the target immune cells or molecules. This targeted approach not only ensures that the therapeutic agents are delivered precisely where needed, but it also reduces the likelihood of off-target effects, which are common in more traditional treatments. This precision is particularly crucial in conditions like autoimmune diseases, where the immune system mistakenly attacks the body's own cells. By targeting only the aberrant immune cells, nanoparticle-based therapies could offer a solution that restores immune balance without broadly suppressing immune function. Additionally, in cancer treatment, these nanoparticles could be used to deliver drugs or therapeutic genes directly into tumor cells, potentially leading to more effective treatment outcomes with fewer side effects. This innovative approach could pave the way for highly personalized medicine, tailored to the unique genetic and molecular landscape of an individual's disease.

5.3.2 Mucosal Immunity

Nanotechnology has been used for formation of inhalable vaccines, which would improve the immunity by means of humoral response and as a result increase efficiency against respiratory pathogens (Qin et al. 2024). Nanotechnology's application extends to enhancing the delivery and effectiveness of vaccines, particularly those administered through inhalation. This innovative approach capitalizes on the natural entry path of many respiratory pathogens, directly introducing the vaccine to the respiratory system. By doing so, it aims to stimulate a more robust and localized immune response where it's most needed. The nanoparticles used in these inhalable vaccines can be designed to mimic the size and shape of respiratory pathogens, thereby training the immune system more effectively to recognize and combat these invaders. This method could potentially lead to a quicker and more potent immune response, offering improved protection against various respiratory diseases. Moreover, the ease of administration of inhalable vaccines, eliminating the need for needles, could significantly increase vaccination rates and accessibility, particularly in regions where medical resources are scarce.

5.3.3 Combination Therapies

The combination of nanoparticles with other immunomodulatory agents, like cytokines and immune checkpoint inhibitors, can potentially give rise to immune-mediated therapies with higher effectiveness for many diseases (Smith et al. 2013). Expanding on this idea, the synergy between nanoparticles and other immunomodulatory agents can be further optimized through precise control over the timing and dosage of the therapy. Nanoparticles can be engineered to release their payload in response to specific stimuli or environmental conditions within the body, such as pH changes or the presence of certain enzymes that are abundant in disease sites. This controlled release mechanism ensures that the combined therapeutic agents are administered at the optimal moment for maximum efficacy, reducing the risk of potential side effects caused by premature or excessive exposure to these agents. Such a targeted and synchronized approach could be especially beneficial in treating complex diseases like cancer, where the tumor microenvironment often evolves in response to therapy. Nanoparticles could be designed to adapt to these changes, continuously delivering the right combination of drugs and immunomodulatory agents in the right amounts, thereby staying one step ahead of the disease's progression. This evolving strategy exemplifies the future of precision medicine, where treatments are not only tailored to the individual but also adaptable to the dynamic nature of diseases.

5.3.4 Personalized Medicine

Given that nanoparticles could be designed in such a way to target individual patients' immune systems, this suggests that personalized and effective immunotherapies could be a reality in the near future (Smith et al. 2013). The concept of personalized medicine using nanoparticles opens a new frontier in healthcare, where treatments are no longer one-size-fits-all but are tailored to the unique genetic makeup and immune response of each individual. This individualization can be achieved by analyzing a patient's specific immune profile, including genetic, proteomic, and metabolic data, and then designing nanoparticles that can precisely interact with their unique immune system. By doing so, therapies become more effective as they are fine-tuned to address the peculiarities of each person's immune response, reducing the chances of adverse reactions and increasing the likelihood of successful treatment outcomes.

Moreover, the use of nanoparticles in personalized medicine extends beyond treatment to prevention and early diagnosis. Nanoparticles can be employed as part of advanced diagnostic tools to detect the early onset of disease by identifying specific biomarkers associated with certain health conditions. For example, in cancer, nanoparticles could be designed to bind to specific cancer cell markers, enabling early detection and the initiation of treatment at a stage where it is likely to be more effective.

The role of artificial intelligence and machine learning in this realm is also pivotal. These technologies can process vast amounts of medical data to identify patterns and correlations that might go unnoticed by the human eye. When combined with nanotechnology, AI can help in predicting how different nanoparticle formulations might interact with an individual's immune system, thereby optimizing the design and efficacy of personalized treatments. This level of personalization also has implications for vaccine development. Nanoparticles could be used to create custom vaccines based on an individual's exposure history or genetic predisposition to certain diseases, offering a bespoke approach to disease prevention. This could be particularly revolutionary in managing diseases with high mutation rates, such as influenza or HIV, where the ability to quickly adapt vaccines to match evolving strains is crucial.

Thus, the marriage of nanotechnology with personalized medicine holds immense promise for revolutionizing healthcare. It offers a future where treatments and preventive measures are not just based on the symptoms or the disease but are specifically designed to work in harmony with each patient's unique biological makeup. As research in this field advances, we are moving closer to a world where healthcare is truly personalized, more effective, and potentially more efficient.

5.4 Conclusion

The convergence of nanotechnology with cellular immunity marks a transformative era in medical science, offering novel therapeutic avenues and reshaping traditional approaches to disease treatment. Nanoparticles, with their ability to target specific immune components, herald a new age of precision in immunotherapy and vaccine development. This advancement promises greater treatment effectiveness, fewer side effects, and the dawn of personalized medicine tailored to individual immune responses. While challenges remain, particularly in understanding long-term effects and ensuring safety, the potential of nanoparticles in revolutionizing healthcare is immense. As research progresses, we edge closer to a future where nanotechnology-enhanced therapies become a cornerstone in the fight against a myriad of diseases, from cancer to chronic inflammatory disorders.

References

Aljabali AA, Obeid MA, Bashatwah RM, Serrano-Aroca Á, Mishra V, Mishra Y, El-Tanani M, Hromić-Jahjefendić A, Kapoor DN, Goyal R, Naikoo GA, Tambuwala MM (2023) Nanomaterials and their impact on the immune system. Int J Mol Sci 24(3):2008. https://doi.org/10.3390/ijms24032008

Brusini R, Varna M, Couvreur P (2020) Advanced nanomedicines for the treatment of inflammatory diseases. Adv Drug Deliv Rev 157:161–178. https://doi.org/10.1016/j.addr.2020.07.010

Dacoba TG, Olivera A, Torres D, Crecente-Campo J, Alonso MJ (2017) Modulating the immune system through nanotechnology. Semin Immunol 34:78–102. https://doi.org/10.1016/j.smim.2017.09.007

Duan X, Chan C, Lin W (2019) Nanoparticle-mediated immunogenic cell death enables and potentiates cancer immunotherapy. Angew Chem Int Ed 58(3):670. https://doi.org/10.1002/anie.201804882

Ernst L, Casals E, Italiani P, Boraschi D, Puntes V (2021) The interactions between nanoparticles and the innate immune system from a nanotechnologist perspective. Nano 11(11):2991. https://doi.org/10.3390/nano11112991

Fadeel B (2019) Hide and seek: nanomaterial interactions with the immune system. Front Immunol 10. https://doi.org/10.3389/fimmu.2019.00133

Jogpal V, Sanduja M, Dutt R, Garg V, Tinku (2022) Advancement of nanomedicines in chronic inflammatory disorders. Inflammopharmacology 30(2):355. https://doi.org/10.1007/s10787-022-00927-x

Keikha R, Daliri K, Jebali A (2021) The use of Nanobiotechnology in immunology and vaccination. Vaccine 9(2):74. https://doi.org/10.3390/vaccines9020074

Kononenko V, Narat M, Drobne D (2015) Nanoparticle interaction with the immune system/Interakcije nanodelcev z imunskim sistemom. Arch Ind Hyg Toxicol 66(2):97–108. https://doi.org/10.1515/aiht-2015-66-2582

Li K, Zhang Z, Mei Y, Li M, Yang Q, Wu Q, Yang H, He L, Liu S (2022) Targeting the innate immune system with nanoparticles for cancer immunotherapy. J Mater Chem B 10(11):1709. https://doi.org/10.1039/d1tb02818a

Liu J, Liu Z, Pang Y, Zhou H (2022) The interaction between nanoparticles and immune system: application in the treatment of inflammatory diseases. J Nanobiotechnol 20(1). https://doi.org/10.1186/s12951-022-01343-7

Lu Y, Huntoon K, Lee DY, Wang Y, Ha JH, Qie Y, Li X, Schrank BR, Dong S, Gallup TD, Kang M, Zhao H, An Y, Yang Z, Li J, Kim BYS, Jiang W (2022) Immunological conversion of solid

tumours using a bispecific nanobioconjugate for cancer immunotherapy. Nat Nanotechnol 17(12):1332. https://doi.org/10.1038/s41565-022-01245-7
Moharana M, Sahu SN, Pattanayak SK, Khan F (2022) A comprehensive study toward the treatment of inflammatory diseases through nanoparticles. In: Advanced nanomaterials for point of care diagnosis and therapy. Elsevier, pp 169–180. https://doi.org/10.1016/B978-0-323-85725-3.00007-6
Qin L, Sun Y, Gao N, Ling G, Zhang P (2024) Nanotechnology of inhalable vaccines for enhancing mucosal immunity. Drug Deliv Transl Res 14(3):597. https://doi.org/10.1007/s13346-023-01431-7
Smith DM, Simon JK, Baker JR (2013) Applications of nanotechnology for immunology. Nat Rev Immunol 13(8):592. https://doi.org/10.1038/nri3488
Zolnik BS, González-Fernández A, Sadrieh N, Dobrovolskaia MA (2010) Minireview: nanoparticles and the immune system. Endocrinology 151(2):458–465. https://doi.org/10.1210/en.2009-1082
Zuo S, Song J, Zhang J, He Z, Sun B, Sun J (2021) Nano-immunotherapy for each stage of cancer cellular immunity: which, why, and what? Theranostics 11(15):7471–7487. https://doi.org/10.7150/thno.59953

The Use of Nanomaterials in Activating the Complement System

6

Vivek P. Chavda, Amol Gholap, Rajashri Bezbaruah, Pankti C. Balar, Damanbhalang Rynjah, Bruce D. Uhal, and Vladimir N. Uversky

6.1 Introduction

The complement system was discovered in the nineteenth century by a Belgian immunologist and microbiologist Jules Jean Baptiste Vincent Bordet (1870–1961). It is a factor of consideration in almost every body action, from the smallest tissue generation and transplantation to the complex diseases, such as autoimmune disease and cancer. It is a crucial part of immune system, which is a crosslink between innate immunity and acquired immunity. It is comprised mainly of proteins

V. P. Chavda (✉)
Department of Pharmaceutics and Pharmaceutical Technology, L.M. College of Pharmacy, Ahmedabad, Gujarat, India
e-mail: Vivek.chavda@lmcp.ac.in

A. Gholap
Department of Pharmaceutics, St. John Institute of Pharmacy and Research, Palghar, Maharashtra, India

R. Bezbaruah
Institute of Pharmacy, Assam Medical College and Hospital, Dibrugarh, Assam, India

P. C. Balar
Pharmacy section, L.M. College of Pharmacy, Ahmedabad, Gujarat, India

D. Rynjah
Girijanada Chowdhury University-Tezpur, Sonitpur, Assam, India

B. D. Uhal
Department of Physiology, College of Human Medicine, Michigan State University, East Lansing, MI, USA
e-mail: uhal@msu.edu

V. N. Uversky
Department of Molecular Medicine and Byrd Alzheimer's Research Institute, Morsani College of Medicine, University of South Florida, Tampa, FL, USA

S. Rehman, E. Al-Suhaimi (eds.), *Nanotechnology Based Microbicides and Immune Stimulators*, https://doi.org/10.1007/978-981-96-0397-8_6

synthesized by liver. It is a crucial factor in activation of the inflammatory cascade in order to eliminate any foreign cell present in the body through the mechanisms of recognition, opsonization, and finally lysis. It works in correlation with immunoglobulins (majorly IgG and IgM) and hence is termed as "*complement*" (Nesargikar et al. 2012). The concretised thought of sticking to conventional methods for the management of various diseases is not acceptable anymore. Nanomaterials hold a vast range of applications in every field of formulation, from a simple moisturizing cream to the treatment and management of various life-threatening diseases, such as psoriasis, cancer, rheumatoid arthritis, and many more. In this chapter the emphasis is on the application of nanomaterials, may it be in the form of carrier or therapeutic moiety, in the management of diseases that can be achieved by the complement system alterations.

6.2 Mechanisms of the Complement System

6.2.1 Background

The complement system (CS) is vital for protection from foreign substances and altered host cells. It is the crucial part of innate immunity and primarily contains a large number of distinct proteins (more than 60 plasma and cellular surface proteins), most of which are produced by liver. Many of these proteins exist in the form of the precursor *zymogens* (inactive precursors that are activated by proteolytic cleavage), which are widely distributed throughout body fluids and tissues without adverse effect. However, at sites of infection, they are locally activated through a triggered-enzyme cascade, interact with each other, and trigger a series of potent inflammatory events that help fight infection. There are three major locations where the complement system operates. Although tissues and plasma are most frequently considered sites of CS action, this system can be functional within the cells as well (Ricklin et al. 2010). This system is required for maintenance of the homeostasis and helps immune cells to combat the pathogens. The complement proteins play important roles in the CS action and work as a cascade for the opsonization of the pathogens and production of a series of inflammatory responses.

There are three different pathways through which the CS is assumed to be activated, such as a classic pathway (CP), lectin pathway (LP), and the alternative pathway (AP). However, these different initiating pathways lead to the common terminal pathway. The human body conducts pathogen surveillance through the AP, which is actually active at a low level permanently throughout the body (Kolev et al. 2014). The AP action on the healthy host cells is inhibited, and they are quite resistant to low levels of AP activation and hence remain protected from the CS attack on them (Shah et al. 2023). The normal homeostasis in the body is maintained through the generation of apoptotic cells, and the three CS pathways are activated on the surface of such cells as a part of the CS activation in the body. The activation of the adaptive components of the immune system is generated by the dying cells. The innate components also react to the same signals, so the elimination of the dying cells is done in a very proper and tightly controlled manner by the CS, with tight regulation of the CS assembly coordinating all these factors involved in the processes (Merle et al. 2015a).

The full level of the CS activation is achieved in cases of infection caused by various pathogens. As a part of the response to the pathogens, the CS induces the activation of the adaptive type of the immune response with the help of inflammation along with the destruction of the pathogens. To this end, the phagocytosis and opsonization processes are used by the CS. Many studies showed that there is requirement for the tight regulation of the CS activation and function, as the overstimulation of this system may lead to severe complications affecting the host cells, tissues, and the organism as a whole. In fact, CS overstimulation may lead to the enhancement of the organism susceptibility toward infections along with various non-infectious diseases *via* chronic inflammation and autoimmunity. Many severe complications, such as induction of the cancer, graft rejection, along with thrombotic microangiopathy are associated with or induced by the CS overstimulation (Merle et al. 2015b; Chavda et al. 2023a, b).

6.2.2 Activation of CS Forms Homeostasis

The complement component 3 (C3) is treated as the central component of the CS, as triggering of the three initiating pathways, CP, LP, and AP, initiates the enzymatic cascades which converge on C3. Activation of C3 ultimately results in the microbial opsonization, the release of chemotactic factors, such as C3a and C5a, and the generation of a membrane attack complex (MAC). The activation of C3 is achieved *via* cleavage by the C4b2a complex (classical *C3 convertase*) that generates the active C3a and C3b components, where C3a plays the role of the inflammation mediator, whereas C3b acts as an opsonin which is able to bind to any surface that is present nearby the site of its production (Hayes and Melrose 2021).

6.2.3 Role of AP in the Complement System

Under normal physiological conditions, the AP is constitutively presented inside the body as a dormant active state. Under these conditions, the basal state of AP is maintained *via* the low level of activity and low concentrations of the corresponding components, and it is needed to protect the body from the harmful pathogens (Noris and Remuzzi 2013). The control of the basal state of AP involves hydrolysis of the internal thioester bond within C3, which is spontaneously hydrolyzed by water in plasma because of the inherent lability of this bond. This spontaneous hydrolysis converts the C3 into the bioactive form C3(H_2O) (Law and Dodds 2008). Although under physiological conditions, the generation of C3(H_2O) occurs at a low but constant rate in blood, the rate of hydrolysis, however, is dramatically increased by the C3 interaction with various biological surfaces. Some artificial surfaces are also known to accelerate the speed of this hydrolysis reaction (Pangburn et al. 1981).

The lipid surfaces, complexes, gas bubbles, along with biomaterial surfaces all have a great influence on this hydrolysis reaction.

The hydrolysis reaction of the thioester bond involves different structural changes in the C3. These changes are required for the exposure of the binding sites, which are needed for the other types of factors worked within the AP, such as Factor B (FB) (Nilsson and Nilsson 2012). The bioactive form of the C3 interacts with the FB, forming a complex, within which the FB is cleaved by the *serine protease* Factor D, into Ba and Bb. The resulting active labile enzymatic complex C3(H_2O) Bb acts as the *AP C3 convertase* (Isenman et al. 1981), which cleaves native C3 molecules into C3a and C3b. The activity of the *C3 convertase* represents a vital part of the normal homeostasis, as it is responsible for the constant production of the C3b molecules, which triggers a positive feedback loop reaction in which each new C3b molecule is potentially capable of forming a new *AP convertase* complex. The C3b is covalently bound, through a thioester domain, to the nearby C3(H_2O)Bb surface that contains hydroxyl groups (Li et al. 2010). The efficiency of the C3b binding varies depending on the position of the hydroxyl group provided on the nearby surfaces. The hydroxy group presented at the sixth position exhibits greater reactivity than that of the average hydroxyl group present in the sugars. Therefore, the sugar component of the pathogen surface is crucial for the activation and reactivity of the CS (Nishida et al. 2006). The activity of C3b molecules on the host cells is regulated with the help of the complement regulators. The complement regulators which are recruited either from the fluid phase or from the membrane, are required for the host cell protection from harm potentially associated with the C3b action (Rodriguez et al. 2015). Since the induction and accumulation of the C3b through AP is very crucial due to its impact on the host, the activities of the complement regulators are essential as well (Winters et al. 2005; Chavda et al. 2023c). The downregulation of the complement system on the host cells is regulated through a set of plasma and membrane-bound inhibitory proteins, which are present in humans and other mammals. Therefore, the activity and location of the complement system is monitored with the help of these proteins (Manderson et al. 2001).

Although the CP and LP are based on the presence of foreign materials, these pathways are also activated under certain conditions, such as tissue ischemia along with reperfusion. The activation of both these pathways is linked to the tissue injury induction.

6.2.4 Clearance of the Apoptotic Cells in Homeostasis

The body homeostasis maintenance involves programmed cell death also known as apoptosis, and many billions of cells die each day in the human body *via* this process. The assistance of the immune system required for the controlling of apoptosis is achieved *via* the constitutive, low levels of the C3b activation in the body, which do not give any danger signal, but rather regulate the clearance of the apoptotic cells from the body (Galvan et al. 2012). Normal cells get protected from the CS action due to the activity of the complement regulators, which are working in synchronization with the membrane-specific activities and actions of the related proteins (Verbovetski et al. 2002). The membrane characteristics of the apoptotic cells are changed as compared to those of normal cells resulting in the expression of complement regulators (Baudino et al. 2014; Clarke et al. 2015). Some of the essential requirements vital for the clearance of the apoptotic cells are the CP initiators, such as the mannose-binding lectin along with the ficolin. These proteins act as initiators interacting with the receptors on the surface of the phagocytic cells (Nauta et al. 2004). Some of the macrophages or immature dendritic cells act as phagocytic cells, which are essential for the immunotolerance and the prevention of the immune responses to the self-antigens (Walport 2001).

6.2.5 Pathogen Elimination by CS

The factors B and D (FB and FD) play essential roles in the clearance of the pathogen from the body. The complement regulators are not present on the pathogens. As a part of the clearance process, the C3b interacts with the FB and FD, which induces the production of the surface-bound *C3 convertase* through AP. The cleavage of the C3 further into the C3a and C3b can also occur during these events (Merle et al. 2015b). The rapid elimination of the pathogens along with immune response is achieved through the activity of proinflammatory factors. One of the highest levels of complement activation is achieved during the process of pathogen recognition. After these events, the complement system induces the anaphylatoxins with the potent inflammatory activity that act with the help of the phagocytes and leukocytes, along with the endothelial cells. All these activities occur at the site of the infections, where platelets also play major roles. The direct killing action of the pathogens originates from the terminal complement component system, while the complement cascade is activated by the two pathways including CP and LP (Harboe and Mollnes 2008). The AP covers more than 70% activity of the terminal complements system during the process of the recognition of the pathogens. AP C3 type of convertase molecules are formed from the C3 molecules with the help of the C3 activation. This type of event is also invoked with the help of the *AP C3 convertase*. The target opsonization is accelerated by this chain reaction, which later enhances the levels of the anaphylatoxins (Lachmann 2009).

6.3 Factors of Nanomaterials Influencing Complement System

There are several different factors impacting the complement system through nanomaterials, including surface chemistry, zeta potential, shape, and size of the particles. Therefore, nanomaterials can be used in a variety of ways for the complement system activation. The classical pathway, including IgG/IgM/C-reactive type of proteins, mediates the type of pathway involved. The mannose-binding type of the lectin or alternative type of the properdin-mediated pathways is also part of this system (Devine et al. 1994; Chavda et al. 2023d). Refer to Fig. 6.1 for the diagrammatic representation of numerous factors that define the action of nanomaterials on CS.

6.3.1 Impact of Nanomaterial Size

There is a direct proportionality of the CS activity with the increase in the particle size supported through the phagocytic action of the cells involved. The increase in the opsonization activity driven through the complement proteins is required for this activity (Madani et al. 2007; Chavda et al. 2023e). The classical pathway of the CS activation is achieved by the nanoparticles having the size of roughly 40 to 250 nm. However, if the size of the nanoparticles increases to 600 nm, the CS activation

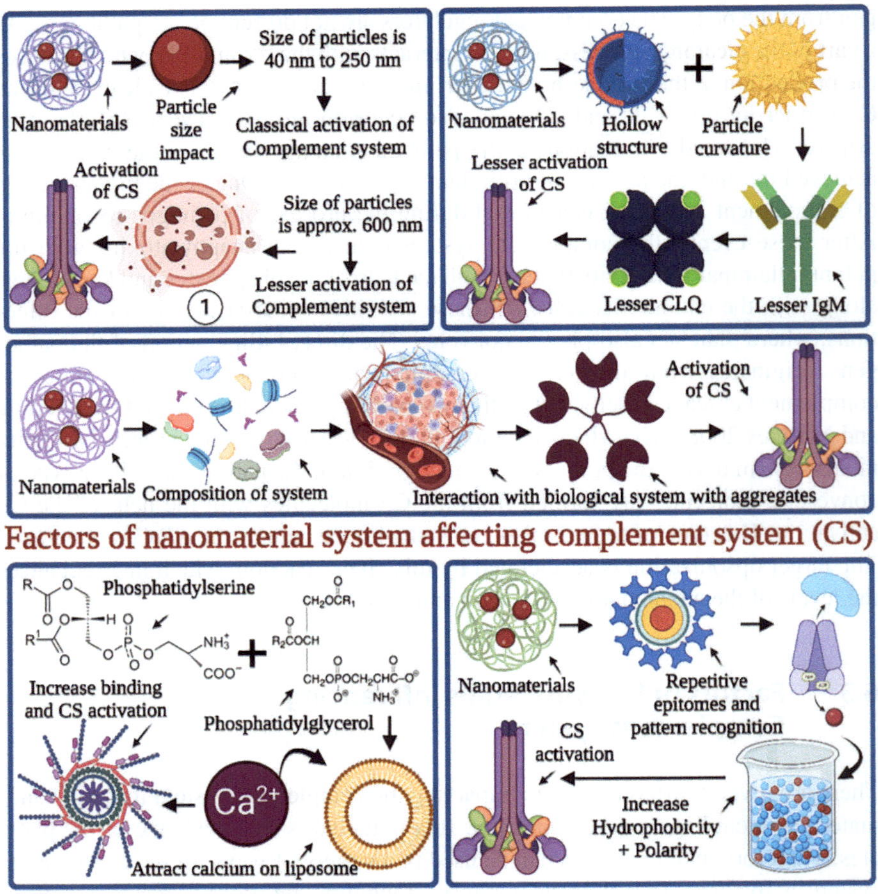

Fig. 6.1 Numerous factors that influence the capability of the nanomaterials to alter the complement system (Created using Biorender.com)

decreases based on the adjustment of the surface areas (Salvadormorales et al. 2006; Sevast'ianov and Tseytlina 1984). The C3b opsonization reaction can be done on the superparamagnetic iron oxide dextran-loaded shell of so-called "nanoworms". The size of these particles is in the range of 200 nm (Salvador-Morales et al. 2008; Hamad et al. 2008). The calcium-based CS system, along with the lectin and classical pathways, are applicable for particles having the size less than 30 nm, owing to the larger size of the C3b which has a surface area of 40 nm^2 (Hamad et al. 2010; Al-Hanbali et al. 2006).

6.3.2 Influence of Nanomaterial Shape and Curvature

In the case of silicon dioxide nanoparticles with sharp curvatures, there is a reduction of the activation of the complement system (Janssen et al. 2006). In the case of peptidoglycan particles with shallow curvatures or that of the sharper curvatures, there is less activation of the CS and the same action is larger in the case of size range of 0.02 to 0.04 nm^{-1} (Pedersen et al. 2010; Malachowski and Hassel 2020; Kouser et al. 2018). The curvature shallower than later figures will not be able to activate the complement cascade due to lesser IgM along with C1q, which are unable to activate CS (Moghimi et al. 2011; Chen et al. 2017; Chavda et al. 2023f).

6.3.3 Nanomaterial Composition for CS

The particles with size less than 30 nm are not capable of activating the CS, and in this case their composition also matters, as they are impacting the interaction of the nanoparticles with the surrounding biological environment (Hamad et al. 2013; Teli et al. 2023). One of the crucial examples of this case is the polyethylene oxide type of polymeric micelles, which are prepared using the polypropylene block type of the copolymer and poloxamer 407, with the size of about 25 nm (Moein Moghimi et al. 2006). These smaller sizes of the micelles are able to significantly activate the CS *via* all three pathways, while in the case of the polyethylene phospholipid micelles the same activation was not induced. There was an increase in the particle's aggregation of micelles due to the activity of poloxamer 407, *via* interaction with chylomicrons along with another lipoprotein. These interactions induced the larger aggregates, which are helpful for the activation of the complement system (Moghimi et al. 2011).

6.3.4 Surface Charge

The surface charge of the nanoparticles, either cationic or anionic, also plays a great role in the CS activation in the body (Szeto and Lavik 2016). Some of the nanomaterials carry anionic charges, as in the example of the liposomes, which are used to attract the calcium ions further resulting in the CS activation *via* the classical pathway (Pham et al. 2011). Many components, such as cardiolipin and phosphatidylserine, along with phosphatidylglycerol contribute to the anionic charge on the liposomes. These components help liposomes to have a great binding potential which also promotes hydrophobic interactions along with hydrogen bonding (Bradley et al. 1999). The alternative pathway induced by the liposomes with sterаylamine gave them cationic charges that helped in the CS activation. The neutral liposomes showed lesser potential for CS activation (Klapper et al. 2014).

6.3.5 Topography Along with the Surface Chemistry

The presence of the repetitive epitopes on the surface of the nanomaterials helps the CS activation *via* pattern recognition mechanisms based on the surface topography (Toda et al. 2008; La-Beck et al. 2021). The attachment of the star-shaped poloxamine-908 to the surface of the nanoparticles resulted in the formation of the repetitive hydrophobicity-polarity pattern (Karmali et al. 2012; Park et al. 2009). Since such type of the patterns can be altered by the modulation of the density of the poloxamine, the docking sites related to the CS pattern recognition also become altered (Hamad et al. 2010; Moghimi et al. 2011). The functional groups are also playing major roles in the activation of the CS pathways, as in the case of high-density amino functional groups along with the hydroxyl functional groups (Moghimi et al. 2001; Simberg et al. 2009). These functional groups are essential for the induction of the nucleophilic attack on the internal thioester bond located within the alpha chain of the nascent C3b (Salvador-Morales et al. 2009; Wang et al. 2016a; Chavda et al. 2023g). Such types of the events increased the efficiency of the CS activation. The cross-linking of the surface polysaccharides also has a great impact on the activation of the CS. In some cases, if the carboxymethyl groups are used for substitution to that of the hydroxyl groups, then there is some decrease in the CS activation (Carreno et al. 1988; Szebeni et al. 2000).

6.4 Clinical Applications

Nanoparticle-induced complement activation leads to the induction of the proinflammatory intermediates, which include anaphylatoxins and also the opsonization of nanoparticles with C3b, which engages with phagocytes (Moghimi et al. 2019). The immune cells are stimulated by the anaphylatoxins (C3a, C4a, and C5a) and generate more inflammatory mediators such as histamine. When the complement activation-related pseudoallergy (CARPA) responses occurred in porcine and canine models, this pattern of inflammatory processes was recognised (Szebeni et al. 2011). It has been shown that some clinically useful nanoparticle-based medications (DaunoXome®, AmBisome®, Amphocil®, Abelcet®, and Doxil®/PLD) elicit hypersensitivity events in individuals that are compatible with CARPA (Kerr and Richards 2012). PEGylated liposomal Doxorubicin (PLD) induces complement in the bloodstreams of those with cancer upon intravenous delivery, and the degree of complement activation (as determined by the generation of s5b-9) was attributed to the production of acute infusion responses (Gabizon and Szebeni 2020). Despite the fact that complement activation triggered by nanoparticles is extensively expressed, clinical evidence of CARPA does not seem to be as typical as one might predict based on studies conducted in vitro (Khadela et al. 2023).

For instance, PLD substantially activates complement in vitro, but in patients, the possibility of acute infusion responses is often below 10%, and they may be minimized through premedications and by lessening the frequency of administration (Gabizon and Szebeni 2020). Nevertheless, the acceptability and pharmacokinetics of medicines provided by nanoparticles can be impacted by unanticipated interactions with circulatory complement proteins (Chavda et al. 2023h).

PEGylation is currently a widely accepted procedure to coat nanomaterials in polyethylene glycol to minimize complement activation, enhance plasma durability, and lengthen circulatory time. All of these factors are crucial for efficient targeted therapy of tumors (Wang et al. 2016b). These methods do not, however, completely eradicate the immunological responses to nanomaterials (Li et al. 2021). In preclinical models, it has been shown by various groups that a systemic approach of the first dosage of PEGylated nanoparticles causes the formation of anti-PEG IgM antibodies, which improve immune detection and removal of the subsequent drug of nanoparticles.

It is essential to highlight that the accelerated blood clearance (ABC) phenomena have not been described in patient reports, and therefore its clinical significance is as of yet unknown. Contrarily, clearance rates in PLD-treated patients fell with subsequent treatment, reaching a maximum of 30% by the course of the third phase (La-Beck and Gabizon 2017). When nanoparticles are delivered systemically, this results in complement-mediated infusion responses, a phenomenon which is typically regarded as unfavorable (Gabizon and Szebeni 2020; Kozma et al. 2019). The uncontrolled complement stimulation might have detrimental and perhaps fatal effects, but controlled complement stimulation by nanomaterials may be desirable for immunization techniques (Kurtovic et al. 2020; Kim et al. 2018).

The pathogen opsonization by complement proteins enhances their absorption by antigen-presenting cells through complement receptors CD21 and CD33 (Chen et al. 2023). When used as intrinsic adjuvants in nanoparticle-based vaccinations, particle-stimulated complement activation components can improve antigen absorption and identification by antigen-presenting cells. It is crucial for T cell co-stimulation and viability because the complement-cleaved products C5a and C3a are produced exclusively at the T cell and APC interfaces (Wang et al. 2019).

The complement CD21 costimulatory receptor and B cell receptors both are engaged by antigens that have been opsonized by complement component C3d, resulting in more effectively activated antibody reactions (Olcina et al. 2019; Chavda et al. 2023i). These complement elements are useful for nanoparticle vaccinations because the targeted stimulation of the complement system promotes the immune reaction to the antigens presented by the nanoparticles (Quach et al. 2018).

Additionally, one study showed that cysteine hydrochloride was integrated into a methoxy poly (ethylene glycol)-block poly (allyl glycidyl ether)-block poly (ε-caprolactone) copolymer to yield a triblock polymer called mPEG5k-PAGE15(NH2)-PCL5k(TPCAH), which links to negatively charged proteins like ovalbumin (OVA) to generate a robust complex through electrostatic association. Additionally, the amino units altered on the lateral chains have the ability to stimulate the complement system in addition to integrating with proteins, which promotes the immune reaction to subunit vaccines (Bezbaruah et al. 2022). The findings demonstrated that the combination of TPCAH@OVA could effectively stimulate the generation of potent anti-OVA-specific antibodies, boost CD4+ T- and CD8+ T-cell stimulation, boost the effectiveness of lymphocyte advancement, and raise the production of several cytokines. Moreover, the TPCAH@OVA's surface is rich with amino groups, which may be used to efficiently trigger the complement system and boost adaptive immune responses. Overall, the findings published to date showed that the triblock copolymer TPCAH may substantially enhance innate and adaptive immunity capacity to resist infections when used as an adjuvant and transporter, thus rendering it a suitable alternative for vaccination purposes (Pan et al. 2019).

The development of a pH-responsive micellar vaccine system (COOH-NPs) utilizing a carboxyl-modified diblock copolymer of poly(2-ethyl-2-oxazoline)-poly(D, L-lactide) (COOH-PEOz-PLA) with hydroxyl groups insulating on the exterior triggered the complement pathway for the increased immune outcomes. The maturing of lymph node-resident dendritic cells, the secretion of antigen-specific IgG, the triggering of antigen-specific CD4+ and CD8+ T-cells, and the quantity of memory T-cell formation in vivo were all considerably raised by COOH-NPs when compared with the standard control nanoparticles (OCH3-NPs). Additionally, COOH-NPs/OVA vaccination in E.G7-OVA tumor-carrier mice not only substantially halted the development of the tumors, but also increased the lifespan of the tumor-carrier animals. These outcomes suggested that COOH-NPs with complement activation potential effectively elevated immunological reactions for the anticancer significance (Li et al. 2019; Chavda et al. 2022).

Complement opsonization can also be used to acquire a therapeutic action, despite the fact that the majority of nanomaterials will retain emphasis on minimizing the harmful effects of the CS. When a wide variety of nanomaterials was tested for uptake in the lung tissue of mice that received injections of nebulized lipopolysaccharide (LPS), it was discovered that albumin nanoparticles which have an agglutinated surface protein had a strong affinity for marginated neutrophils in the lung tissue of the LPS-induced mice, but not in the case of ferritin nanocages (a near-crystalline arrangement of surface proteins). Furthermore, it was shown that near-crystalline protein nanoparticles are deficient in the C3 opsonization necessary for agglutinated surface protein nanoparticle absorption in marginated neutrophils. Perhaps most strikingly, the agglutinated surface protein-liposomes (liposomes with aggregated exterior proteins) led the marginated neutrophils to demarginate (depart their presence in the lung capillaries) and depart to the spleen, in addition to having an affinity for them. The acute respiratory disease syndrome-like phenotypes that resulted had a profound therapeutic impact. As a result, complement-opsonized nanoparticles functioned like a kind of bait, diverting marginated neutrophils out of their pro-inflammatory function and lessening the severity of an acute inflammatory circumstance (Wang and Brenner 2021). Therefore, it is potentially feasible to use nanoparticles' propensity to promote complement activation to increase their effectiveness as antigen transporters for immunizations (Quach et al. 2018; Hatterer et al. 2019).

6.5 Limitations

It has been shown in several studies over the past decade that complement protein production and stimulation hasten the progression of carcinomas in both animal models and human patients. It is also feasible that this stimulation might have additionally reduced the therapeutic efficiency of nanoparticle-based medications, given that nanoparticles are frequently administered to cancer patients to stimulate complement.

Considering the C5a-mediated engagement of immune-suppressive cells and a possible function for complement in endothelial cell retraction, increasing complement activation by both metastatic and non-metastatic malignant cell-derived exosomes could serve as an integral component in encouraging tumor progression (Kerr and Richards 2012; Chavda et al. 2023j). The subsequent process may encourage the propagation of cancer cells in the circulatory system and accelerate the dissemination of metastatic disease. This assumption is further supported by the fact that exosomes obtained from cancer cell lines are noticeably less rigid than those derived from non-malignant cell lines (Lachmann 2018; Chavda et al. 2023k). It has been shown that polymer nanoparticles that can stimulate the complement system can accelerate tumor development in a C5aR1-dependent way. This is likely because they can release C5a and attract and trigger proinflammatory macrophages and Tregs (Macor et al. 2018). The cytokine release in tumor-associated macrophages (TAM) and CD8+ T cells was compromised, and tumor antigen-specific immune reactions were also marginalized, implying that this accelerated tumor development was linked to the inhibition of anticancer immune function (Pio et al. 2019).

The prolonged inflammatory reactions brought on by complement fission byproducts and the absorption of nanoparticles by lymphocytes through complement receptors may counteract the curative effect of the carrier (Moghimi and Farhangrazi 2014). Complement activation linked to these blank nanoparticles has the potential to accelerate tumor development, whereas drug-loaded nanoparticles did not have the same impact. It is speculated that the anticancer medicinal product that is incorporated within the nanoparticle reduces the negative effects of the carrier by preventing both tumor cells and TAMs from ingesting the nanoparticles. Therefore, the tumor-enhancing capability of nanoparticle-induced complement activation for cytotoxic chemotherapy might not be fully comprehended. However, nanoparticle-induced complement activation might potentially reduce the efficiency of other medications, such as immunotherapeutics, when they are administered by nanoparticles and fail to have the effect of directly destroying cancerous cells (Nam et al. 2019). Another factor to consider is that complement stimulation in the blood, which arises after an intravenous medicine infusion containing nanoparticles, is momentary and probably will not endure long enough to have a long-term effect on tumor growth. Although it is unknown if nanoparticles that build up in the tumor cells cause continuous activation of complement, persistent inflammation and complement stimulation can regardless accelerate up the growth of tumors.

The primary problem with research on the intricate interconnections between nanomaterials and innate immunity is that studies conducted in vitro and in vivo experiments do not accurately reflect how biologically relevant nanomedicines operate with the immune systems of patients with cancer (Teunissen et al. 2021). The most common in vivo models employed to show that treatments, particularly nanomedicines, are effective against cancer are xenograft tumor prototypes. However many of these models rely on immunocompromised mice. Better choices for analyzing the intricate interactions among the tumor immunologic context and nanomedicine would be modified genetically and syngeneic tumor systems using immunocompetent mice (Hare et al. 2017). Preclinical toxicity outcomes are significantly influenced by the type of animal species, sensitivity, and pattern of nanoparticle clearance in preclinical studies of nanomedicines. Rats and non-human primates are two common preclinical models that may be resistant to the complement activation and cytokine-storm stimulation caused by nanoparticles (Stančič et al. 2021; Vessillier et al. 2015). In certain situations, it may be appropriate to augment in vivo research with in vitro tests that use human blood.

6.6 Toxicity

Numerous therapeutically authorised drug delivery methods exhibit some toxicity as a result of immune response induction and stimulation of the complement system (DeLoughery 2019). Many therapeutically used nanomaterials have been linked to the development of allergic C5a and signs of complement activation, including anaphylaxis, dyspnea, hypotension, facial swelling, back pain, headaches, tachypnea, chills, hypertension or hypotension, and chest discomfort. The mechanics and clinical instances of complement stimulation by drug-delivering vehicles were studied by a variety of researchers. Liposomal doxorubicin (Doxil), liposomal daunorubicin (DaunoXome™, Gilead Sciences, San Dimos, CA), and Taxol™, Bristol-Myers Squibb, New York, NY (paclitaxel emulsion in Cremophor EL) (Muñoz-Cano et al. 2017) have all been shown to cause adverse impacts post-intravenous infusion in a substantial number of patients. When administered intravenously, dextran iron oxide (ferumoxides, Feridex™) showed a similar pattern of complement-dependent adverse effects, which compelled its producer to halt the medicine from production. Furthermore, the majority of these outcomes might be attributable to complement stimulation by the formulation's lipid materials or, in the instance of Feridex™, the dextran-coated iron oxide (Shah and Dobrovolskaia 2018).

The complement system may substantially increase the capability of blood to clot and stimulate the inflammatory process, both of which increase coagulation. There have been reports of unregulated complement stimulation for a number of nanomedicines, particularly reported "stealth" nanoparticles like liposomes (Stater et al. 2021). The complement activation-related pseudo-allergy (CARPA) and hypersensitivity responses have been linked to complement stimulation; hence low complement activation is preferred for safety considerations. When the immune function recognizes nanomedicines, it causes CARPA, which can be fatal since it activates the complement cascade (Hannon et al. 2019). Additionally, not all individuals who had complement activation exhibited in vivo hypersensitive responses. Despite most patients experiencing cessation of CARPA within a few minutes to several hours of ceasing the IV infusion, a small percentage of these individuals may experience a potentially fatal reaction, with significant interindividual heterogeneity in complement reaction to the particular stimulator (Szebeni and Storm 2015). Only 11% (out of 46) of the observed blood contraindications in vivo were brought on by hematological issues. Leukocyte stimulation, hemolysis, and changes in the total blood count were also reported effects. The RBC membrane disruption that causes the release of the iron-containing protein hemoglobin into the circulation is described as hemolysis. Anemia and impaired oxygen-carrying ability also can impact the brain and various organs and be potentially fatal when there is a therapeutically significant decrease in erythrocyte count. The immune system or other blood cell sections, including coagulation, may be affected by hemolysis within the vascular system (De La Harpe et al. 2019).

6.7 Conclusions

For more than 20 years from the time of the first nanoparticle discovery, nanomaterials represent a focal point for researchers in different fields in which the efficacy of the nanoparticles is repeatedly proven. One of the vivid illustrations of applicability of nanoparticles is given by their ability to alter the CS, which is an underlying cause of many diseases. Although CS is initiated via three pathways, CP, LP, and AP, they all culminate in the common terminal pathway. The CS is an important system that has multiple functions ranging from homeostasis maintenance, to clearance of the apoptotic cells and pathogen elimination among others discussed above. The potential of the nanoparticles to activate, control, and regulate CS can be modulated itself by various factors, such as size, shape, curvature, and composition of nanoparticles. With all the perks they possess, nanoparticles have their limitations and might show toxicity as well. However, based on the balanced comparison of all the pros and cons, the advantages outweigh the disadvantages. In conclusion, we can confidently state that the applications of nanomaterials represent a vast field and yet need to be further elaborated for the further benefit of mankind.

References

Al-Hanbali O, Rutt KJ, Sarker DK, Hunter AC, Moghimi SM (2006) Concentration dependent structural ordering of poloxamine 908 on polystyrene nanoparticles and their modulatory role on complement consumption. J Nanosci Nanotechnol 6(9):3126–3133

Baudino L, Sardini A, Ruseva MM, Fossati-Jimack L, Cook HT, Scott D et al (2014) C3 opsonization regulates endocytic handling of apoptotic cells resulting in enhanced T-cell responses to cargo-derived antigens. Proc Natl Acad Sci USA 111(4):1503–1508

Bezbaruah R, Chavda VP, Nongrang L, Alom S, Deka K, Kalita T et al (2022) Nanoparticle-based delivery systems for vaccines vaccines. Vaccines (Basel) 10(11):1946

Bradley AJ, Brooks DE, Norris-Jones R, Devine DV (1999) C1q binding to liposomes is surface charge dependent and is inhibited by peptides consisting of residues 14–26 of the human C1qA chain in a sequence independent manner. Biochim Biophys Acta Biomembr 1418(1):19–30

Carreno MP, Labarre D, Jozefowicz M, Kazatchkine MD (1988) The ability of sephadex to activate human complement is suppressed in specifically substituted functional sephadex derivatives. Mol Immunol 25(2):165–171

Chavda VP, Vihol D, Mehta B, Shah D, Patel M, Vora LK et al (2022) Phytochemical-loaded liposomes for anticancer therapy: an updated review. Nanomedicine 17(8):547–568

Chavda VP, Nalla LV, Balar P, Bezbaruah R, Apostolopoulos V, Singla RK et al (2023a) Advanced phytochemical-based nanocarrier systems for the treatment of breast cancer. Cancers 15(4):1023

Chavda VP, Khadela A, Shah Y, Postwala H, Balar P, Vora L (2023b) Current status of cancer nanotheranostics: emerging strategies for cancer management. Nano 7(4):368–379

Chavda V, Dange K, Joglekar M (2023c) Mass spectrometry, protein interaction and amalgamation of bioinformatics. In: Chavda V, Anand K, Apostolopoulos V (eds) Bioinformatics tools for pharmaceutical drug product development [Internet], 1st edn. Wiley, pp 77–93. [cited 2023 Feb 27]. Available from: https://onlinelibrary.wiley.com/doi/10.1002/9781119865728.ch5

Chavda VP, Balar PC, Patel SB (2023d) Nanotheranostics-based management of head and neck cancer. Nano 7(2):202–209

Chavda VP, Balar PC, Patel SB (2023e) Interventional nanotheranostics in hepatocellular carcinoma. Nano 7(2):128–141

Chavda VP, Valu DD, Parikh PK, Tiwari N, Chhipa AS, Shukla S et al (2023f) Conventional and novel diagnostic tools for the diagnosis of emerging SARS-CoV-2 variants. Vaccine 11(2):374

Chavda VP, Vuppu S, Mishra T, Balar P (2023g) The emergence of hybrid variants of SARS-CoV-2: towards hybrid immunity. Vaccine 11(4):764

Chavda VP, Vihol D, Patel A, Redwan EM, Uversky VN (2023h) Introduction to bioinformatics, AI, and ML for pharmaceuticals. In: Chavda V, Anand K, Apostolopoulos V, editors. Bioinformatics tools for pharmaceutical drug product development [Internet]. 1st ed. Wiley; 2023 [cited 2023 Mar 5]. p. 1–18. Available from: https://onlinelibrary.wiley.com/doi/10.1002/9781119865728.ch1

Chavda VP, Balar PC, Teli D, Davidson M, Bojarska J, Apostolopoulos V (2023i) Antibody–biopolymer conjugates in oncology: a review. Molecules 28(6):2605

Chavda VP, Pandya A, Kumar L, Raval N, Vora LK, Pulakkat S et al (2023j Apr) Exosome nanovesicles: a potential carrier for therapeutic delivery. Nano Today 49:101771

Chavda VP, Sugandhi VV, Pardeshi CV, Patil RJ, Joshi M, Patel B et al (2023k) Engineered exosomes for cancer theranostics: next-generation tumor targeting. J Drug Delivery Sci Technol 85:104579

Chen F, Wang G, Griffin JI, Brenneman B, Banda NK, Holers VM et al (2017) Complement proteins bind to nanoparticle protein corona and undergo dynamic exchange in vivo. Nat Nanotechnol 12(4):387–393

Chen ST, Oliveira TY, Gazumyan A, Cipolla M, Nussenzweig MC (2023) Article B cell receptor signaling in germinal centers prolongs survival and primes B cells for selection in brief B cell receptor signaling in germinal centers prolongs survival and primes B cells for selection. Immunity 56:547–561

Clarke EV, Weist BM, Walsh CM, Tenner AJ (2015) Complement protein C1q bound to apoptotic cells suppresses human macrophage and dendritic cell-mediated Th17 and Th1 T cell subset proliferation. J Leukoc Biol 97(1):147–160

De La Harpe KM, Kondiah PPD, Choonara YE, Marimuthu T, Du Toit LC, Pillay V (2019) The hemocompatibility of nanoparticles: a review of cell–nanoparticle interactions and hemostasis. Cells 8(10):1209

DeLoughery TG (2019) Safety of oral and intravenous iron. Acta Haematol 142(1):8–12
Devine DV, Wong K, Serrano K, Chonn A, Cullis PR (1994) Liposome—complement interactions in rat serum: implications for liposome survival studies. Biochim Biophys Acta Biomembr 1191(1):43–51
Gabizon A, Szebeni J (2020) Complement activation: a potential threat on the safety of poly(ethylene glycol)-coated nanomedicines. ACS Nano 14(7):7682–7688
Galvan MD, Greenlee-Wacker MC, Bohlson SS (2012) C1q and phagocytosis: the perfect complement to a good meal. J Leukoc Biol 92(3):489–497
Hamad I, Christy Hunter A, Rutt KJ, Liu Z, Dai H, Moein MS (2008) Complement activation by PEGylated single-walled carbon nanotubes is independent of C1q and alternative pathway turnover. Mol Immunol 45(14):3797–3803
Hamad I, Al-Hanbali O, Hunter AC, Rutt KJ, Andresen TL, Moghimi SM (2010) Distinct polymer architecture mediates switching of complement activation pathways at the nanosphere−serum interface: implications for stealth nanoparticle engineering. ACS Nano 4(11):6629–6638
Hamad I, Hunter AC, Moghimi SM (2013) Complement monitoring of Pluronic 127 gel and micelles: suppression of copolymer-mediated complement activation by elevated serum levels of HDL, LDL, and apolipoproteins AI and B-100. J Control Release 170(2):167–174
Hannon G, Lysaght J, Liptrott NJ, Prina-Mello A (2019) Immunotoxicity considerations for next generation cancer nanomedicines. Adv Sci 6(19):1900133
Harboe M, Mollnes TE (2008) The alternative complement pathway revisited. J Cell Mol Med 12(4):1074–1084
Hare JI, Lammers T, Ashford MB, Puri S, Storm G, Barry ST (2017) Challenges and strategies in anti-cancer nanomedicine development: an industry perspective. Adv Drug Deliv Rev 108:25–38
Hatterer E, Barba L, Noraz N, Daubeuf B, Aubry-Lachainaye JP, von der Weid B et al (2019) Co-engaging CD47 and CD19 with a bispecific antibody abrogates B-cell receptor/CD19 association leading to impaired B-cell proliferation. MAbs 11(2):322–334
Hayes AJ, Melrose J (2021 Aug) Neural tissue homeostasis and repair is regulated via CS and DS proteoglycan motifs. Front Cell Dev Biol 2(9):696640
Isenman DE, Kells DIC, Cooper NR, Mueller-Eberhard HJ, Pangburn MK (1981) Nucleophilic modification of human complement protein C3: correlation of conformational changes with acquisition of C3b-like functional properties. Biochemistry 20(15):4458–4467
Janssen BJC, Christodoulidou A, McCarthy A, Lambris JD, Gros P (2006) Structure of C3b reveals conformational changes that underlie complement activity. Nature 444(7116):213–216
Karmali PP, Chao Y, Park JH, Sailor MJ, Ruoslahti E, Esener SC et al (2012) Different effect of hydrogelation on antifouling and circulation properties of dextran–iron oxide nanoparticles. Mol Pharm 9(3):539–545
Kerr H, Richards A (2012) Complement-mediated injury and protection of endothelium: lessons from atypical haemolytic uraemic syndrome. Immunobiology 217(2):195–203
Khadela A, Popat S, Ajabiya J, Valu D, Savale S, Chavda VP (2023) AI, ML and other bioinformatics tools for preclinical and clinical development of drug products. In: Chavda V, Anand K, Apostolopoulos V (eds) Bioinformatics tools for pharmaceutical drug product development [Internet]. 1st ed. Wiley; 2023 [cited 2023 Mar 5]. pp 255–284. Available from: https://onlinelibrary.wiley.com/doi/10.1002/9781119865728.ch12
Kim YJ, Kim KH, Ko EJ, Kim MC, Lee YN, Jung YJ et al (2018) Complement C3 plays a key role in inducing humoral and cellular immune responses to influenza virus strain-specific hemagglutinin-based or cross-protective M2 extracellular domain-based vaccination. J Virol 92(20):e00969-18.
Klapper Y, Hamad OA, Teramura Y, Leneweit G, Nienhaus GU, Ricklin D et al (2014) Mediation of a non-proteolytic activation of complement component C3 by phospholipid vesicles. Biomaterials 35(11):3688–3696
Kolev M, Friec GL, Kemper C (2014) Complement — tapping into new sites and effector systems. Nat Rev Immunol 14(12):811–820

Kouser L, Paudyal B, Kaur A, Stenbeck G, Jones LA, Abozaid SM et al (2018 Feb) Human properdin opsonizes nanoparticles and triggers a potent pro-inflammatory response by macrophages without involving complement activation. Front Immunol 12(9):131
Kozma GT, Mészáros T, Vashegyi I, Fülöp T, Örfi E, Dézsi L et al (2019) Pseudo-anaphylaxis to polyethylene glycol (PEG)-coated liposomes: roles of anti-PEG IgM and complement activation in a porcine model of human infusion reactions. ACS Nano 13(8):9315–9324
Kurtovic L, Boyle MJ, Opi DH, Kennedy AT, Tham WH, Reiling L et al (2020) Complement in malaria immunity and vaccines. Immunol Rev 293(1):38–56
La-Beck NM, Gabizon AA (2017) Nanoparticle interactions with the immune system: clinical implications for liposome-based cancer chemotherapy. Front Immunol 8(APR):416
La-Beck NM, Islam MR, Markiewski MM (2021) Nanoparticle-induced complement activation: implications for cancer nanomedicine. Front Immunol 8(11):603039
Lachmann PJ (2009) The amplification loop of the complement pathways. In: Advances in immunology [Internet]. Elsevier, pp 115–149. [cited 2023 Apr 25]. Available from: https://linkinghub.elsevier.com/retrieve/pii/S0065277608040042
Lachmann PJ (2018) Looking back on the alternative complement pathway. Immunobiology 223(8–9):519–523
Law SKA, Dodds AW (2008) The internal thioester and the covalent binding properties of the complement proteins C3 and C4. Protein Sci 6(2):263–274
Li K, Gor J, Perkins SJ (2010) Self-association and domain rearrangements between complement C3 and C3u provide insight into the activation mechanism of C3. Biochem J 431(1):63–72
Li C, Lu Y, Chen Q, Hu H, Zhao X, Qiao M et al (2019) Tailored polymers with complement activation ability to improve antitumor immunity. Mol Pharm 16(6):2648–2660
Li M, Jiang S, Simon J, Paßlick D, Frey ML, Wagner M et al (2021) Brush conformation of polyethylene glycol determines the stealth effect of nanocarriers in the low protein adsorption regime. Nano Lett 21(4):1591–1598
Macor P, Capolla S, Tedesco F (2018) Complement as a biological tool to control tumor growth. Front Immunol 9(SEP):2203
Madani F, Bessodes M, Lakrouf A, Vauthier C, Scherman D, Chaumeil JC (2007 Feb) PEGylation of microspheres for therapeutic embolization: preparation, characterization and biological performance evaluation. Biomaterials 28(6):1198–1208
Malachowski T, Hassel A (2020) Engineering nanoparticles to overcome immunological barriers for enhanced drug delivery. Eng Regener 1:35–50
Manderson AP, Pickering MC, Botto M, Walport MJ, Parish CR (2001) Continual low-level activation of the classical complement pathway. J Exp Med 194(6):747–756
Merle NS, Noe R, Halbwachs-Mecarelli L, Fremeaux-Bacchi V, Roumenina LT (2015a) Complement system Part II: Role in immunity. Front Immunol [Internet]. 2015 May 26 [Cited 2023 Apr 25];6. Available from: http://www.frontiersin.org/Molecular_Innate_Immunity/10.3389/fimmu.2015.00257/abstract
Merle NS, Church SE, Fremeaux-Bacchi V, Roumenina LT (2015b) Complement system Part I â€“ Molecular mechanisms of activation and regulation. Front Immunol [Internet]. 2015 June 2 [Cited 2023 Apr 25];6. Available from: http://journal.frontiersin.org/Article/10.3389/fimmu.2015.00262/abstract
Moein Moghimi S, Hamad I, Andresen TL, Jørgensen K, Szebeni J, Moein Moghimi S et al (2006) Methylation of the phosphate oxygen moiety of phospholipid-methoxy(polyethylene glycol) conjugate prevents PEGylated liposome-mediated complement activation and anaphylatoxin production. FASEB J 20(14):2591–2593
Moghimi SM, Farhangrazi ZS (2014) Just so stories: the random acts of anti-cancer nanomedicine performance. Nanomedicine 10(8):1661–1666
Moghimi SM, Hunter AC, Murray JC (2001) Long-circulating and target-specific nanoparticles: theory to practice. Pharmacol Rev 53(2):283–318
Moghimi SM, Andersen AJ, Ahmadvand D, Wibroe PP, Andresen TL, Hunter AC (2011) Material properties in complement activation. Adv Drug Deliv Rev 63(12):1000–1007

Moghimi SM, Simberg D, Skotland T, Yaghmur A, Hunter AC (2019) Special section on drug delivery technologies-minireview the interplay between blood proteins, complement, and macrophages on nanomedicine performance and responses. J Pharmacol Exp Ther 370:581–592

Muñoz-Cano R, Pascal M, Araujo G, Goikoetxea MJ, Valero AL, Picado C et al (2017) Mechanisms, cofactors, and augmenting factors involved in anaphylaxis. Front Immunol 8:1193

Nam J, Son S, Park KS, Zou W, Shea LD, Moon JJ (2019) Cancer nanomedicine for combination cancer immunotherapy. Nat Rev Mater 4(6):398–414

Nauta AJ, Castellano G, Xu W, Woltman AM, Borrias MC, Daha MR et al (2004) Opsonization with C1q and mannose-binding lectin targets apoptotic cells to dendritic cells. J Immunol 173(5):3044–3050

Nesargikar P, Spiller B, Chavez R (2012) The complement system: history, pathways, cascade and inhibitors. Eur J Microbiol Immunol 2(2):103–111

Nilsson B, Nilsson EK (2012) The tick-over theory revisited: Is C3 a contact-activated protein? Immunobiology 217(11):1106–1110

Nishida N, Walz T, Springer TA (2006) Structural transitions of complement component C3 and its activation products. Proc Natl Acad Sci USA 103(52):19737–19742

Noris M, Remuzzi G (2013 Nov) Overview of complement activation and regulation. Semin Nephrol 33(6):479–492

Olcina MM, Kim RK, Melemenidis S, Graves EE, Giaccia AJ (2019) The tumour microenvironment links complement system dysregulation and hypoxic signalling. Br J Radiol 92(1093):20180069

Pan Y, Qi Y, Shao N, Tadle AC, Huang Y (2019) Amino-modified polymer nanoparticles as adjuvants to activate the complement system and to improve vaccine efficacy in vivo. Biomacromolecules 20(9):3575–3583

Pangburn MK, Schreiber RD, Müller-Eberhard HJ (1981) Formation of the initial C3 convertase of the alternative complement pathway. Acquisition of C3b-like activities by spontaneous hydrolysis of the putative thioester in native C3. J Exp Med 154(3):856–867

Park JH, Von Maltzahn G, Zhang L, Derfus AM, Simberg D, Harris TJ et al (2009) Systematic surface engineering of magnetic nanoworms for in vivo tumor targeting. Small 5(6):694–700

Pedersen MB, Zhou X, Larsen EKU, Sørensen US, Kjems J, Nygaard JV et al (2010) Curvature of synthetic and natural surfaces is an important target feature in classical pathway complement activation. J Immunol 184(4):1931–1945

Pham CTN, Mitchell LM, Huang JL, Lubniewski CM, Schall OF, Killgore JK et al (2011) Variable antibody-dependent activation of complement by functionalized phospholipid nanoparticle surfaces. J Biol Chem 286(1):123–130

Pio R, Ajona D, Ortiz-Espinosa S, Mantovani A, Lambris JD (2019) Complementing the cancer-immunity cycle. Front Immunol 10(APR):774

Quach QH, Kong RLX, Kah JCY (2018) Complement activation by PEGylated gold nanoparticles. Bioconjug Chem 29(4):976–981

Ricklin D, Hajishengallis G, Yang K, Lambris JD (2010) Complement: a key system for immune surveillance and homeostasis. Nat Immunol 11(9):785–797

Rodriguez E, Nan R, Li K, Gor J, Perkins SJ (2015) A revised mechanism for the activation of complement C3 to C3b. J Biol Chem 290(4):2334–2350

Salvadormorales C, Flahaut E, Sim E, Sloan J, Hgreen M, Sim R (2006) Complement activation and protein adsorption by carbon nanotubes. Mol Immunol 43(3):193–201

Salvador-Morales C, Basiuk EV, Basiuk VA, Green MLH, Sim RB (2008) Effects of covalent functionalization on the biocompatibility characteristics of multi-walled carbon nanotubes. J Nanosci Nanotechnol 8(5):2347–2356

Salvador-Morales C, Zhang L, Langer R, Farokhzad OC (2009) Immunocompatibility properties of lipid–polymer hybrid nanoparticles with heterogeneous surface functional groups. Biomaterials 30(12):2231–2240

Sevast'ianov VI, Tseytlina EA (1984) The activation of the complement system by polymer materials and their blood compatibility. J Biomed Mater Res 18(9):969–978

Shah A, Dobrovolskaia MA (2018) Immunological effects of iron oxide nanoparticles and iron-based complex drug formulations: therapeutic benefits, toxicity, mechanistic insights, and translational considerations. Nanomedicine 14(3):977–990

Shah H, Chavda V, Soniwala MM (2023) Applications of bioinformatics tools in medicinal biology and biotechnology. In: Chavda V, Anand K, Apostolopoulos V (eds) Bioinformatics tools for pharmaceutical drug product development [Internet], 1st edn. Wiley, pp 95–116. [Cited 2023 Mar 5]. Available from: https://onlinelibrary.wiley.com/doi/10.1002/9781119865728.ch6

Simberg D, Park JH, Karmali PP, Zhang WM, Merkulov S, McCrae K et al (2009) Differential proteomics analysis of the surface heterogeneity of dextran iron oxide nanoparticles and the implications for their in vivo clearance. Biomaterials 30(23–24):3926–3933

Stančič B, Qvarfordt B, Berglund MM, Brenden N, Sydow Bäckman M, Fransson M et al (2021) The blood endothelial cell chamber—an innovative system to study immune responses in drug development. Int Immunopharmacol 90:107237

Stater EP, Sonay AY, Hart C, Grimm J (2021) The ancillary effects of nanoparticles and their implications for nanomedicine. Nat Nanotechnol 16(11):1180–1194

Szebeni J, Storm G (2015) Complement activation as a bioequivalence issue relevant to the development of generic liposomes and other nanoparticulate drugs. Biochem Biophys Res Commun 468(3):490–497

Szebeni J, Baranyi L, Savay S, Lutz HU, Jelezarova E, Bunger R et al (2000) The role of complement activation in hypersensitivity to pegylated liposomal doxorubicin (Doxil®). J Liposome Res 10(4):467–481

Szebeni J, Muggia F, Gabizon A, Barenholz Y (2011) Activation of complement by therapeutic liposomes and other lipid excipient-based therapeutic products: prediction and prevention. Adv Drug Deliv Rev 63(12):1020–1030

Szeto GL, Lavik EB (2016) Materials design at the interface of nanoparticles and innate immunity. J Mater Chem B 4(9):1610–1618

Teli D, Balar P, Patel K, Sharma A, Chavda V, Vora L (2023) Molnupiravir: a versatile prodrug against SARS-CoV-2 variants. Meta 13(2):309

Teunissen AJP, Burnett ME, Prévot G, Klein ED, Bivona D, Mulder WJM (2021) Embracing nanomaterials' interactions with the innate immune system. Wiley Interdiscip Rev Nanomed Nanobiotechnol 13(6):e1719

Toda M, Kitazawa T, Hirata I, Hirano Y, Iwata H (2008) Complement activation on surfaces carrying amino groups. Biomaterials 29(4):407–417

Verbovetski I, Bychkov H, Trahtemberg U, Shapira I, Hareuveni M, Ben-Tal O et al (2002) Opsonization of apoptotic cells by autologous iC3b facilitates clearance by immature dendritic cells, down-regulates DR and CD86, and up-regulates CC chemokine receptor 7. J Exp Med 196(12):1553–1561

Vessillier S, Eastwood D, Fox B, Sathish J, Sethu S, Dougall T et al (2015) Cytokine release assays for the prediction of therapeutic mAb safety in first-in man trials — whole blood cytokine release assays are poorly predictive for TGN1412 cytokine storm. J Immunol Methods 424:43–52

Walport MJ (2001) Complement. Second of two parts. N Engl J Med 344(15):1140–1144

Wang Z, Brenner JS (2021) The nano-war against complement proteins. AAPS J 23(5):1–8

Wang G, Chen F, Banda NK, Holers VM, Wu L, Moghimi SM, et al (2016a) Activation of human complement system by dextran-coated iron oxide nanoparticles is not affected by dextran/Fe ratio, hydroxyl modifications, and crosslinking. Front Immunol [Internet]. 2016 Oct 10 [cited 2023 Apr 26];7. Available from: http://journal.frontiersin.org/article/10.3389/fimmu.2016.00418/full

Wang G, Chen Y, Wang P, Wang Y, Hong H, Li Y et al (2016b) Preferential tumor accumulation and desirable interstitial penetration of poly(lactic-co-glycolic acid) nanoparticles with dual coating of chitosan oligosaccharide and polyethylene glycol-poly(D,L-lactic acid). Acta Biomater 1(29):248–260

Wang Y, Zhang H, He YW (2019) The complement receptors C3aR and C5aR are a new class of immune checkpoint receptor in cancer immunotherapy. Front Immunol 19(10):1574

Winters MS, Spellman DS, Lambris JD (2005) Solvent accessibility of native and hydrolyzed human complement protein 3 analyzed by hydrogen/deuterium exchange and mass spectrometry. J Immunol 174(6):3469–3474

Role of Nanoparticle Formulation for the Combination Delivery of Multiple Antigens

7

Ganesh Lahane, Trupti Ghatage,
Jegadheeswari Venkadakrishnan, Twisha Dube,
Dishank Arondekar, Rakesh Kumar, Arti Dhar,
and Audesh Bhat

7.1 Introduction

In modern medicine, the development and application of vaccines have played an indispensable role in safeguarding public health (Han et al. 2018). Vaccines are a cornerstone in preventing and controlling infectious diseases, saving countless lives every year. Vaccines have traditionally relied on weakened or inactivated pathogens to stimulate the immune system's response, protecting against specific diseases (Srivastava and Liu 2003; Seder and Hill 2000; Anderson and May 1985). However, in recent years, the field of vaccinology has undergone a paradigm shift, focusing on novel strategies to enhance the efficacy and versatility of immunization. One promising approach in this pursuit is the development of nanoparticle formulations for the delivery of multiple antigens (Lung et al. 2020; Gregory et al. 2013). Nanoparticles (NPs) are minuscule structures, typically ranging in size from 1 to 100 nanometers (nm), that offer unique advantages in the field of vaccine development (Shinde et al. 2012). Their size allows them to penetrate cellular barriers, interact with immune cells, and enhance antigen delivery, making them ideal candidates for improving the efficacy of vaccines. Moreover, NPs can encapsulate

G. Lahane · T. Ghatage · J. Venkadakrishnan · T. Dube · D. Arondekar · A. Dhar
Department of Pharmacy, Birla Institute of Technology and Sciences Pilani, Hyderabad, Telangana, India

R. Kumar
Department of Biotechnology, Shri Mata Vaishno Devi University, Katra, Jammu and Kashmir, India

A. Bhat (✉)
Centre for Molecular Biology, Central University of Jammu, Jammu, Jammu and Kashmir, India
e-mail: abhat@cujammu.ac.in

S. Rehman, E. Al-Suhaimi (eds.), *Nanotechnology Based Microbicides and Immune Stimulators*, https://doi.org/10.1007/978-981-96-0397-8_7

multiple antigens, creating what is known as a multivalent or polyvalent vaccine (Gregory et al. 2013; Bhatia and Bhatia 2016). This innovation holds tremendous potential for addressing complex diseases that require a multifaceted immune response.

As infectious diseases continue to evolve and adapt, the development of vaccines that can combat multiple antigens simultaneously is becoming increasingly essential. Diseases like HIV (human immunodeficiency virus), tuberculosis, and malaria are characterized by their ability to mutate and diversify, challenging the conventional vaccine development paradigm (Collins et al. 2002). In the past, these pathogens have eluded single-antigen vaccines by developing resistance or evading immune responses. However, by harnessing the power of nanotechnology, we have the opportunity to create a new generation of vaccines capable of targeting multiple antigens from the same pathogen or even different pathogens in a single formulation (Peek et al. 2008; Zdrojewicz et al. 2015).

The ability to deliver multiple antigens through NP-based vaccines opens up new possibilities for enhancing immunogenicity, expanding coverage, and addressing emerging health threats. This approach can revolutionize vaccine design, not only for infectious diseases but also for cancer immunotherapies and autoimmunity management (Nasirmoghadas et al. 2021; Park et al. 2018). Nanoparticles have gained tremendous attention in the past few decades for their unique properties and versatile applications in biomedicine (Nikzamir et al. 2021). These properties include their high surface area, tunable size and shape, biocompatibility, and the ability to encapsulate, protect, and deliver a variety of cargo, including antigens (Nikzamir et al. 2021; De et al. 2008). These features make NPs ideal candidates for use in designing next-generation vaccines and immunotherapies, where the co-delivery of multiple antigens is a paramount requirement (De et al. 2008). Nanoparticle-based formulations are widely used in the delivery of immunostimulatory or immunomodulatory molecules in combination with other immunotherapies (Leleux and Roy 2013; Feng et al. 2019). Nanoparticles are used as vehicles for antigens and adjuvants, protect them from systemic toxicity, and increase their half-life with bioavailability (Shao et al. 2015). It altogether contributes to stronger stimulation of immunity. The targeted and efficient delivery of molecules is a successful innovation in the development of NP formulations. Another critical advantage of NP-based formulations is their ability to incorporate adjuvants, molecules that enhance the immune response. Adjuvants play a pivotal role in promoting the maturation of antigen-presenting cells (APCs), increasing cytokine secretion, and improving antigen persistence (Kheirollahpour et al. 2020; Zhao et al. 2023). By co-delivering antigens and adjuvants within the same NP system, these formulations can amplify the immune response while minimizing potential side effects associated with free adjuvant administration (Awate et al. 2013).

In this chapter, we will discuss the nanoparticle-based formulations for the delivery of antigens. We will delve into the various types of NP-based formulations, their principle and applications, the suitability of the material, and their current status. Furthermore, we will examine the challenges and prospects of this burgeoning field, shedding light on how NP-based formulations may revolutionize the development of vaccines and immunotherapies. Through this exploration, we hope to provide valuable insights into the promising role of NP formulation for the combination delivery of multiple antigens.

7.2 Nanoparticle Formulation for Antigen Delivery

Historically, traditional vaccines have predominantly relied on single-antigen formulations. While effective for many diseases, this approach may fall short when dealing with pathogens that exhibit antigenic variation or diseases with multiple strains. For instance, the influenza virus constantly mutates, requiring an annual update to the seasonal flu vaccine (Hay et al. 2001). Similarly, human HIV presents a significant challenge due to its genetic diversity (Hu et al. 2013). Addressing these issues necessitates the development of vaccines that can simultaneously deliver multiple antigens, enabling the immune system to recognize a broader spectrum of epitopes by continuous release of antigens. Traditional vaccines, while undeniably effective, are not without limitations as described in Fig. 7.1. Studies have found that inclusion of antigens in NPs enhances T-cell and B-cell responses (Shao et al. 2015; Nembrini et al. 2011). The delivery of peptide antigens through single-walled carbon nanotubes enhanced IgG responses (Villa et al. 2011). Nanoparticle formulation has gained popularity in delivering vaccines and antigens for the treatment of

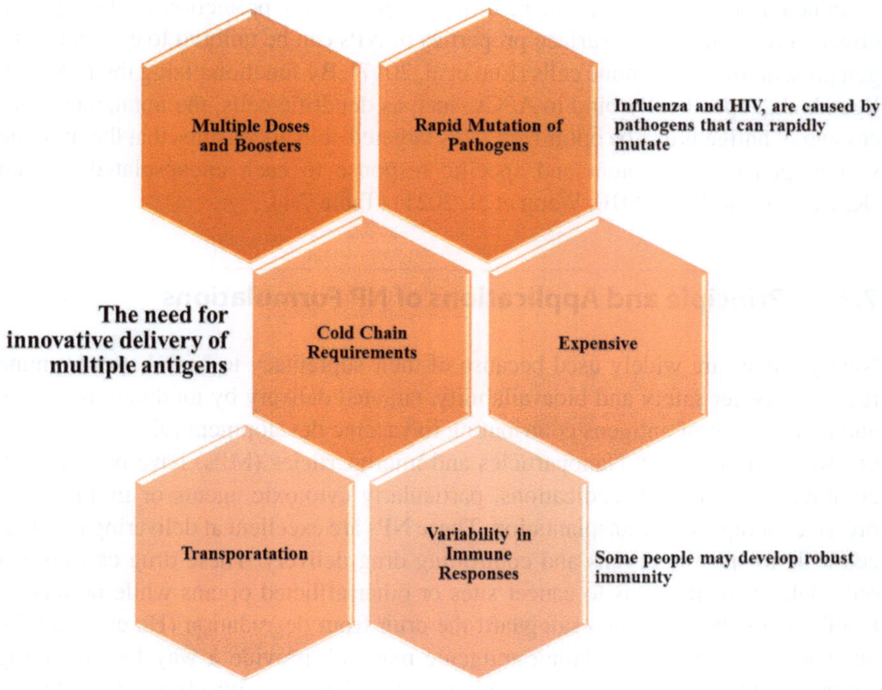

Fig. 7.1 The need for innovative delivery of multiple antigens

cancer, preventing infectious diseases, and various other therapeutic applications (Chesson and Zloza 2017). Nano-formulations are benefiting in delivering the antigens/vaccine specific to the site and targeted delivery owing to having better therapeutic potential (Patra et al. 2018). Antigens which are formulated as a nanoparticle have added advantages their degradation due to extracellular protease is minimal which further increases the bioavailability (Such et al. 2015). The antigen encapsulated as nano-capsules and NPs warrants the transportation of it into the cellular components increasing the vaccine immunogenicity effect in comparison to the vaccine alone (Taki and Smooker 2015). Recently, silica, liposomes, and synthetic polymers are alluring interest in the development of vaccine carriers (Taki and Smooker 2015). The materials chosen for the encapsulation are in such a way that the composites are site-specific, disintegrate easily, and stimulate localized response that not only releases the antigen instantaneously but also enhances prolonged immune system exposure (Gregory et al. 2013).

Simultaneous delivery of multiple antigens has multiple effects, such as stimulating multiple dendritic cell targets and prolonging stimulation by continuous release of antigens (Klippstein and Pozo 2010), offering several advantages for optimizing immunotherapies. First and foremost, it can encapsulate different antigens within a single carrier, preserving their stability and bioactivity. By doing so, these NPs-based formulations ensure that each antigen reaches its target in a coordinated manner, preventing premature degradation and maximizing the immune response (Ho et al. 2021). Additionally, the controlled release of antigens from NPs can lead to sustained immune stimulation, resulting in prolonged protection or therapeutic effects. Furthermore, the surface properties of NPs can be tailored to enhance antigen presentation to immune cells (Luo et al. 2017). By functionalizing the NPs with ligands that specifically bind to APCs, such as dendritic cells, the uptake and processing of antigens can be optimized. This targeted delivery ensures that the immune system generates a potent and specific response to each encapsulated antigen (Klippstein and Pozo 2010; Wang et al. 2021) (Table 7.1).

7.3 Principle and Applications of NP Formulations

Nanoparticles are widely used because of their supremacy to have better immune response, better safety and bioavailability, targeted delivery by modulation of size, and conjugation of antigens contributing to vaccine development (Kim et al. 2019; Oyewumi et al. 2010). Nanoparticles and microparticles (MPs) have been widely employed to transport medications, particularly cytotoxic agents or immunosuppressive therapies for transplantation. These NPs are excellent at delivering pharmaceuticals to specific organs and controlling drug delivery. These drug carriers not only deliver medications to cancer sites or other afflicted organs while protecting healthy cells, but they also safeguard the drug from degradation (He et al. 2011). Nanoparticles, by encapsulating antigenic material, provide a way for delivering antigens that would otherwise disintegrate quickly after injection or produce a localized immune response (Gregory et al. 2013). The antigen can be exposed on

Table 7.1 Advantages of nanoparticle formulation for the combination delivery of multiple antigens

S. no.	Advantages	Descriptions
1.	NPs as versatile vaccine carriers	NPs, as the name suggests, are tiny particles with dimensions typically in the range of 1–100 nanometers. These minuscule carriers offer a multitude of advantages when it comes to vaccine design and delivery.
2.	Enhanced antigen presentation	NPs can be engineered to display antigens in a highly organized and repetitive manner. This arrangement mimics the presentation of pathogens to the immune system, leading to more robust and targeted immune responses.
3.	Protection of antigens	Antigens encapsulated within NPs are shielded from degradation and immune clearance, allowing for prolonged exposure to immune cells and better chances of generating a strong and sustained immune response.
4.	Co-delivery of multiple antigens	One of the most significant advantages of nanoparticle-based vaccine formulations is their ability to carry multiple antigens simultaneously. This is a game-changer for diseases caused by complex pathogens with multiple antigens, such as the human papillomavirus (HPV) and the hepatitis B virus.
5.	Tailored release profiles	Researchers can engineer NPs to release antigens and adjuvants at specific rates, optimizing the timing and duration of immune stimulation.
6.	Adjuvant incorporation	NPs can incorporate adjuvants, substances that enhance the immune response to antigens. This eliminates the need for separate adjuvant administration and ensures precise co-delivery.
7.	Reduced dosing frequency	By offering sustained antigen exposure and enhanced immune responses, nanoparticle-based vaccines have the potential to reduce the need for multiple booster shots.
8.	Improved stability	Some NPs can be lyophilized (freeze-dried) for long-term stability, eliminating the cold chain requirements associated with traditional vaccines and simplifying distribution logistics.

the surface or inside the NPs in this sort of nanoparticle. One essential feature of these vaccines is that, although being constituted of a virus protein, they self-assemble without encapsulating viral RNA, which means they will not reproduce and are not infectious. Vaccinations made using virus-like NPs were among the first to hit the market, specifically against the hepatitis B virus in 1986, followed by vaccinations against the human papillomavirus in 2006 and hepatitis E in 2011 (Yadav et al. 2019). Formulating NPs for the combination delivery of multiple antigens is a complex process that involves several principles to ensure the successful development of effective vaccines or immunotherapies. The key principles involved in NP formulation are shown in flow chart format in Fig. 7.2.

Various physiochemical properties are determining factors for the delivery and efficacy of nanovaccines. The size of the nanovaccine with respect to the surface area of the antigen increases the possibility of antigen uptake and interaction with the targeted cells in comparison to conventional vaccines (Mumper et al. 2003; Wendorf et al. 2006). Studies suggested that NPs of diameter from 20 to 200 nm are

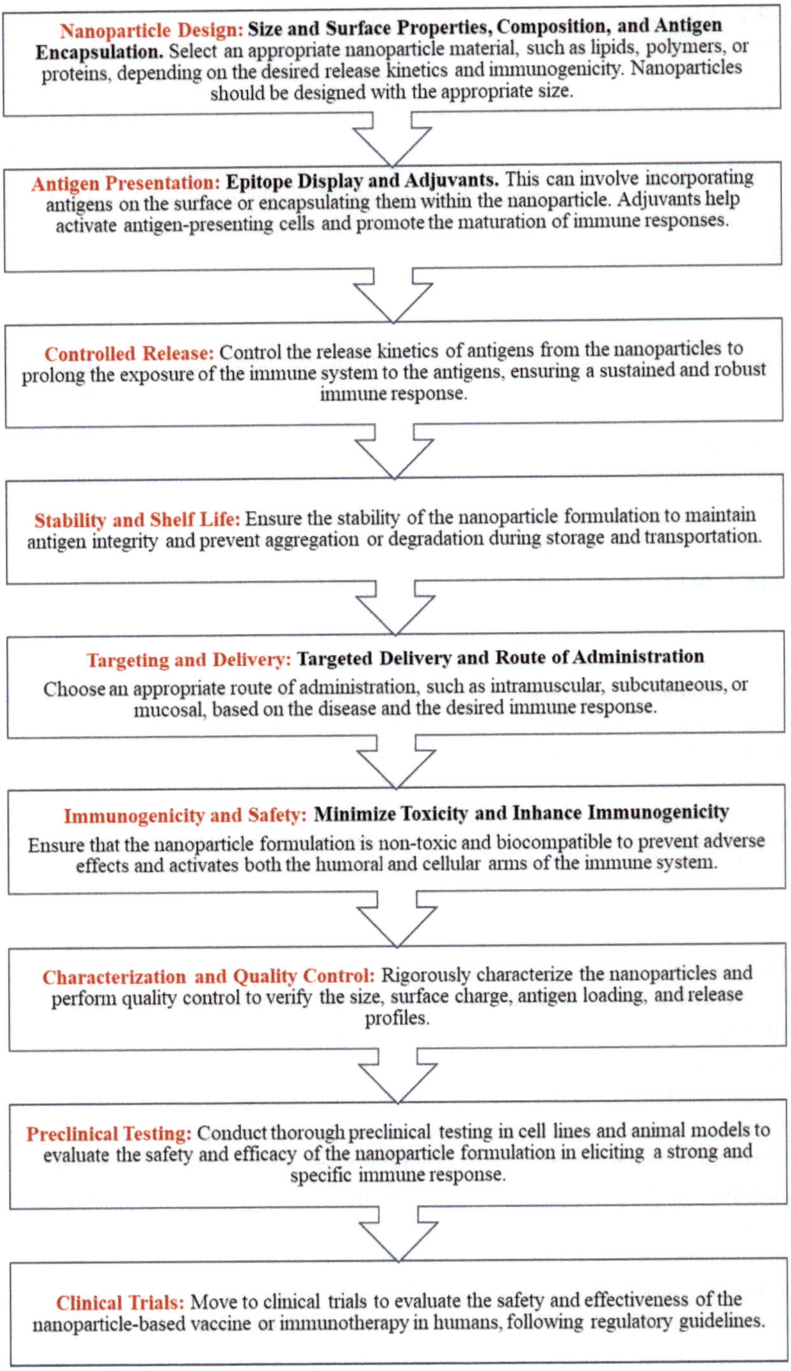

Fig. 7.2 Principle of combination delivery of multiple antigens in nanoparticle formulations

entrapped by the process of endocytosis into APCs inducing adaptive immune responses through T cells whereas NPs ranging 0.5–5 μm are entrapped by phagocytosis further activating humoral immune responses (Falcone et al. 2006; Xiang et al. 2006). Apart from the size of the NP, various other factors which influence the internalization and further transportation of the antigen in the targeted region is the surface charge of the nanovaccine. The cationic-charged NPs are easily entrapped by APCs owing to the cross-presentation through a lysosomal exchange (Gao et al. 2019). Apart from size and surface charges, the matrix system that encapsulates the antigen is a very salient feature in NP-based antigen delivery (Table 7.2).

Vaccine-like particles (VLPs) are nanoscale structures made up of assembled viral proteins that lack viral genetic material and are hence non-infectious (Bai et al. 2008). Because of the hollow inside their structure, VLPs are being investigated as carriers or delivery systems for bio- and nanomaterials such as medications, vaccines, quantum dots, and imaging chemicals. VLPs have been employed in vaccinations to protect against chronic inflammatory disorders such as hypertension, Alzheimer's disease, and rheumatoid arthritis, as well as drug addiction vaccines (Jennings and Bachmann 2009; Buonaguro et al. 2011). Merck's Gardasil® and GlaxoSmithKline's Cervarix® are two commercially available vaccinations that have been designed to prevent HPV-related malignancies developed based on VLP technology (Szarewski 2010). Nano-emulsion adjuvants are oil-in-water emulsions made up of solvents and surfactants. MF59, which is made up of squalene oil, polysorbate 80 (Tween 80), and sorbitan trioleate (Span 85). MF59 is a clinical vaccination adjuvant for influenza that has been delivered intramuscularly to tens of millions of healthy adults, elderly people, and children throughout Europe (Della Cioppa et al. 2012). The nano-emulsion $W_{805}EC$, which is made of soybean oil rather than squalene, has been shown in animal models and, more recently, in humans to improve vaccine antigen targeting and to safely induce potent mucosal, humoral, and cellular immune responses after intranasal administration (Stanberry et al. 2012).

Liposomes and liposome-derived nanovesicles such as archaeosomes and virosomes have emerged as key vaccine carrier systems, and interest in liposome-based vaccinations has grown significantly (Schwendener 2014). Vaccines produced using CAF01, an adjuvant comprised of the immunomodulator glycolipid trehalose dibehenate, a synthetic immune-stimulating mycobacterial cord factor, and the cationic membrane-forming chemical dimethyl dioctadecylammonium, are reported so far (Schwendener 2014; Christensen et al. 2011).

7.4 Suitability of the Material

The suitability of materials for NP formulation in the context of combination delivery of multiple antigens is a critical consideration in the field of vaccine development and immunotherapy. These materials play a pivotal role in ensuring the stability, efficacy, and safety of multi-antigen delivery systems (Peltonen et al. 2010). Several key factors must be taken into account when selecting materials for

Table 7.2 Nanoparticle-based antigen delivery with their size, antigen, and applications

Nano-formulations	Matrix/ expression system	Size	Antigen (pathogen)	Applications	References
Virus-like particles	Baculovirus (Sf9, Sf21, Hi5)	55–60 nm (HPV)	Major capsid protein, L1 (HPV);	It is a vaccine targeting HPV types 6, 11, 16, 18	Villa et al. (2005)
	Mammalian cells	80–120 nm (H1N1)	Gag precursor protein, pr45 (HIV);	It produced CTL activity against V3 peptide-treated target cells.	Kang et al. (1999)
	Yeast		Haemagglutinin (H1N1); Nicotinamide (H1N1) Matrix protein M1 (H1N1)	Safe and effective vaccination against the deadly influenza virus	Quan et al. (2007)
Nanoemulsions	MF59	20 nm to 200 nm	H9N2, H3N2	Nanoemulsion encapsulating antigen to prevent influenza	Zhao et al. (2020)
	$W_{805}EC$	400 nm	NE-Ag	Induce potent mucosal, humoral, and cellular immune responses	Myc et al. (2013)
Liposomes (non-viral lipids)	MPLA;	50–500 nm	R32NS1	Vaccination against the prevention of malaria	Alving et al. (1986)
	Phospholipid S100 and cholesterol		Cholera toxin;	Vaccine for cholera toxin	Alving et al. (1986)
			CtUBE fusion peptide (*H. pylori*);	Protected significant gastritis caused due to *H. pylori*	Zhao et al. (2007)
			KWC *Y. pestis*	Stimulate immunity in the respiratory tract	Baca-Estrada et al. (2000)

(continued)

Table 7.2 (continued)

Nano-formulations	Matrix/ expression system	Size	Antigen (pathogen)	Applications	References
Synthetic	Silica;	5–470 nm	Hepatitis B	Vaccine for prevention of hepatitis B	Bharali et al. (2008)
Polymeric	Poly(lactic-*co*-glycolic acid) (PLGA)	100–200 nm	Docetaxel	Used to target cancers including breast, ovarian, and non-small cell lung cancer	Bharali et al. (2008)
			TetHc (tetanus)	Can be a replacement for conventional tetanus	Gupta et al. (1996)
	Chitosan		Dtxd (Diptheria)	For preventing diphtheria	Schipper et al. (2017)

NP formulation in this context such as biocompatibility, encapsulation efficiency, controlled release, stability, size and surface properties, immunogenicity, scalability, targeting capabilities, adjuvant properties, and regulatory approval (Peltonen et al. 2010; Peltonen and Hirvonen 2010) (Fig. 7.2). For example, a significant drawback of polymeric micelles is the relative instability in blood upon systematic administration, which leads to rapid dissociation and burst release. By using the derivatization of polymers, we can introduce various properties in these polymers, such as increasing the hydrophobicity of the core and introducing hydrogen-bond interaction in the core, promoting electrostatic interactions, and crosslinking the micelle core by thermal or photo-induced polymerization (Zhang et al. 2013). PEGylated and Poly (lactic-co-glycolic acid) (PLGA) is preferred over the conventional polymer for multiple antigen delivery as they have shown to have improved drug stability, half-life, enhancing the biodistribution and pharmacokinetics (Chen et al. 2021; Prakash 2023). The judicious choice of materials can significantly impact the success of multi-antigen delivery systems, ultimately advancing the field of vaccine development and immunotherapy. Nanoparticle-based formulation and their suitable materials are shown in Table 7.3.

Table 7.3 Nanoparticle-based formulation and their suitable materials

Formulations	Materials used	Advantages	Limitations	References
Polymeric NPs	Poly(d,l-lactide-co-glycolide)(PLG), poly(d,l-lactic-coglycolic acid)(PLGA), poly(g-glutamic acid) (g-PGA), poly(ethylene glycol) (PEG), polystyrene.	1. Versatile in terms of material Choices. 2. Controlled release of antigens. 3. Extended circulation time. 4. Protection of antigens from Degradation.	1. Possible toxicity of some polymers. 2. Controlled release may be challenging. 3. Potential aggregation and stability Issues.	Zhao et al. (2014)
Liposome	Distearoyl phosphatidylcholine (DSPC), dipalmytoyl phosphatidylcholine (DPPC), and dimyristoyl phosphatidylcholine (DMPC) and cholesterol.	1. Biocompatible and biodegradable. 2. Facilitates co-encapsulation of Hydrophilic and hydrophobic antigens. 3. Enhanced antigen protection.	1. Limited payload capacity. 2. Potential stability issues in certain Physiological conditions.	Anderson and Omri (2004)
Chitosan NPs	Trimethyl chitosan, dimethyl ethyl chitosan, diethyl methyl chitosan, triethyl chitosan, chitosan-thioglycolic acid, chitosan-cysteine, chitosan-glutathione, and chitosan-thioethylamidine, carboxylated chitosan, amphiphilic chitosan, PEGylated chitosan and lactose-modified chitosan.	1. Chitosan NPs can enhance the mechanical strength of biodegradable plastics. 2. Chitosan is a biodegradable, biocompatible polymer regarded as safe for human dietary use and approved for wound dressing applications.	1. Low solubility in neutral and alkaline pH and low mechanical resistance. 2. Chitosan may also interfere with the absorption of fat-soluble vitamins such as vitamins A, D, E, and K, as well as calcium and magnesium.	Zhu et al. (2012), Mohammed et al. (2017), Bernkop-Schnürch et al. (2003)

Nanosponge	Alkyloxycarbonyl cyclodextrins, methyl-β-cyclodextrin, hydroxypropylβ-cyclodextrins, ethylcellulose, poly valerolactone, eudrgit RS100, acrylic polymers, poly (valerolactone allylvalerolactone), poly (valerolactone-allylvalerolactone oxepanedione), ethyl cellulose, polyvinyl alcohol carbonyldiimidazoles, carboxylic acid di anhydrides, diaryl carbonates, dichloromethane, diisocyanates, diphenylcarbonate, epichloridine, gluteraldehyde, pyromellitic anhydride, 2,2-bis (acrylamido)acetic acid.	1. Non-irritating, non-mutagenic and non-toxic. 2. Nanosponges help to remove the toxic and venom substances from the body. 3. Nanosponges drug delivery system minimizes side effect. 4. Increase formulation stability and enhance the flexibility of the formulation.	1. Size limitation for drugs, as only small molecules can be loaded into the microscopic pores of nanosponges (molecular weight, 100 to 400 kDa). 2. Dose dumping may happen. 3. The loading ability of nanosponges rests on degree of crystallization.	Jilsha and Viswanad (2013), Singh et al. (2016)
Inorganic NPs (gold and silica NPs).	PEG, poly(vinyl alcohol) (PVAlc).	1. High stability and durability. 2. Precise control over particle size and surface properties. 3. Tunable surface functionalization.	1. Limited biocompatibility. 2. Potential toxicity depends on material and size. 3. Limited payload capacity for some Materials.	Paul and Sharma (2020)
Protein-based (albumin NPs, ferritin NPs).	PEG, polylactide (PLA) Polylactide-co-glycolide (PLGA) Poly epsilon-caprolactone (PCL) Poly Isobutylecyanoacrylate(PIBCA) Polyacrylate (Eudragit).	1. Biocompatible and biodegradable. 2. Low immunogenicity in many cases. 3. Controlled release of antigens. 4. Suitable for encapsulating biological Antigens.	1. Limited stability under certain conditions. 2. May require modification for antigen stability.	Kaltbeitzel and Wich (2023, Hong et al. (2020), Verma et al. (2018)

7.5 Current Status and Future Prospects

In this book chapter, we reviewed the role and status of NP formulations for the combination delivery of multiple antigens. As we discussed earlier challenges and applications, few experimental studies and clinical trials were still being conducted to study these NPs and combat the various drawbacks. Fifty-eight nanoparticle treatments and imaging agents have been licensed for clinical use by the FDA in the United States or the European Medicines Agency (EMA) in the European Union. The clinically approved NP formulations have been widely used in various types of cancer, arthritis, pneumonia, amyloidosis, hepatitis, and fibrosis (Anselmo and Mitragotri 2019; Abdellatif and Alsowinea 2021). Some of the approved drugs are listed in Fig. 7.3.

Despite all the potential benefits of NPs, just a few NP-based medications have obtained clinical approval. This is due to various difficulties faced during the manufacturing stages of NP research and manufacture. Researchers have witnessed stability and scale-up problems with liposomal and polymeric NP delivery. There are toxicity concerns, too, due to the accumulation of NP in the cells on continuous exposure. The components used to manufacture NPs also pose problems in the biological system (Gregory et al. 2013).

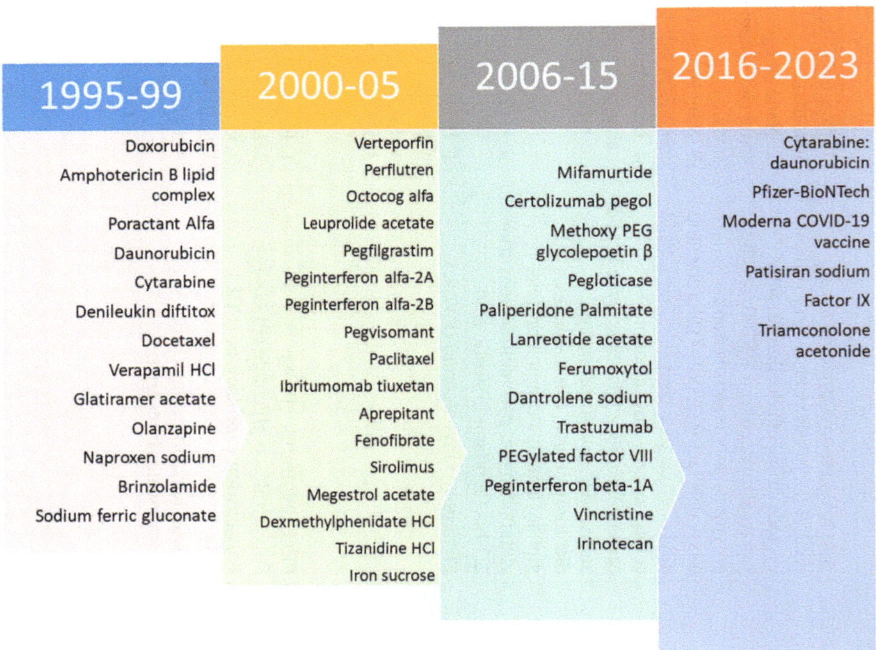

Fig. 7.3 Current status of nanoparticle formulations. (Anselmo and Mitragotri 2019; Abdellatif and Alsowinea 2021)

Bridging the gap between innovative NP technologies and their translation into practical vaccines for human use is a substantial challenge in this field. Firstly, achieving the precise co-encapsulation of different antigens within NPs with consistent efficacy is a complex task. Variability in antigen loading can lead to uneven immune responses, making it crucial to develop reliable encapsulation techniques. Additionally, ensuring the stability of these NPs during storage and transportation is a challenge, as antigens may degrade or aggregate over time. Controlling the release kinetics of multiple antigens from the NPs to mimic the desired immune response can also be challenging, especially when antigens have distinct release profiles. The interactions between different antigens within the NPs and their effects on immune cells require comprehensive investigation. Moreover, safety concerns related to potential toxicity and immune reactions to the NPs themselves must be carefully addressed. Furthermore, regulatory hurdles and the approval process for novel NP-based vaccine formulations can be time-consuming and resource-intensive (Diaz-Arévalo and Zeng 2020; Desai 2012).

Despite the challenges, nanoparticle formulation for combination antigen delivery offers promising opportunities in the field of vaccinology and immunotherapy. One key advantage is the ability to enhance the overall immunogenicity of vaccines. By co-delivering multiple antigens within NPs, it is possible to induce a more robust and diverse immune response, which can be particularly advantageous in the context of infectious diseases or cancer (Park et al. 2013). This approach opens up opportunities for the development of next-generation vaccines that provide broader protection against a range of pathogens (Ho et al. 2021; Ward and Wilson 2020). Nanoparticle-based vaccines also offer opportunities for improved antigen stability and controlled release. These NPs can protect antigens from degradation, thereby extending shelf life and simplifying storage requirements (Bezbaruah et al. 2022). Precise control over the release kinetics of antigens allows for the design of vaccines tailored to elicit specific immune responses, such as rapid or sustained antibody production or T-cell activation. Furthermore, the versatility of NP platforms enables the co-delivery of adjuvants and immunomodulators, enhancing the overall vaccine effectiveness (Akagi et al. 2012; Grego et al. 2020).

In the realm of biomedical research and vaccine development, the integration of NP formulations for the combination delivery of multiple antigens represents a pioneering frontier with profound implications for the future. As we look ahead, several exciting perspectives emerge. Nanoparticle-based antigen delivery is poised to revolutionize vaccine design. These platforms allow for the precise co-delivery of diverse antigens, enhancing the immune system's ability to mount a comprehensive response. This opens doors to novel vaccines for complex diseases like HIV, tuberculosis, and cancer (Mishra et al. 2017). The adaptability of NP formulations lends itself to personalized medicine. Tailoring antigen combinations to an individual's unique immune profile could usher in an era of highly effective, patient-specific immunotherapies. Nanoparticles can be engineered to target specific cells or tissues, minimizing off-target effects (Nguyen 2011). This precision holds tremendous promise for treating autoimmune disorders, allergies, and chronic infections. By co-delivering antigens with immune-stimulating molecules, NPs can potentiate vaccine efficacy. Future research will likely focus on optimizing these adjuvant formulations to boost immune responses further. Nanoparticle-based antigen delivery isn't limited to vaccines. It has potential applications in drug delivery, gene therapy, and diagnostics, expanding its impact across the entire spectrum of healthcare.

7.6 Conclusion

In conclusion, the use of nanoparticle formulations for the combination delivery of multiple antigens represents a groundbreaking approach with significant potential to revolutionize vaccine development and immunotherapy. By harnessing the unique properties of NPs, we can create safer, more effective, and more versatile vaccines that are better equipped to meet the evolving challenges of infectious diseases and other health threats. Continued research and collaboration in this field will undoubtedly lead to breakthroughs that will benefit global health and improve our ability to combat a wide range of diseases.

Acknowledgments A.B. is thankful to ICMR for the financial support (grant nos. RBMH/CAR 1312018-19 and 67/9/2020-DDI/BMS).

References

Abdellatif AA, Alsowinea AF (2021) Approved and marketed nanoparticles for disease targeting and applications in COVID-19. Nanotechnol Rev 10(1):1941–1977

Akagi T, Baba M, Akashi M (2012) Biodegradable nanoparticles as vaccine adjuvants and delivery systems: regulation of immune responses by nanoparticle-based vaccine. Polymer Nanomed:31–64

Alving CR, Richards RL, Moss J, Alving LI, Clements JD, Shiba T et al (1986) Effectiveness of liposomes as potential carriers of vaccines: applications to cholera toxin and human malaria sporozoite antigen. Vaccine 4(3):166–172

Anderson RM, May RM (1985) Vaccination and herd immunity to infectious diseases. Nature 318(6044):323–329

Anderson M, Omri A (2004) The effect of different lipid components on the in vitro stability and release kinetics of liposome formulations. Drug Deliv 11(1):33–39

Anselmo AC, Mitragotri S (2019) Nanoparticles in the clinic: an update. Bioeng Transl edicine 4(3):e10143

Awate S, Babiuk LA, Mutwiri G (2013) Mechanisms of action of adjuvants. Front Immunol 4:114

Baca-Estrada ME, Foldvari M, Snider M, Harding K, Kournikakis B, Babiuk LA et al (2000) Intranasal immunization with liposome-formulated Yersinia pestis vaccine enhances mucosal immune responses. Vaccine 18(21):2203–2211

Bai B, Hu Q, Hu H, Zhou P, Shi Z, Meng J et al (2008) Virus-like particles of SARS-like coronavirus formed by membrane proteins from different origins demonstrate stimulating activity in human dendritic cells. PLoS One 3(7):e2685

Bernkop-Schnürch A, Hornof M, Zoidl T (2003) Thiolated polymers – thiomers: synthesis and in vitro evaluation of chitosan–2-iminothiolane conjugates. Int J Pharm 260(2):229–237

Bezbaruah R, Chavda VP, Nongrang L, Alom S, Deka K, Kalita T et al (2022) Nanoparticle-based delivery systems for vaccines. Vaccine 10(11):1946

Bharali DJ, Pradhan V, Elkin G, Qi W, Hutson A, Mousa SA et al (2008) Novel nanoparticles for the delivery of recombinant hepatitis B vaccine. Nanomedicine 4(4):311–317

Bhatia S, Bhatia S (2016) Nanoparticles types, classification, characterization, fabrication methods and drug delivery applications. In: Natural polymer drug delivery systems: nanoparticles, plants, and algae, pp 33–93

Buonaguro L, Tagliamonte M, Tornesello ML, Buonaguro FM (2011) Developments in virus-like particle-based vaccines for infectious diseases and cancer. Expert Rev Vaccines 10(11):1569–1583

Chen B-M, Cheng T-L, Roffler SR (2021) Polyethylene glycol immunogenicity: theoretical, clinical, and practical aspects of anti-polyethylene glycol antibodies. ACS Nano 15(9):14022–14048

Chesson CB, Zloza A (2017) Nanoparticles: augmenting tumor antigen presentation for vaccine and immunotherapy treatments of cancer. Nanomedicine 12(23):2693–2706

Christensen D, Korsholm KS, Andersen P, Agger EM (2011) Cationic liposomes as vaccine adjuvants. Expert Rev Vaccines 10(4):513–521

Collins KR, Quiñones-Mateu ME, Toossi Z, Arts EJ (2002) Impact of tuberculosis on HIV-1 replication, diversity, and disease progression. AIDS Rev 4(3):165–176

De M, Ghosh PS, Rotello VM (2008) Applications of nanoparticles in biology. Adv Mater 20(22):4225–4241

Della Cioppa G, Nicolay U, Lindert K, Leroux-Roels G, Clement F, Castellino F et al (2012) Superior immunogenicity of seasonal influenza vaccines containing full dose of MF59® adjuvant: results from a dose-finding clinical trial in older adults. Hum Vaccin Immunother 8(2):216–227

Desai N (2012) Challenges in development of nanoparticle-based therapeutics. AAPS J 14(2):282–295

Diaz-Arévalo D, Zeng M (2020) Nanoparticle-based vaccines: opportunities and limitations. In: Nanopharmaceuticals. Elsevier, pp 135–150

Falcone S, Cocucci E, Podini P, Kirchhausen T, Clementi E, Meldolesi J (2006) Macropinocytosis: regulated coordination of endocytic and exocytic membrane traffic events. J Cell Sci 119(22):4758–4769

Feng X, Xu W, Li Z, Song W, Ding J, Chen X (2019) Immunomodulatory nanosystems. Adv Sci 6(17):1900101

Gao S, Yang D, Fang Y, Lin X, Jin X, Wang Q et al (2019) Engineering nanoparticles for targeted remodeling of the tumor microenvironment to improve cancer immunotherapy. Theranostics 9(1):126

Grego EA, Siddoway AC, Uz M, Liu L, Christiansen JC, Ross KA et al (2020) Polymeric nanoparticle-based vaccine adjuvants and delivery vehicles. In: Nanoparticles for rational vaccine design, pp 29–76

Gregory AE, Titball R, Williamson D (2013) Vaccine delivery using nanoparticles. Front Cell Infect Microbiol 3:13

Gupta RK, Chang A-C, Griffin P, Rivera R, Siber GR (1996) In vivo distribution of radioactivity in mice after injection of biodegradable polymer microspheres containing 14C-labeled tetanus toxoid. Vaccine 14(15):1412–1416

Han J, Zhao D, Li D, Wang X, Jin Z, Zhao K (2018) Polymer-based nanomaterials and applications for vaccines and drugs. Polymers 10(1):31

Hay AJ, Gregory V, Douglas AR, Lin YP (2001) The evolution of human influenza viruses. Philos Trans R Soc Lond B 356(1416):1861

He Q, Gao Y, Zhang L, Zhang Z, Gao F, Ji X et al (2011) A pH-responsive mesoporous silica nanoparticles-based multi-drug delivery system for overcoming multi-drug resistance. Biomaterials 32(30):7711–7720

Ho W, Gao M, Li F, Li Z, Zhang XQ, Xu X (2021) Next-generation vaccines: nanoparticle-mediated dna and mrna delivery. Adv Healthc Mater 10(8):2001812

Hong S, Choi DW, Kim HN, Park CG, Lee W, Park HH (2020) Protein-based nanoparticles as drug delivery systems. Pharmaceutics 12(7):604

Hu DJ, Pieniazek D, Mastro TD (2013) The genetic diversity and global molecular epidemiology of HIV. In: AIDS and other manifestation of HIV infection

Jennings GT, Bachmann MF (2009) Immunodrugs: therapeutic VLP-based vaccines for chronic diseases. Annu Rev Pharmacol Toxicol 49:303–326

Jilsha G, Viswanad V (2013) Nanosponges: a novel approach of drug delivery system. Int J Pharm Sci Rev Res 19(2):119–123

Kaltbeitzel J, Wich PR (2023) Protein-based nanoparticles: from drug delivery to imaging, nanocatalysis and protein therapy. Angew Chem Int Ed 62:e202216097

Kang CY, Luo L, Wainberg MA, Li Y (1999) Development of HIV/AIDS vaccine using chimeric gag-env virus-like particles. Biol Chem 380(3):353–364

Kheirollahpour M, Mehrabi M, Dounighi NM, Mohammadi M, Masoudi A (2020) Nanoparticles and vaccine development. Pharm Nanotechnol 8(1):6–21

Kim CG, Kye Y-C, Yun C-H (2019) The role of nanovaccine in cross-presentation of antigen-presenting cells for the activation of CD8+ T cell responses. Pharmaceutics 11(11):612

Klippstein R, Pozo D (2010) Nanotechnology-based manipulation of dendritic cells for enhanced immunotherapy strategies. Nanomedicine 6(4):523–529

Leleux J, Roy K (2013) Micro and nanoparticle-based delivery systems for vaccine immunotherapy: an immunological and materials perspective. Adv Healthc Mater 2(1):72–94

Lung P, Yang J, Li Q (2020) Nanoparticle formulated vaccines: opportunities and challenges. Nanoscale 12(10):5746–5763

Luo M, Samandi LZ, Wang Z, Chen ZJ, Gao J (2017) Synthetic nanovaccines for immunotherapy. J Control Release 263:200–210

Mishra SS, Banode KB, Belgamwar VS (2017) Nanotechnology: a tool for targeted drug delivery. Nanotechnol Appl Pharm Technol:113–137

Mohammed MA, Syeda JT, Wasan KM, Wasan EK (2017) An overview of chitosan nanoparticles and its application in non-parenteral drug delivery. Pharmaceutics 9(4):53

Mumper RJ, Cui Z, Oyewumi MO (2003) Nanotemplate engineering of cell specific nanoparticles. J Dispers Sci Technol 24(3–4):569–588

Myc A, Kukowska-Latallo JF, Smith DM, Passmore C, Pham T, Wong P et al (2013) Nanoemulsion nasal adjuvant W805EC induces dendritic cell engulfment of antigen-primed epithelial cells. Vaccine 31(7):1072–1079

Nasirmoghadas P, Mousakhani A, Behzad F, Beheshtkhoo N, Hassanzadeh A, Nikoo M et al (2021) Nanoparticles in cancer immunotherapies: an innovative strategy. Biotechnol Prog 37(2):e3070

Nembrini C, Stano A, Dane KY, Ballester M, Van Der Vlies AJ, Marsland BJ et al (2011) Nanoparticle conjugation of antigen enhances cytotoxic T-cell responses in pulmonary vaccination. Proc Natl Acad Sci 108(44):E989–EE97

Nguyen KT (2011) Targeted nanoparticles for cancer therapy: promises and challenge. J Nanomed Nanotechnol 2(05)

Nikzamir M, Akbarzadeh A, Panahi Y (2021) An overview on nanoparticles used in biomedicine and their cytotoxicity. J Drug Deliv Sci Technol 61:102316

Oyewumi MO, Kumar A, Cui Z (2010) Nano-microparticles as immune adjuvants: correlating particle sizes and the resultant immune responses. Expert Rev Vaccines 9(9):1095–1107

Park Y-M, Lee SJ, Kim YS, Lee MH, Cha GS, Jung ID et al (2013) Nanoparticle-based vaccine delivery for cancer immunotherapy. Immune Netw 13(5):177–183

Park W, Heo Y-J, Han DK (2018) New opportunities for nanoparticles in cancer immunotherapy. Biomater Res 22:1–10

Patra JK, Das G, Fraceto LF, Campos EVR, Rodriguez-Torres MP, Acosta-Torres LS et al (2018) Nano based drug delivery systems: recent developments and future prospects. J Nanobiotechnol 16(1):1–33

Paul W, Sharma CP (2020) Inorganic nanoparticles for targeted drug delivery. In: Biointegration of medical implant materials, pp 333–373

Peek LJ, Middaugh CR, Berkland C (2008) Nanotechnology in vaccine delivery. Adv Drug Deliv Rev 60(8):915–928

Peltonen L, Hirvonen J (2010) Pharmaceutical nanocrystals by nanomilling: critical process parameters, particle fracturing and stabilization methods. J Pharm Pharmacol 62(11):1569–1579
Peltonen L, Valo H, Kolakovic R, Laaksonen T, Hirvonen J (2010) Electrospraying, spray drying and related techniques for production and formulation of drug nanoparticles. Expert Opin Drug Deliv 7(6):705–719
Prakash S (2023) Nano-based drug delivery system for therapeutics: a comprehensive review. Biomed Phys Eng Exp 9:052002
Quan F-S, Huang C, Compans RW, Kang S-M (2007) Virus-like particle vaccine induces protective immunity against homologous and heterologous strains of influenza virus. J Virol 81(7):3514–3524
Schipper P, van der Maaden K, Groeneveld V, Ruigrok M, Romeijn S, Uleman S et al (2017) Diphtheria toxoid and N-trimethyl chitosan layer-by-layer coated pH-sensitive microneedles induce potent immune responses upon dermal vaccination in mice. J Control Release 262:28–36
Schwendener RA (2014) Liposomes as vaccine delivery systems: a review of the recent advances. Ther Adv Vaccines 2(6):159–182
Seder RA, Hill AV (2000) Vaccines against intracellular infections requiring cellular immunity. Nature 406(6797):793–798
Shao K, Singha S, Clemente-Casares X, Tsai S, Yang Y, Santamaria P (2015) Nanoparticle-based immunotherapy for cancer. ACS Nano 9(1):16–30
Shinde NC, Keskar NJ, Argade PD (2012) Nanoparticles: advances in drug delivery systems. Res J Pharm Biol Chem Sci 3(1):922–929
Singh D, Soni G, Prajapati S (2016) Recent advances in nanosponges as drug delivery system: a review. Eur J Pharm Med Res 3(10):364–371
Srivastava IK, Liu MA (2003) Gene vaccines. Ann Intern Med 138(7):550–559
Stanberry L, Simon J, Johnson C, Robinson P, Morry J, Flack M et al (2012) Safety and immunogenicity of a novel nanoemulsion mucosal adjuvant W805EC combined with approved seasonal influenza antigens. Vaccine 30(2):307–316
Such GK, Yan Y, Johnston AP, Gunawan ST, Caruso F (2015) Interfacing materials science and biology for drug carrier design. Adv Mater 27(14):2278–2297
Szarewski A (2010) HPV vaccine: Cervarix. Expert Opin Biol Ther 10(3):477–487
Taki A, Smooker P (2015) Small wonders—the use of nanoparticles for delivering antigen. Vaccine 3(3):638–661
Verma D, Gulati N, Kaul S, Mukherjee S, Nagaich U (2018) Protein based nanostructures for drug delivery. J Pharm 2018:1
Villa LL, Costa RL, Petta CA, Andrade RP, Ault KA, Giuliano AR et al (2005) Prophylactic quadrivalent human papillomavirus (types 6, 11, 16, and 18) L1 virus-like particle vaccine in young women: a randomised double-blind placebo-controlled multicentre phase II efficacy trial. Lancet Oncol 6(5):271–278
Villa CH, Dao T, Ahearn I, Fehrenbacher N, Casey E, Rey DA et al (2011) Single-walled carbon nanotubes deliver peptide antigen into dendritic cells and enhance IgG responses to tumor-associated antigens. ACS Nano 5(7):5300–5311
Wang F, Ullah A, Fan X, Xu Z, Zong R, Wang X et al (2021) Delivery of nanoparticle antigens to antigen-presenting cells: from extracellular specific targeting to intracellular responsive presentation. J Control Release 333:107–128
Ward AB, Wilson IA (2020) Innovations in structure-based antigen design and immune monitoring for next generation vaccines. Curr Opin Immunol 65:50–56
Wendorf J, Singh M, Chesko J, Kazzaz J, Soewanan E, Ugozzoli M et al (2006) A practical approach to the use of nanoparticles for vaccine delivery. J Pharm Sci 95(12):2738–2750
Xiang SD, Scholzen A, Minigo G, David C, Apostolopoulos V, Mottram PL et al (2006) Pathogen recognition and development of particulate vaccines: does size matter? Methods 40(1):1–9
Yadav R, Zhai L, Tumban E (2019) Virus-like particle-based L2 vaccines against HPVs: where are we today? Viruses 12(1):18
Zdrojewicz Z, Waracki M, Bugaj B, Pypno D, Cabała K (2015) Medical applications of nanotechnology. Adv Hyg Exp Med 69:1196–1204

Zhang Y, Chan HF, Leong KW (2013) Advanced materials and processing for drug delivery: the past and the future. Adv Drug Deliv Rev 65(1):104–120

Zhao W, Wu W, Xu X (2007) Oral vaccination with liposome-encapsulated recombinant fusion peptide of urease B epitope and cholera toxin B subunit affords prophylactic and therapeutic effects against H. pylori infection in BALB/c mice. Vaccine 25(44):7664–7673

Zhao L, Seth A, Wibowo N, Zhao C-X, Mitter N, Yu C et al (2014) Nanoparticle vaccines. Vaccine 32(3):327–337

Zhao L, Zhu Z, Ma L, Li Y (2020) O/W nanoemulsion as an adjuvant for an inactivated H3N2 influenza vaccine: based on particle properties and mode of carrying. Int J Nanomedicine Volume 15:2071–2083

Zhao T, Cai Y, Jiang Y, He X, Wei Y, Yu Y et al (2023) Vaccine adjuvants: mechanisms and platforms. Signal Transduct Target Ther 8(1):283

Zhu X, Su M, Tang S, Wang L, Liang X, Meng F et al (2012) Synthesis of thiolated chitosan and preparation nanoparticles with sodium alginate for ocular drug delivery. Mol Vis 18:1973

Passive Nano Particulate Chemotherapeutic Gene Targeting of Immune Cells

8

Odate K. Tadros

8.1 Introduction "Nanoparticles Era"

Cancer is a major disease that causes death worldwide despite all the progress in developing new therapy strategies. Surgery, radiation, and the use of chemotherapeutic agents considered the frequently used therapies that face major challenges due to severe toxicity and risk of development of severe side effects. To improve the therapeutic outcomes and lower treatment risks, targeted therapy has been developed. Yet, ineffective drug delivery was an issue as multidrug resistance happens due to failure of drug delivery at the desired target site. Resistance mechanism development in cancer cells and relapse susceptibility limits effective cancer treatment. Nanomaterials are used for cancer treatment due to their high specificity and minute size which allow delivery of anticancer drugs to the target tumor site by either active or passive means (Tiwari et al. 2023).

Figure 8.1 shows different types of nanoparticles that can be used in cancer treatment (Elumalai et al. 2024).

Compared to conventional drugs, Nanoparticle drug delivery methods have better tumor targeting, biocompatibility, stability, and pharmacokinetics. They significantly lower systemic toxicity (Verma et al. 2023).

Passive targeting method uses tumor microenvironment, permeability, and retention effect. On the other hand, active targeting uses a ligand-coated nanoparticles.

Nano-oncology is a branch of biomedical research and engineering that studies cancer treatment and diagnosis using nanotechnology.

Nanotechnology can be effectively used to early diagnose cancer using tumor imaging to find specific biomarkers. For cancer therapy, nanotechnology applies photo-induced nano-sensitizers, reverses multidrug resistance, and enables efficient

O. K. Tadros (✉)
Department of Allied Health Sciences, Al-Balqa Applied University, Salt, Jordan
e-mail: o.tadros@bau.edu.jo

S. Rehman, E. Al-Suhaimi (eds.), *Nanotechnology Based Microbicides and Immune Stimulators*, https://doi.org/10.1007/978-981-96-0397-8_8

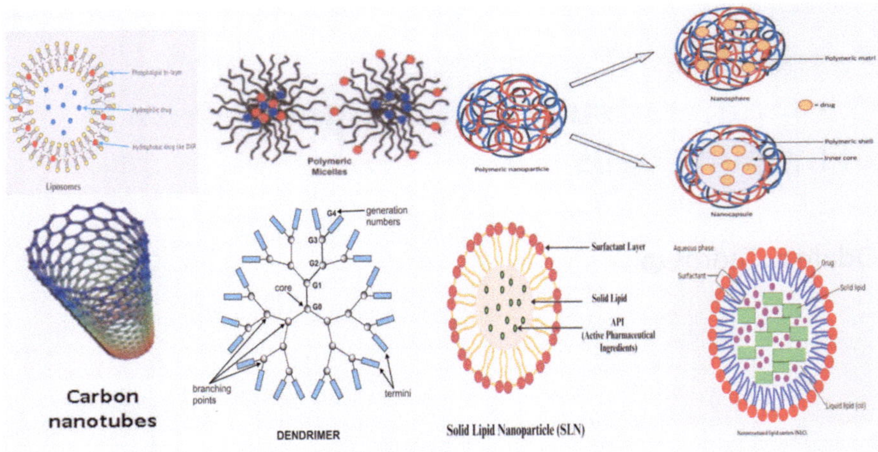

Fig. 8.1 Different types of nanoparticles are used in cancer treatment

delivery of drugs for therapeutic applications. Peptides, ligands, metals, polymeric materials, and semiconductors have bulky sizes that limit their use for cancer detection and therapy. Medical applications and assay have been done by linking these biomolecules covalently to active nanoparticles.

To conquer the depression in bone marrow, ototoxicity, poor stability, and nephrotoxicity of conventional drugs, several nano-formulations have been designed (Elumalai et al. 2024).

Nanoparticles improve drug delivery systems, allowing controlled and targeted release of medications that will reduce side effects and enhance the efficacy of cancer therapy. Nanoparticles encapsulate drugs to protect them from degradation while delivering them directly to the target site, allowing controlled manner of drug release that will maximize the therapeutic effects (Rajeeva et al. 2014) and improving patient outcomes. Healthcare professionals can personalize treatment by controlling dosage and timing of drug release (Gashtasbi 2015). Additionally, nanoparticles penetrate blood-brain barrier, allowing therapeutic drugs to be delivered directly to the brain for neurological disorders like Alzheimer's disease (Elumalai et al. 2024).

Intravenous nanoparticles reach the blood and eradicate foreign materials through their interaction with the surroundings. Nanoparticles are recognized by phagocytic cells as external agents, that are being taken as soon as they undergo opsonization in blood stream. Barriers as antigen expression and permeability and tumor diverse surrounding prevent uniform delivery of drugs and nanoparticles to tumor. Advancements should be made to nanotechnology to conquer body clearance and increase accumulation in specific tumor. The development in drug delivery and significance of nano-technological modes of targeting cancer cells to reverse multidrug resistance is shown in Fig. 8.2 (Tiwari et al. 2023).

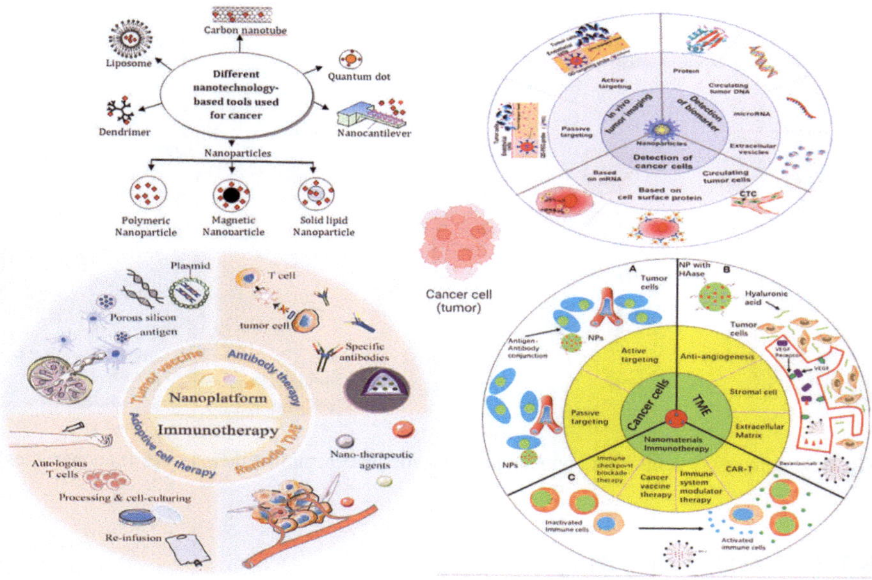

Fig. 8.2 Cancer diagnosis and treatment using nanotechnology with different types of nanotechnological tools: dendrimers, liposome, micelles, and polymers are utilized for drug nanoparticles targeted delivery that may be active or passive

Nano-particles-based delivery system (Fig. 8.3 summarizes delivery systems):

- Nanoparticles-based tumor-targeted delivery systems, ICD (immunogenic cell death) is delivered to induce cytotoxic agents to specific sites that have tumor, NPs concentrate in the interstitial space of tumor, and this will change TME or trigger ICD to generate immune response and cause tumor cells to die.
- NPs release ICD-inducing cytotoxic drugs, inducing cell surface CRT translocation, ATP will be released, and HMGB1 expression will eventually increase. All of that will drive tumor antigens procedure and lead to the production of cytotoxic T cells production to cause death to tumor cells.
- On the other hand, nanoparticle systems will change TME by targeting TAM. Target TAM: polarizing macrophages to M1 phenotype, then stimulate inflammatory response, to suppress tumor growth.
- Target immunosuppressive molecules: NPs act on TME immunosuppressive molecules as TGF-β and IDO, to increase infiltration of CD8+ T cells and NK cells, which will inhibit tumor growth.
- Target immunocytes: NPs directly act on immune cells as cytotoxic T cells and macrophages to produce pro-inflammatory cytokines (IFN-γ and IL-2), which will increase antitumor immune response and suppress tumor growth (Qiyan et al. 2022).

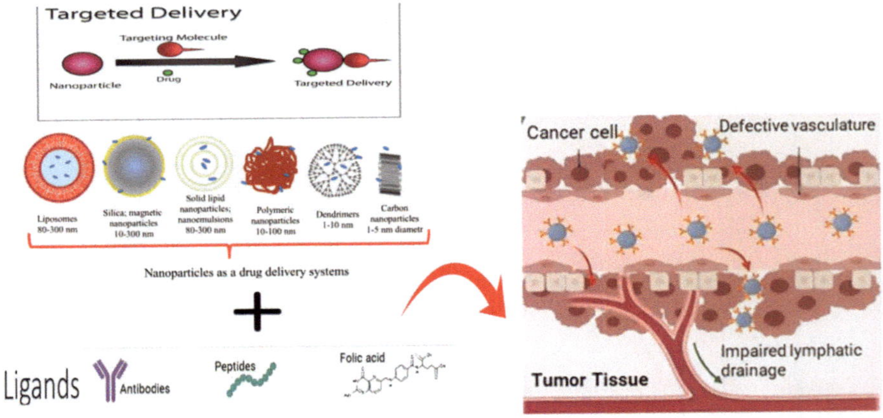

Fig. 8.3 delivery systems based on nanoparticles for cancer immunotherapy. NP (nanoparticles)

8.1.1 Passive Targeting

Pharmacological factors as well as physico-chemical factors control drug administration using a drug carrier system which involves passive targeting at the targeted region while taking into account nanoparticle features (chemistry and size), tumor vasculature, increased permeability and retention effect (EPR), and tumor microenvironment (TME). Angiogenesis takes place as tumor tissues seek nutrition and oxygen. Tumor cells emit chemicals that signal the creation of new blood vessels in the host tissues. FENDRR, a long non-coding RNA (LncRNA), is one example; it increases the proliferation and angiogenesis of human retinal endothelial cells in the context of elevated glucose levels (Shi et al. 2019). The distances between angiogenic blood arteries are greater than those between normal cells. Inadequate lymph removal is caused by poor vasculature, resulting in the EPR effect. This vascular embolism permits nanoparticles to spread via extravascular areas and collect in tumor tissues. Nanoparticles' size should not exceed 100 nm (Ramamurthi et al. 2022). Covalent linkage of nanoparticles and PEG can also be employed for passive targeting, which helps desorption and aggregation of the ligand when released in the bloodstream (Gao et al. 2022). According to studies, elevated concentrations of vascular mediators such as bradykinins, prostaglandins, nitric oxide, basic fibroblast growth factors, and platelet-derived growth factors are responsible for abnormal vasculature during angiogenesis. Compared to free medication administration, nanoparticles improve drug effectiveness and accumulation. The distribution of nanoparticles into tumor tissue is regulated by circulation, surface properties, half-life, size, and angiogenesis level.

Also, passive targeting is aided by the tumor cell's surrounding microenvironment. Tumor cells require a fast metabolic rate to proliferate quickly. The body's supply of nutrient and oxygen usually will not meet tumor tissue requirements. Hypoxic glycolysis (metabolism) will be used as extra energy source tumor cells, and this will create an acidic environment. Liposomes (drug delivery nanoparticles source) are designed to remain at 7.4 pH level but degraded in acidic surroundings to release the active drug.

Targeting MMPs (matrix metalloproteinases; an enzyme produced by cancer cells that affect angiogenesis and metastasis) through tumor-activated prodrug therapy has significantly improved cancer treatment efficacy. Doxorubicin (a potent anti-neoplastic agent) was observed to be cleaved by the enzyme (Tiwari et al. 2023). Figure 8.4 illustrates liposomal drug delivery.

Nanoparticles heterogeneous localization inside tumor cell is still a major drawback of passive targeting. Covalent bonds are a powerful mode of cancer therapy passive drug targeting. Polymer-drug conjugates are chemically bonded drug molecules to polymers' side chains. These conjugates enter the blood as inactive form of prodrugs (type of medications that is inert and becomes active at site of action by regular metabolic processes, as enzymatic breakdown of drug-polymer link.

The administration of paclitaxel via covalent conjugation with PEG-dihydrolipoic acid via a disulfide connection (MeO-PEG2k-DHLA) has demonstrated anti-tumor effect against the cancer cell lines BEL-7402, 4 T1, MCF-7, and A549 (Wang et al. 2019).

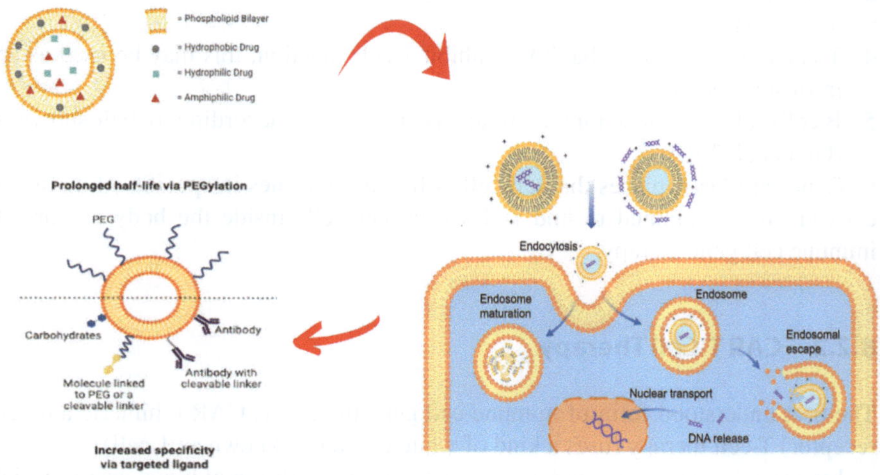

Fig. 8.4 Liposome-based drug delivery (loading, endocytosis, and improved specificity)

8.2 Chemotherapeutic Gene Targeting of Immune Cells (Gun et al. 2019)

Immunotherapy is the use of immune system to kill cancer cells as part of cancer treatment. This can be achieved by different ways, one of which is immune cell gene therapy that is also known as adoptive cell transfer (ACT).

Immunotherapy concept was developed based on the evidence that tumor growth can be retarded by immune system manipulation.

8.2.1 Immune Cells in Cancer

1. Macrophages: cancer metastasis is enhanced by *tumor-associated macrophages* (TAMs) through different mechanisms as prompting tumor growth, inducing angiogenesis, and enhancing tumor invasion and migration. Data show association between the number of TAM present in tumor microenvironment (TME) and poor prognosis for many types of cancer. Controlling this number may help fight cancer. Figure 8.5a and 8.5b shows barriers to immunotherapy from TME.

Abbreviations: APC, antigen presenting cell; DC, dendritic cell; TAM, tumor-associated macrophage; MDSC, myeloid-derived suppressor cell; ECM, extracellular matrix; CAF, cancer-associated fibroblast.

2. Neutrophils: they have been activated in tumor cells, indicating host inflammation stage. Modulating neutrophils to anti-tumor phenotypes will help fight cancer.
3. NK cells: they reflect cytotoxicity which means low level of NK cells indicates tumor control.
4. T cells: it was shown that TME inhibit T-cell function, this may be used as an intrinsic checkpoint.
5. B cells: effects can be tumor stimulants or inhibitors according to their subtypes (Gun et al. 2019).

Genes are DNA pieces that tell cells what to do. Genes in specific white blood cells are reprogrammed to find and kill cancer cells inside the body as part of immune cell gene therapy.

8.2.2 CAR T-Cell Therapy

The best understood form of immune cell gene therapy is CAR (chimeric antigen receptor) T-cell therapy (uses a kind of white blood cell known as T cell).

First step is "apheresis" and is done by filtering T cells out of blood using special machine, then putting the blood back. The extracted T cells are sent to a lab to change genes (allowing them to make proteins called CARs). CARs are better

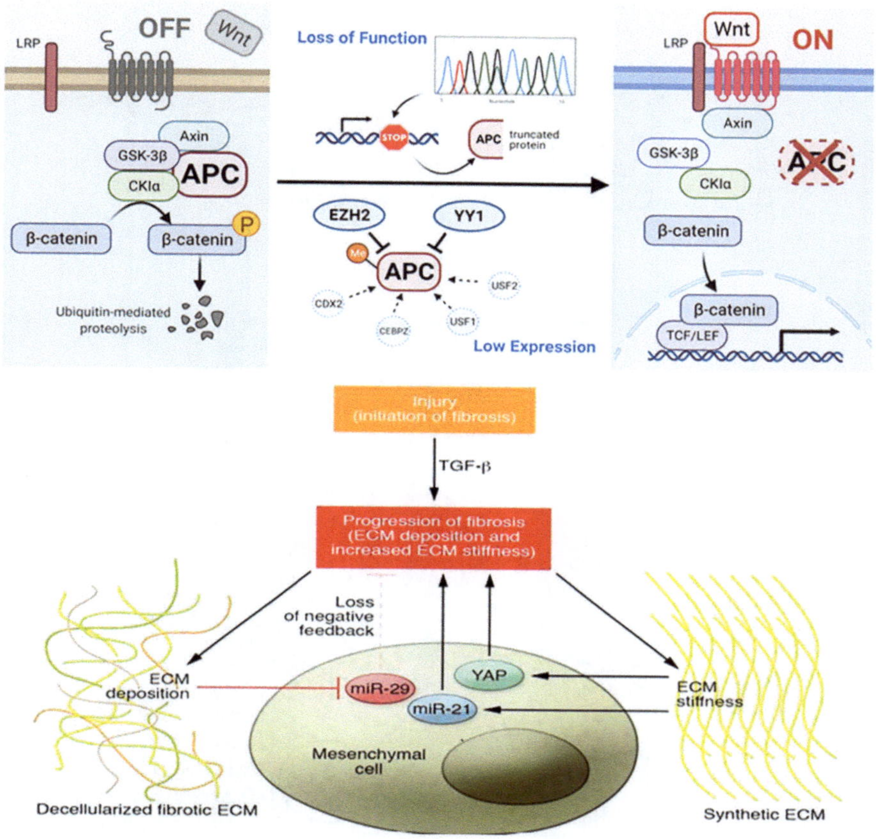

Fig. 8.5a Several barriers to cancer immunotherapy from TME. (1) Inhibition of APC function. (2) The formation of an immunosuppressive environment

explained as inserting keys on T cells that will fit locks on cancer cells. Then the T cells are frozen and sent to your doctor.

The processed T cells are given back through IV route (direct to the blood) where they will travel to find locks and terminate cancer cells.

CAR T cells will grow and divide many times inside your body and may fight cancer cells for months or even years.

8.2.3 TCR Therapy

T-cell receptor (TCR) is similar to CAR T-cell therapy, but these T cells can find hidden "locks" inside cancer cells.

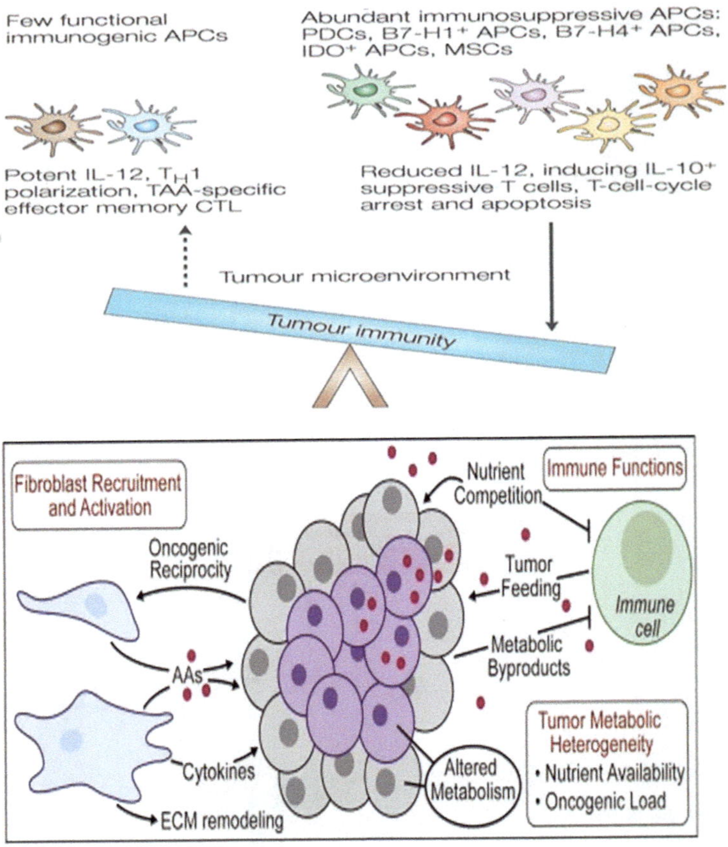

Fig. 8.5b Several barriers to cancer immunotherapy from TME. (3) Fibrosis of the extracellular matrix. (4) Abnormal cellular metabolism

8.2.4 TIL Therapy

Tumor-infiltrating lymphocytes (TILs) are white blood cells inside a tumor, which may indicate the efforts of immune system to kill cancer cells.

TILs are removed from tumor and multiplied on the lab and then large numbers of them are put in the body to help fight cancer cells (Webmd.com).

Figure 8.6 shows advantages vs disadvantages of adoptive cell therapy.

Advances in immunotherapy have opened promising models of therapy that include monoclonal antibodies, cellular therapies, small molecules, proteins, and peptides. Yet, the benefit is limited due to delivery problems that include poor target specificity, nonspecific phagocytic uptake, poor tumor tissue permeation, and many others. Nanoparticles (NPs) have offered the solution to the above problems as they have good transport properties and improved distribution. Advancements have been made to nanoparticle systems to develop and anticipate expansion in NPs possibilities. Engineered NPs development for cancer therapy offers the promise of efficient

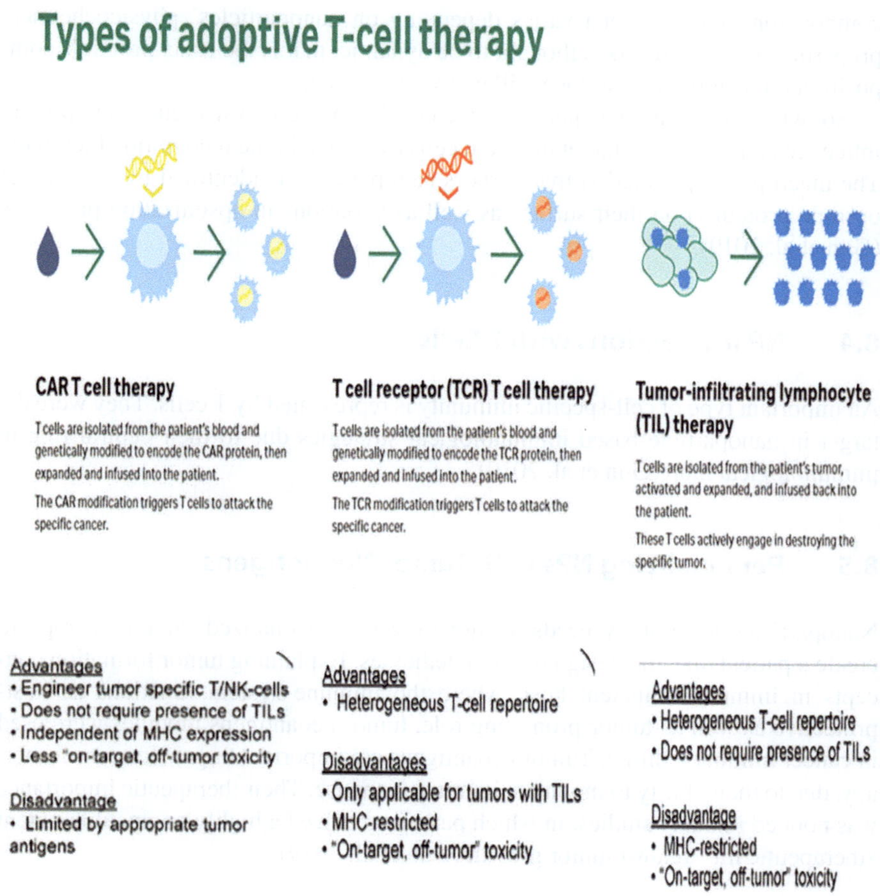

Fig. 8.6 Advantages and disadvantages of adoptive cell transfer

and specific delivery of TME (tumor microenvironment). Several nano-therapeutic agents have been approved for cancer treatment by the FDA. FDA-approved treatments depend on delivery of drugs aimed at malignant tumor, without any consequence on the TME of immune cells. Nano-therapeutic agents focus on shifting toward changing immune cells activity for cancer therapeutic drug treatment (Gun et al. 2019).

8.3 Nanoparticle Use for Immunotherapy

Keeping in mind that the interaction between nanoparticles and immune cells is affected by physicochemical properties, that is when nanoparticles get exposed to blood or lymph fluid, deposition and adsorption of plasma proteins occurs rapidly on Nanoparticles surface leads to the formation of "protein corona". The composition of this corona varies depending on nanoparticles' physicochemical properties. Protein corona is thought to be dynamic, that is the exact protein's composition on nanoparticle surface will evolve with time.

So, when designing nanoparticles, the developing corona will either temporally influence cell activity or target different cells over time in the recognition hierarchy. The macrophage population that harbors nanoparticles is identified by the type of opsonic protein(s) on their surface as well as variations in opsonization processes (Gun et al. 2019).

8.4 NP Interactions with T Cells

An important type of cell-specific immunity is represented by T cells. They were the target in nanoparticle-based immunological strategies due to their central role in immunological map (Gun et al. 2019).

8.5 Personalizing NPs with Tumor Neoantigens

Nanoparticles technology needs to move toward personalized immunotherapy to create a patient-specific design of nanomedicines. Explaining tumor formation concepts in immunocompetent host, where the immune system functions as host-protective as well as tumor-promoting role, tumor neoantigens modules were used in cancer immune-editing. Tumor neoantigens are important targets for cancer therapy, due to their ability to modulate immune response. Their therapeutic importance was noticed in many studies, in which patients received a highly compatible potent "therapeutic hit" against tumor growth (Gun et al. 2019).

8.6 Testing NP Immunotherapeutic Potential

Controlled kinetics were achieved by using degradable nanoparticles that tend to achieve therapeutic delivery method. It is important to consider that any induced immune system change raises concern about adverse effects or even toxicity. That is why the development of nano-immunotherapy should be done with a balance between toxicity and therapeutic safety (Gun et al. 2019). Figure 8.7 presents a summary of nano-immunotherapy.

8.7 Investigative Models for Evaluating Nano-immunotherapy

Suitable models of the TME are needed to develop and test immunotherapeutic NPs. Murine models are usually used to study nanoparticles in vivo, they are considered as a powerful system in determining NP organ's uptake of nanoparticles and its distribution.

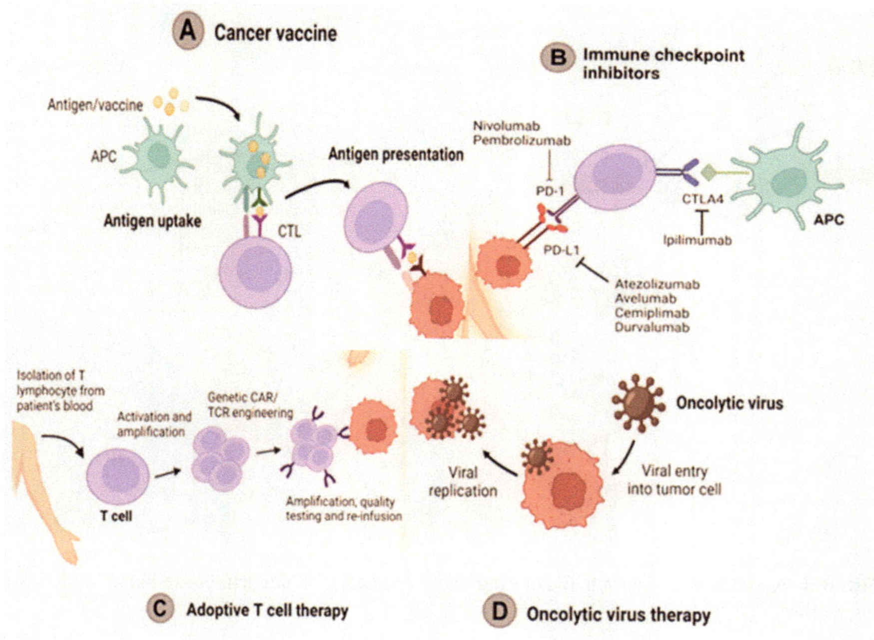

Fig. 8.7 Recent developments in nanoparticle (NP)-mediated immunotherapy may target antigen-presenting cells (APCs) or the adaptive arm of immunity

Microfluidic models of the cancer TME represent a vital intermediate step to link in vitro screenings, studies on animals, and trials on clinical patients. This module has been used successfully to test different immunotherapies efficacy (Yu et al. 2022).

8.8 Dendritic Cells (DCs) in Skin Draining Lymph Nodes

Therapeutic vaccine nanoformulations that contain small nanomaterials allow an efficient passive targeting transport to endogenous DCs in draining lymph nodes (dLNs), promoting presentation of antigens by activated DCs to naïve T cells and subsequently priming antigen-specific anti-tumor effector T cells (Yu et al. 2022). Figure 8.8 shows the hypothesis of dendrimer-mediated interaction.

Vaccines that are synthesized from mixed adjuvants and antigens might reduce anti-tumor immunity and induce immune tolerance. Exogenous nanoparticles generally tend to be uptaken by DCs using endocytosis, acidic endosome compartments will trap them (Mellman and Steinman 2001).

Using nano-formulation for immunization allows efficacious elimination of different types of tumors and has been first tried on humans in 2023 with a great

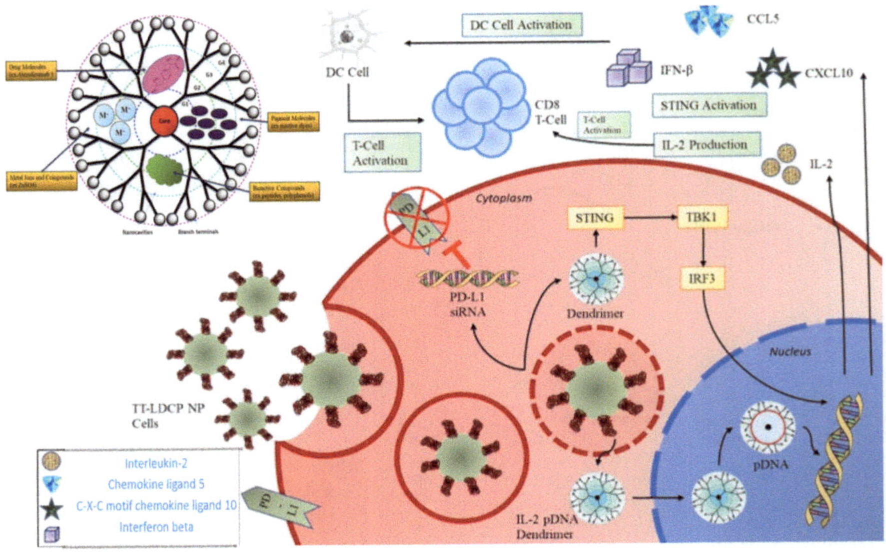

Fig. 8.8 A schematic diagram illustrating the hypothesis of dendrimer-mediated multivalent interaction

potential in a personalized vaccine formulation. Nano-STING agonists will improve in-vivo stability of nanoparticles that gives an attractive approach for cancer (Yu et al. 2022).

8.9 DCs at Tumor Sites

As opposed to therapeutic vaccination in well-defined TAAs (tumor-associated antigens) or TSAs (tumor-specific antigens), ISV (in situ vaccination) in TME that documented the endogenous antigens from the fading cancer cells, which could be exhibited to tumor-infiltrating DCs to produce adaptable T-cell immune response.

ISV system overcomes the idea of difficulty finding specific tumor antigens. Many ISV methods that aim induction of ICD (immunogenic cell death) or activation of DCs are being studied as agonists of TLR, CD40 receptor and STING, oncolytic viruses, Fms-like tyrosine kinase 3 ligand (FLT3L), and radiotherapy.

Coating nanoformulations with erythrocyte membranes will for sure prolong circulation time by alleviating immune cells clearance.

Targeting DCs or TME offered implications for nanoformulation vaccination use in cancer immunotherapy that include:

(1) Adaptable manipulation of surface characteristics and elasticity. (2) Avoiding taking up free antigens by immature DCs. (3) Lengthened retention of antigen in DCs, to improve efficacy (Yu et al. 2022).

In conclusion, research must persist to improve the efficacy of nano-particles-based immunotherapies and rationalize their usefulness. The presence of combined immunotherapies should take into consideration different sets of nanoparticles modules to achieve outstanding results for patients. These anticipations of NPs should consider the correlation between the physicochemical characters of NPs and the resultant protein corona, as well as specific immune-cell subsets and patient-specific immunity to improve therapeutic efficacy.

Author Disclaimer *The author of this chapter would like to clarify that while every effort has been made to provide accurate and up-to-date information, this text is intended for general informational purposes only. The author shall not be held responsible for any consequences arising from the use of information contained within this chapter.*

Conflict of Interest *The author declares no conflicts of interest regarding the publication of this chapter.*

References

Elumalai K., S. Srinivasan, A. Shanmugam, Review of the efficacy of nanoparticle-based drug delivery systems for cancer treatment, Biomed Technol, 5, 2024, 109–122. https://doi.org/10.1016/j.bmt.2023.09.001. (ISSN 2949-723X. https://www.sciencedirect.com/science/article/pii/S2949723X23000533)

Gao Y, Qiu W, Liang M, Ma X, Ye M, Xue P, Kang Y, Deng J, Xu Z (2022) Active targeting redox-responsive mannosylated prodrug nanocolloids promote tumor recognition and cell internalization for enhanced colon cancer chemotherapy. Acta Biomater 147:299–313. [CrossRef]

Gashtasbi GP (2015) Nanotoxicology and challenges of translation. Nanomedicine (N. Y., NY, U. S.) 10:3121–3129

Gun SY, Lee SWL, Sieow JL, Wong SC (2019) Targeting immune cells for cancer therapy. Redox Biol 25:101174. https://doi.org/10.1016/j.redox.2019.101174. Epub 2019 Mar 20. PMID: 30917934; PMCID: PMC6859550

Mellman I, Steinman RM (2001) Dendritic cells: specialized and regulated antigen processing machines. Cell 106:255

Qiyan L, Yulin L, Zihua H, Yajie G, Qingjiao L (2022) Triggering immune system with nanomaterials for cancer immunotherapy. Front Bioeng Biotechnol 10. https://doi.org/10.3389/fbioe.2022.878524. https://www.frontiersin.org/articles/10.3389/fbioe.2022.878524. ISSN 2296-4185

Rajeeva BB, Menz R, Zheng Y (2014) Towards rational design of multifunctional theranostic nanoparticles: what barriers do we need to overcome. Nanomedicine (N. Y., NY, U. S.) 9:1767–1770

Ramamurthi P, Zhao Z, Burke E, Steinmetz NF, Müllner M (2022) Tuning the hydrophilic–hydrophobic balance of molecular polymer bottlebrushes enhances their tumor homing properties. Adv Healthc Mater 11:2200163. [CrossRef]

Shi Y, Chen C, Xu Y, Liu Y, Zhang H, Liu Y (2019) LncRNA FENDRR promotes high-glucose-induced proliferation and angiogenesis of human retinal endothelial cells. Biosci Biotechnol Biochem 83:869–875. [CrossRef]

Tiwari H, Rai N, Singh S, Gupta P, Verma A, Singh AK, Kajal SP, Singh SK, Gautam V (2023) Recent advances in nanomaterials-based targeted drug delivery for preclinical cancer diagnosis and therapeutics. Bioengineering (Basel) 10(7):760. https://doi.org/10.3390/bioengineering10070760. PMID: 37508788; PMCID: PMC10376516

Verma S, Singh A, Kukreti G, Bharkatiya M, Dobhal K, Parashar T, Suyal J, Jakhmola V (2023) Nanoparticles – a booming drug delivery system in chemotherapy. Biomed Pharmacol J 16(3):1785–1790
Wang Z, Ling L, Xia Q, Li X (2019) Disulfide-crosslinked reduction-responsive prodrug micelles for on-demand paclitaxel release. J Drug Deliv Sci Technol 53:101168. [CrossRef]
Yu M, Yang W, Yue W, Chen Y (2022) Targeted cancer immunotherapy: nanoformulation engineering and clinical translation. Adv Sci 9:2204335. https://doi.org/10.1002/advs.202204335

Nanoscale Diagnostics for Antigen-Specific Cells and Monitoring the Progress of an Immunotherapy

9

Areej Fatima, Maleeha Azhar, Aroosa Younis Nadeem, Adeeb Shehzad, and Ebtesam A. Al-Suhaimi

9.1 Introduction

Immunology is a diverse discipline that occupies a prominent position in biomedical research. It focuses on comprehending the complexities of the immune system, which is a magnificent assemblage of cells, tissues, and components responsible for protecting the body against diseases and a range of infections (Abbas et al. 2019). Immunology employs innate as well as adaptive immunity at its core. Innate immunity is the primary, nonspecific defense against pathogens, consisting of physical barriers such as the epidermis and cellular components including macrophages and neutrophils. Integrated innate immunity is a defense system that offers instantaneous protection but lacks specificity. In contrast, adaptive immunity exhibits a remarkably specialized and highly precise response. The response involves the activation of distinct immune cell populations, namely T and B cells, which possess the ability to identify antigens, typically located on the external surface of pathogens, by binding to their corresponding receptors (Thaiss et al.

Authors Areej Fatima, Maleeha Azhar and Aroosa Younis Nadeem have equally contributed to this chapter.

A. Fatima · M. Azhar · A. Y. Nadeem
Department of Biomedical Sciences, School of Mechanical and Manufacturing Engineering (SMME), National University of Sciences and Technology (NUST), Islamabad, Pakistan

A. Shehzad (✉)
Biodiversity Unit, Research Center, Dhofar University, Salalah, Oman
e-mail: ashehzad@du.edu.om

E. A. Al-Suhaimi
Vice presidency for Scientific Research and Innovation, Imam Abdulrahman bin Faisal University (IAU), Dammam, Saudi Arabia

S. Rehman, E. Al-Suhaimi (eds.), *Nanotechnology Based Microbicides and Immune Stimulators*, https://doi.org/10.1007/978-981-96-0397-8_9

2016). The identification of pathogens triggers the generation of antibodies via B cells and the activation of T cells, which either immediately eliminate pathogenic cells or coordinate a more specific immune response. Immunological memory, a fundamental characteristic of adaptive immunity, serves to ensure persistent protection. T cells play a crucial role in cell-mediated immunity, and B cells, which are responsible for antibody-mediated immunity, constitute the backbone of the adaptive immune system (Hildner et al. 2008). Macrophages, serving as both phagocytes and antigen-presenting cells, play a crucial role in pathogen elimination and immune activation (Mosser and Edwards 2008). These immune cells execute a coordinated reaction in response to cytokines, signaling molecules (O'Shea and Paul 2010). Immunological research has significantly broadened our comprehension of the immune system, allowing us to decipher complex phenomena such as antigen presentation by specialized cells like dendritic cells, as well as the intricate regulatory roles of cytokines (O'Shea and Paul 2010). The subject of immunology has undergone significant progress in the past two decades. The traditional categorization of the immune response into two discrete domains, namely the innate and adaptive elements, which work together to respond to external antigens or perform self-nonself marginalization, has undergone a more complex evolution. The progress and application of innovative technologies have resulted in new findings that have established a new domain in which the autoimmune system establishes communication, not only among immune components but also with commensal microorganisms and other essential systems, such as the endocrine and nervous systems (Miller et al. 2009). The advancements have compelled immunologists to revise the immunological framework responsible for providing immunity, hence rendering the investigation of the immune system particularly appealing. Moreover, due to these advancements, there has been a growing inclination toward enhancing the understanding, regulation, and modulation of the immune response in both states of well-being and pathological conditions.

Immunotherapy encompasses the utilization of specific immune system constituents, for instance, antibodies, cells, and cytokines, in the therapeutic management of several malignancies and autoimmune disorders. Additionally, immunotherapy involves manipulating the immune system via vaccinations, which are utilized as a preventive and therapeutic measure against infectious and allergic disorders. The utilization of microbes or their constituents in vaccines for immunotherapy has a longstanding history dating back millennia. Over the years, soluble compounds, including poly- and monoclonal antibodies and cytokines, have been employed extensively. However, in recent times, cellular immunotherapy has arisen as a novel approach in medical practices. Immunotherapy is a versatile therapeutic approach that finds use in a wide range of medical conditions, including infections, macular degeneration, autoimmune disorders, allergic reactions, and, notably, the field of oncology, where its utilization is extensive (Couzin-Frankel 2013; Kelly 2018).

9.2 Types of Immunotherapies

9.2.1 Serotherapy

"Serotherapy" was pioneered by Dr. Roux and Dr. Yersin, who administered anti-diphtheria serum as a therapeutic intervention for multiple pediatric patients (Varadé et al. 2021).

Serum comprises diverse proteins, notably antibodies, pivotal in mounting an immune response against infections. Serotherapy has been utilized in the past for the treatment of disorders, and has more recently been employed in the field of immunotherapy. Many patients diagnosed with humoral immunodeficiencies are effectively undergoing treatment to mitigate their susceptibility to contracting infectious pathogens. The application of immunoglobulins for therapeutic purposes has been on the rise lately. These have demonstrated their effectiveness in addressing numerous illnesses, such as COVID-19, that stem from SARS-Cov-2 infection, autoimmune disorders, and Kawasaki syndrome in children. With time, the use of immunoglobulins has widened, and they are now employed in treating an even more extensive range of disorders. The observed positive impacts appear to be facilitated through various immunological processes, such as virus neutralization, suppression of inflammatory cells, and stimulation of immune regulators (Perez et al. 2017).

9.2.2 Monoclonal Antibodies (mAbs)

The groundbreaking discovery of monoclonal antibodies (mAbs) by esteemed scientists C. Milstein and G. Köhler in 1975, who were subsequently awarded the Nobel Prize in Physiology/Medicine in 1984, brought about a profound transformation in medicine and immunology. The number of monoclonal antibodies (mAbs) licensed for human therapy has steadily grown since 1975 (Köhler and Milstein 1975). Additionally, numerous mAbs are currently undergoing clinical trials. mAbs have greater adaptability due to a different mechanism of action. mAbs can undergo the process of neutralizing or inhibiting soluble components. For instance, the inhibition of cytokines (specifically TNF-α) and growth factors (such as vascular endothelium growth factor) hinders the manifestation of their respective effects, namely inflammatory and angiogenic effects, respectively (Lonberg 2008).

The activation of the complement cascade through the classical pathway by IgG/IgM antibodies results in the subsequent killing of the target cell. Natural killer (NK) cells can enhance the efficacy of monoclonal antibody (mAb)-mediated death of target cells. Upon attaching to a target cell, the monoclonal antibody (mAb) can bind to Fc receptors on the surface of natural killer (NK) cells. This interaction subsequently initiates the release of granzymes and perforin by the NK cells, thereby promoting the death of the targeted cell (Rogers et al. 2014).

Inducing cellular death by apoptosis by certain monoclonal antibodies (mAbs) targeting specific membrane molecules is another mode of immunotherapy. The inhibition of stimulating signals as antibodies can inhibit specific membrane-associated receptors, hence preventing cellular activation and proliferation. As carriers of radioactive materials, pro-drugs, enzymes, and poisons, mAbs can concentrate specific molecules around target cells, making them a significantly more discerning therapeutic approach compared to conventional chemotherapy or radiotherapy (Beck et al. 2017).

The recent advancement in cancer therapy has been facilitated by discovering many inhibitory compounds that can be effectively neutralized by monoclonal antibodies (mAbs). This neutralization process subsequently triggers the activation and multiplication of T cells with antitumoral properties. Molecules such as cytotoxic T-lymphocyte-associated protein 4 (CTLA-4) and programmed cell death protein 1 (PD-1), along with its receptor programmed death-ligand 1 (PD-L1), function as key players in controlling the activity of immunological cells. Nevertheless, it is feasible to reinvigorate the immune responses against tumors by employing monoclonal antibodies (mAbs) to obstruct specific molecules. This mode of immunotherapy can be achieved by targeting a single molecule or utilizing a combination of antibodies, such as those targeting CTLA-4 and PD1 (Fritz and Lenardo 2019).

9.2.3 Recombinant Proteins

Recombinant cytokines refer to a class of proteins that play a crucial function in regulating immune responses. Interleukins, such as IL-2, IL-7, IL-12, interferon-alpha, and interferon-beta, are representative cytokines employed in immunotherapy. A comprehensive array of recombinant proteins is currently used in human therapeutic applications, with interleukin 2 (IL-2), GM-CSF, and IFN. These substances have the potential to be administered to enhance the immune response against malignancies or transmissible diseases. Another cytokine that exhibits adjuvant antitumoral effects is granulocyte-macrophage colony-stimulating factor (GM-CSF). GM-CSF stimulates the development of granulocytes and monocytes from myeloid progenitor cells in the bone marrow. Various cytokines, including IL-5, IL-7, IL-12, IL-15, IL-18, and IL-21, are being evaluated in numerous clinical trials. Specific cytokines, either in isolation or in conjunction with others, may hold promise for future use in antitumoral therapy (Ruderman and Pope 2005).

9.2.4 Microorganism-Based Therapies

Bacteria

Recently, there has been a notable upsurge in the utilization of microbes in therapeutic applications, which can be attributed to the enhanced understanding of the human microbiome. Certain gastrointestinal diseases, as in those caused by *Clostridium difficile*, can be effectively treated with the transplantation of intestinal bacteria obtained from individuals in good health, commonly known as fecal microbiota transplantation (FMT). Numerous additional attempts to employ microorganisms to treat inflammatory diseases, such as Crohn's disease, ulcerative colitis, and many others, have possessed a slight success, emphasizing that this kind of treatment exhibits a significantly higher level of intricacy than previously predicted. Therefore, it is imperative to conduct additional investigations to ascertain the viability of employing this method for curative immunotherapy. Scientists are currently conducting studies on genetically modified or synthetic bacteria, such as those derived from *Salmonella enterica*, Listeria monocytogenes, or *Lactobacillus lactis.* However, thus far, the observed outcomes have been constrained in their scope (Decker and Safdar 2009).

Oncolytic Viruses (OVs)

The utilization of bacteria in the context of antitumoral therapy has been predominantly limited, whereas the application of therapeutic viruses is experiencing a notable surge. Virus-based therapy was initially introduced during the 1990s, primarily employing adenovirus as the vehicle for delivery. However, its practical implementation within healthcare systems has become more prevalent in recent years (Davola and Mossman 2019). Oncologic viruses can selectively target tumor cells, stimulate immunogenic cell death (ICD), and elicit a host immune response against the tumor. Genetic engineering techniques can modify viral sequences, resulting in an increased propensity of the virus to infect specific cells and augmenting its ability to infiltrate tissues and target tumors. The investigation of combinations involving additional components such as immunomodulators, drugs, and cytokines is underway. These combinations aim to inhibit the immune response against viruses and improve the cytotoxicity against tumors (Zheng et al. 2019) (Fig. 9.1).

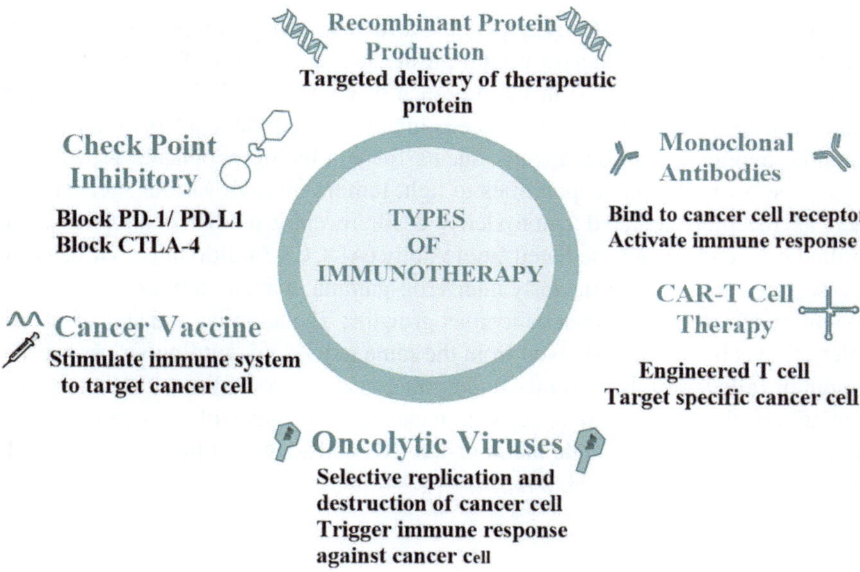

Fig. 9.1 Types of immunotherapies and their target sites

9.2.5 Vaccines

Vaccines have emerged as a significant asset in the realm of immunotherapy, namely within the domain of cancer treatments. Cancer vaccines elicit an immune response in the patient, thereby facilitating the recognition and subsequent targeting of cancerous cells. These vaccines target specific antigens found on cancer cells, inducing the immune system to perceive them as exogenous pathogens. There exist two main classes of cancer vaccines, namely preventative and therapeutic. Preventive vaccines, such as the human papillomavirus (HPV) vaccine, can potentially mitigate the risk of acquiring infections that can potentially progress into malignant disorders. In contrast, therapeutic cancer vaccinations are provided to people already affected by cancer. Tumor-specific antigens or immune-enhancing drugs can be included in these therapies to augment the body's immunological response against the tumor. Immunotherapy vaccines can enhance the body's ability to combat cancer and other diseases by harnessing the body's natural defense mechanisms, offering patients more precise and less toxic treatments (Kim et al. 2021).

Various viruses, including genital herpes, hepatitis B, Epstein-Barr, and human papilloma, as well as the bacterium Helicobacter pylori, have been linked to the development of uterine and liver cancers, Burkitt's lymphoma, and oral, genital, and stomach cancers, respectively. Hence, the vaccination process against these pathogens provides a barrier against infection and the onset of cancer (Gutowska-Owsiak and Ogg 2017).

9.2.6 Lymphokine-Activated Killer (LAK) Cells

Lymphokine-activated killer (LAK) cells, a diverse group comprising NKT, NK, and T cells, can be produced through an in vitro cultivation of peripheral blood mononuclear cells (PBMCs) in the presence of interleukin-2 (IL-2). The inherent characteristics of natural killer (NK) cells manifest remarkable adaptability in their potential application as therapeutic agents for combating neoplastic growth. NK cells possess exceptional capabilities to fight tumors through various mechanisms, such as receptor-mediated cytotoxicity, death receptor-mediated apoptosis, and antibody-dependent cell-mediated cytotoxicity (ADCC). Additionally, NK cells can also secrete cytokines, particularly interferon-gamma, that aid in boosting the adaptive immune response against cancerous growths. The adoptive transfer of natural killer (NK) cells, whether derived from the same individual (autologous) or a different donor (allogenic), is presently undergoing evaluation in clinical trials targeting hematological disorders and solid malignancies. Other types of cells employed for tumor immunotherapy include memory-like NK cells, NK cell lines, and stem cell-derived NK cells (Davis et al. 2015).

9.2.7 Tils

Since the beginning of the 1980s, Dr. Rosenberg has diligently endeavored to substantiate and enhance the productive utilization of tumor-infiltrating lymphocytes (TILs). The experimental procedure commences with a surgical intervention, wherein tumor-infiltrating lymphocytes (TILs) are extracted. Subsequently, the isolated TILs are subjected to a series of manipulations, beginning with their activation within a controlled laboratory environment, facilitated by adding specific signaling molecules known as cytokines. This activation process is accompanied by the expansion of the TIL population, promoting their proliferation and amplification. Ultimately, the augmented TILs are reintroduced into the original patient, completing the therapeutic cycle. Since its inception, this therapeutic intervention has undergone significant advancements, resulting in a notable enhancement of optimal responses. The initial success rate, less than 30%, has now escalated to 50–75% in some instances (Chen et al. 2007; Hu et al. 2020).

9.2.8 Chimeric Antigen Receptor (CAR)

CAR T-cell therapy (chimeric antigen receptor) is a pioneering type of immunotherapy used to treat specific forms of cancer. It entails the modification of a patient's T cells for them to express a synthetic receptor (CAR) that targets a particular antigen identified in cancer cells. Once put back into the patient, these modified CAR T cells can recognize and destroy cancer cells containing the specific antigen. T lymphocytes have undergone the process of synthesizing a chimeric antigen receptor (CAR) derived from a specific monoclonal antibody that recognizes and binds to a specific surface molecule. The altered T cells possess the ability to identify tumor cells independently of antigen processing or presentation by antigen-presenting cells. In addition, the chimeric antigen receptor (CAR) encompasses the essential constituents necessary for intracellular signaling, as well as the stimulation of cytotoxic T cells and helper cells (Cho et al. 2018).

9.2.9 Dendritic Cells

The utilization of dendritic cell-based immunotherapy has emerged as a highly promising and effective method within the realm of cancer treatment. Numerous research investigations have provided evidence supporting the capacity of dendritic cells (DCs) to function as effective antigen-presenting cells, hence facilitating the initiation and augmentation of immune responses against tumors. The process of DC-based immunotherapy entails the extraction of a patient's dendritic cells (DCs), which are subsequently loaded with tumor-specific antigens or peptides. These antigen-laden DCs are then reintroduced into the patient's body through reinfusion. This procedure has been demonstrated to elicit potent cytotoxic T-cell responses, promoting the identification, and eradicating malignant cells (Gardner et al. 2020).

9.3 Significance of Nanotechnology in Immunotherapy

Nanotechnology is crucial to immunotherapy, a rapidly emerging method for treating cancer and autoimmune diseases by manipulating the immune system. Nanotechnology manipulates materials at the nanoscale (1–100 nanometers) to generate new features and applications (Demetzos and Demetzos 2016). Nanotechnology has various immunotherapy applications.

9.3.1 Drug Delivery

The drug delivery field involves using nanoparticles specifically designed to transport and administer therapeutic substances, like peptides, antibodies, or small molecules, to specific cells or tissues within the body. The utilization of targeted drug delivery systems serves to mitigate the potential impact of the drugs on healthy tissues, limiting the incidence of adverse effects while simultaneously augmenting the treatment's effectiveness. A recent study involved the utilization of hyaluronic acid, a polysaccharide present in the extracellular matrix, as a ligand-appended component in various nanocarriers. These nanocarriers have demonstrated encouraging outcomes in enhancing the antitumor efficacy against melanoma stem-like cells, breast cancer cells, and pulmonary adenocarcinoma cells. Additionally, they have shown potential in facilitating the delivery of drugs into the vitreous humor for retinal gene therapy (Dourado 2020).

Polymeric micelles are nanostructures composed of amphiphilic block copolymers that self-assemble into a core-shell configuration in an aquatic environment. Hydrophobic drugs such as camptothecin, docetaxel, and paclitaxel can be encapsulated within the hydrophobic core of the system. Simultaneously, the hydrophilic shell of the system imparts water solubility and stabilizes the body. Polymeric micelles typically have a size below 100 nm and possess a limited distribution, preventing rapid renal elimination. Consequently, these micelles can accumulate in tumor tissues by utilizing the enhanced permeability and retention (EPR) effect. After topical administration, polymeric micelles are employed to achieve targeted delivery to the posterior ocular tissues through the transscleral pathway (Kuperkar et al. 2022).

9.3.2 Immune Modulation

The immune system can be modulated by designing specific nanoparticles. For example, nanoparticles can be equipped with antigens or immunomodulatory molecules to initiate or inhibit targeted immune responses. This system holds significant value in the treatment of autoimmune disorders and in augmenting the efficacy of cancer immunotherapies. Curcumin has the potential to serve as an immunomodulatory agent, too (Getts et al. 2015). Immunomodulatory nano systems offer a promising strategy for combatting viral infections by enhancing target-specific drug delivery and promoting strong viral-specific immune responses (Khatun et al. 2023). EGCG was tested in both regular and nanoparticle formulations, which shows beneficial effects in a psoriasis-like model, indicating potential for large-scale testing and further exploration of nano formulations to improve EGCG pharmacokinetics (Moudgil and Venkatesha 2023). Immunomodulatory nano-preparations hold promise for regulating local immune responses and inducing antigen-specific immune tolerance in rheumatoid arthritis by delivering self-antigens, immunomodulators, or cytokines. These nano-preparations provide insights into developing new immunotherapies for Rheumatoid arthritis (Li et al. 2023a).

9.3.3 Enhanced Antigen Presentation

Nanoparticles help immune cells recognize and fight infections and cancer cells by improving antigen presentation. This can be accomplished by designing nanoparticles resembling pathogens or tumor cells, stimulating a more robust immune response. The co-cultivation of biodegradable nanoparticles with dendritic cells isolated from bone marrow (BMDCs) led to an increased display of ovalbumin-derived peptides. This augmentation was measured using B3Z cells, a hybridoma of CD8+ T cells (Zom et al. 2019).

9.3.4 Vaccine Development

Nanotechnology is paramount in advancing vaccinations, particularly within the realm of viral illnesses. The utilization of nanoparticles has demonstrated enhanced efficacy in the delivery of antigens, hence augmenting the immune response and potentially facilitating the production of innovative vaccination formulations. The immune system can effectively induce cellular and humoral immune responses due to the comparable size of NPs and infections. ISCOMs, also known as immune-stimulating complexes, are spherical structures with a radius of around 40 nm. These complexes are composed of many components, including phospholipids, saponin adjuvant Quil A, cholesterol, and protein antigens. The spherical particles can effectively capture the antigen through apolar interactions (Yeolekar 1996; Stelzner 2018).

9.3.5 Imaging and Diagnosis

Nanoparticles can be engineered as contrast agents for several imaging modalities, including magnetic resonance imaging (MRI) and computed tomography (CT) scans. This facilitates the early detection of diseases and enables the tracking of immunotherapy advancements. The researchers employed a study wherein oleic acid-coated FeO nanoparticles were enclosed within oleic acid-conjugated chitosan (oleyl-chitosan). This encapsulation aimed to investigate the accumulation of these nanoparticles in tumor cells, utilizing the enhanced permeability and retention (EPR) effect in an in vivo state for analytical purposes facilitated by magnetic resonance imaging (MRI) and near-infrared techniques (Patra et al. 2018; Sharma et al. 2022).

9.3.6 Personalized Medicine

Nanotechnology allows personalized therapy based on immunological and health profiles. This can improve treatment efficacy and adaptability. Theragnostic, which combines therapeutic and diagnostic capabilities, may boost personalized treatment. A theranostic NP delivery system assists siRNA therapy. The entry of negatively charged small interfering RNA (siRNA) into the cytosol of target cells poses a challenge. However, this obstacle can be addressed by using theranostic nanoparticles (NPs) in siRNA therapy. Iron oxide nanoparticles (NPs) have the potential to serve as a theranostic nanoplatform due to their ability to be imaged noninvasively using magnetic resonance imaging (MRI) and the capacity for drug loading (Revia et al. 2019; Ryu et al. 2012).

9.3.7 Combination Therapies

Using nanoparticles presents a promising approach for simultaneously delivering multiple therapeutic agents. The utilization of combination therapies in cancer immunotherapy, such as the integration of checkpoint inhibitors, cytokines, and antigens, holds significant promise in augmenting the immune response and ultimately enhancing treatment efficacy. Multidrug resistance (MDR) poses a significant impediment to the efficacy of cancer therapies, thereby playing a pivotal role in cancer metastasis and recurrence (Mokhtari et al. 2017). Combination therapy has been widely recognized as a highly effective strategy for addressing multidrug resistance (MDR) in various disease contexts. The combination effect of drugs can be affected in vivo due to the varying pharmacokinetic (PK) profiles, particularly when there is a significant disparity in the physicochemical properties of the combined substances, such as macromolecules and small drugs. To tackle this problem, extensive research has been conducted using nanotechnology-based codelivery practices (Zhang et al. 2017).

9.3.8 Immune and Disease Monitoring

Real-time immune response monitoring by nanoscale sensors and technologies can optimize immunotherapy treatments. MultiCas-NP, a more multiplexed approach, targets apoptotic signaling caspases (caspase-3, -8, and -9). A series of dark quenchers using mesoporous silica NPs quenches several colors in this nanoprobe. To detect caspase-3, -8, and -9 simultaneously, the nano quencher device was coupled with caspase-specific substrate peptides and dyes (Ryu et al. 2014).

9.3.9 Biological Barriers

Nanoparticles can surmount biological barriers, including the blood-brain barrier, which can restrict the targeted transportation of therapeutic agents to tissues of the brain. Identifying novel therapeutic approaches presents promising prospects for addressing diseases that manifest in anatomically complex regions. In vitro, investigations have demonstrated that using polymeric nanoparticles notably improves drug delivery to the brain. This enhanced delivery mechanism has been observed to reduce oxidative stress, inflammation, and plaque load. Specifically, the delivery of curcumin, a potential therapeutic agent for Alzheimer's disease, has been significantly improved through these nanoparticles (Zhang et al. 2017). Furthermore, the internalization of doxorubicin into human glioma cells is highly efficient when facilitated by these nanoparticles. This efficient internalization has subsequently led to a cytotoxic effect on cancer cells. Many studies have highlighted the significance of nanomedicine in managing complex brain diseases like Alzheimer's and glioma. One of the studies elaborated on the role of nanoparticles in drug delivery but did not extensively delve into the blood-brain barrier (BBB) (Anwar et al. 2023). In contrast, another study specifically focused on lipoprotein-based nano drug-delivery systems, emphasizing their potential to overcome the BBB's challenges for brain disease therapy (Wang et al. 2023). Similarly, another study provided a comprehensive review of strategies to customize drug delivery nano-vehicles for efficient brain tumor treatments, which indirectly addressed BBB challenges (Miao et al. 2023). Vector-associated nanoparticles stand out as the most explicitly dedicated to addressing the BBB's role in brain disease treatment (Lu et al. 2023).

Nanotechnology presents a versatile and promising array of tools that have the potential to significantly enhance the efficacy of immunotherapy in the treatment of diverse diseases. Utilizing this technology facilitates accurate drug administration, modulation of the immune system, and the advancement of individualized therapeutic approaches. Consequently, this progress culminates in providing treatments that are more specific in their action and exhibit reduced toxicity. The ongoing improvement in nanotechnology and immunotherapy holds great promise for developing novel applications and significant advancements.

9.4 Nanoscale Diagnostic Methods

Nanoparticles exhibit distinctive optical and physicochemical characteristics that have the potential to revolutionize various aspects of biomedicine, specifically in the fields of diagnostics, therapy, and imaging. Recent developments in biomolecular diagnostics have primarily centered around enhancing nanoparticle-based platforms for detecting and analyzing DNA, RNA, and proteins at the single-molecule level. Nanoparticles have been found to enhance the analytical capability compared to conventional techniques, exhibiting superior resolution and medium-high throughput. Furthermore, there has been a notable focus on investigating single nucleotide polymorphism (SNP) detection, gene expression profiles, and the characterization of biomarkers using colorimetric, spectrometric, or electrochemical strategies. In recent years, there has been a growing recognition of the significance of nanoscale diagnostic methods in medical research. These methods have garnered considerable attention and interest due to their inherent capabilities for improving sensitivity and accuracy in detecting and diagnosing various diseases and conditions (Wang et al. 2018).

9.4.1 Nanoparticles for Imaging and Detection

Integration of nanoparticles into conventional techniques has significantly enhanced the field of molecular imaging. This advancement has facilitated the development of non-invasive imaging procedures, effectively illuminating specific areas of interest. Consequently, researchers can now investigate and comprehend intricate biological processes at cellular and molecular levels. Various imaging modalities exhibit a notable drawback in terms of their low sensitivity. However, this concern can be addressed using probes, such as contrast agents designed explicitly for X-ray and magnetic resonance imaging and radiolabeled molecules intended for nuclear medicine applications (Ou et al. 2020).

Moreover, it is worth noting that nanoparticles possess the remarkable ability to serve as carriers for targeted delivery of contrast agents, thereby surmounting the inherent constraints associated with conventional modalities. Using iron oxide-based nanoparticles (NPs) presents a promising avenue for the non-invasive imaging of biological systems by applying magnetic resonance (MR) imaging techniques. These nanoparticles are carriers of contrasting agents, enabling enhanced visualization and characterization of targeted tissues or structures. The utilization of colloidal superparamagnetic iron oxide (SPIO) has been investigated as a contrast agent for magnetic resonance imaging (MRI) due to its capacity to lower T2 proton relaxation periods in particular tissues. The enhancement of MRI resolution to the extent of single cells is achieved by modifying the surface of colloidal nanoparticles with recognition elements, such as ligands or antibodies, which specifically target the tissue. This enables the precise identification of molecular components associated with significant cellular features and processes (Du et al. 2019).

9.4.2 Quantum Dots

Quantum dots (QDs) are nanostructures composed of a semiconductor core enveloped by an outer semiconductor shell, exhibiting a characteristic diameter ranging from 2 to 10 nanometers. Due to their remarkable characteristics, such as tunable nanoscale dimensions, exceptional photostability, extensive absorption spectra, and precise emission bands, quantum dots (QDs) have been effectively employed as fluorescent markers to visualize a diverse array of biological structures and processes optically (Sun et al. 2021). These encompass but are not limited to DNA, minute organelles, malignant tumors, intercellular interactions, and cellular signaling mechanisms. Quantum dots are pivotal in optical imaging, specifically in the domains of in vitro, in vivo, and ex vivo applications. Significant advancements have been made in surface modification, biocompatibility, and targeting specificity, resulting in noteworthy enhancements in the in vivo utilization of quantum dots. Furthermore, it is worth noting that quantum dots (QDs) can be tailored for precise target recognition by applying various coatings such as oligonucleotides, antibodies, or streptavidin (Sun et al. 2021).

9.4.3 Nanoparticle-Based Bio-Bar Codes

The utilization of bio-bar codes based on nanoparticles is a prominent approach in the field of diagnosis, facilitating the detection of biomolecules with exceptional sensitivity and the ability to detect multiple targets simultaneously. This technology holds immense significance in timely disease identification and tracking specific health markers. This groundbreaking methodology employs a novel strategy wherein distinct amalgamations of magnetic nanoparticles (MNPs) and gold nanoparticles (AuNPs) endowed with precise DNA sequences are meticulously crafted to locate specific biomarkers, including proteins or nucleic acids, selectively. Upon detecting these biomarkers within a patient's sample, an intriguing phenomenon occurs whereby they exhibit an affinity toward the specific DNA sequences located on the AuNPs. This interaction leads to the formation of a unique and discernible "bar code" for each target. The subsequent utilization of magnetic separation isolates the magnetic nanoparticles (MNPs) that bear the attached gold nanoparticles (AuNPs). Consequently, the encoded information can be deciphered, thereby facilitating the accurate determination and measurement of numerous biomarkers concurrently. The methodology above exhibits an extraordinary level of sensitivity, frequently attaining detection limits in the picomolar or femtomolar range. As a result, it holds significant value in the realm of early disease identification and personalized medicine. By enabling the acquisition of a comprehensive array of biomarkers, this approach facilitates the purchase of crucial diagnostic insights (Kim et al. 2008).

9.4.4 Nanoporous Silica Chips

Nanoporous silica chips have been recognized as highly groundbreaking tools in diagnostic applications. Utilizing silica-based structures with nanoscale pores presented a notable platform for diverse diagnostic applications. The distinctive characteristics of these entities provided meticulous regulation of fluid transportation, immobilization of biomolecules, and amplification of signals, rendering them significant in detecting several diseases and pathogens. The chips exhibited a distinctive feature that displayed a network of pores at the nanoscale level. This resulted in a significantly large surface area, enabling the capture and concentration of specific biomolecules. Upon applying a biological sample onto the chip, the molecules present within the sample are attracted toward the nanopores, resulting in their specific immobilization. Many signaling techniques, including fluorescence or electrochemical signals, commonly achieve the detection and quantification of specific binding interactions. This procedure improved the diagnostic assays' accuracy and precision, facilitating the identification of disease indicators, infections, or genetic material. Moreover, the potential for shrinking exhibited by these chips presents an opportunity to create portable and expeditious diagnostic instruments, offering significant improvements in healthcare and biotechnology (Fontana et al. 2021).

9.4.5 Multiplexed Dendrimers

Multiplexed dendrimers evolved as a viable diagnostic tool, as they can detect numerous biomarkers simultaneously within a single specimen. The dendritic nanoparticles possess a complex branching architecture and can be modified with diverse recognition molecules, including DNA probes, antibodies, or aptamers, that exhibit specificity toward distinct analytes. The capacity to do multiplexing facilitates the identification of several targets, a particularly significant aspect of intricate disorders such as cancer and infectious diseases. In these cases, multiple biomarkers can offer a more thorough diagnostic profile. Research investigations have demonstrated the promise of multiplexed dendrimers in enhancing diagnostic accuracy and broadening our comprehension of disease pathways. Through the utilization of their multifunctionality, dendrimers play a significant role in the progression of personalized medicine and the early identification of diseases, ultimately leading to the improvement of healthcare and treatment modalities (Shields IV et al. 2020).

9.5 Techniques for Immunotherapy Monitoring

9.5.1 Imaging

In the last ten years, cancer treatment has seen a significant advancement through cancer immunotherapy, specifically using immune checkpoint inhibitors. These inhibitors work by activating the body's immune system to fight advanced cancers. However, because immunotherapy operates differently from traditional treatments, it can lead to unusual effects on how tumors respond, progress, and even how the drugs affect the body.

In this context, medical imaging plays a crucial role. It helps us objectively understand and measure these unique responses of the immune system to cancer and the side effects of the treatment. Additionally, there's exciting potential in a field known as molecular imaging, which enables us to visualize the specific molecules targeted by immune checkpoint inhibitors. This can provide even more insight into how these treatments work and help us develop better strategies for combating cancer (Carter et al. 2018).

MRI

Magnetic resonance imaging (MRI) stands as a widely accessible clinical imaging modality, boasting numerous strengths and the potential to tackle various clinical challenges associated with assessing the response to immunotherapy. As previously discussed, it effectively overcomes the limitations inherent in nuclear medicine imaging techniques. An MRI scan utilizes a range of sequences to investigate the anatomical, microstructural, physiological, and functional aspects of tissues. This imaging technique can provide remarkable contrast for soft tissues with a high level of spatial precision throughout the entirety of the human body.

Through the integration of several sequences within a multiparametric framework, an MRI scan has the capability to generate a comprehensive depiction of the tumor microenvironment. Moreover, several of these sequencing elements have the potential to provide quantitative maps that offer direct or indirect insights into the underlying biological processes and changes occurring inside the tumor microenvironment after therapy. This capacity of MRI makes it a valuable tool for the evaluation and understanding of immunotherapeutic responses in the clinical context (Carter et al. 2018).

Magnetic resonance imaging (MRI) distinguishes itself as an ideal tool for functional and molecular imaging due to its ability to harness various measured parameters, which create image contrast by relying on the local chemical environment. This unique characteristic enables MRI to provide valuable insights into essential biological properties of tissues. For instance, it leverages the spin–lattice relaxation (T1) and spin–spin relaxation (T2) times of protons (1H) within water and fat in tissues, shedding light on their underlying composition and properties.

Furthermore, MR spectroscopy (MRS) operates on the principle that nuclei within molecules resonate at distinct frequencies in a magnetic field, a frequency inherently tied to the chemical structure of the molecule. This principle allows MRS to non-invasively identify and quantify individual tissue metabolites, offering a deeper understanding of tissue composition (Lau et al. 2022).

Over the years, a diverse array of MRI techniques has been developed to delve into the intricacies of the tumor immune microenvironment. These techniques play a pivotal role in monitoring treatment responses to cancer immunotherapy and assessing immune-related adverse events. This expansive toolkit encompasses approaches that can visualize microstructural and functional changes within the tumor microenvironment, as well as molecular imaging methods adept at detecting cellular and biochemical alterations within tumors (Lau et al. 2022). This versatility empowers MRI to provide invaluable contributions to the field of oncology and immunotherapy.

An MRI scan is presently employed as a routine clinical tool to observe the anatomical modifications that occur subsequent to immunotherapy. It is incorporated in various Response Evaluation Criteria in Solid Tumors (RECIST) guidelines, such as iRECIST and iRANO, which specifically address immune-related assessments and the evaluation of immunotherapy response in central nervous system tumors, such as glioblastoma (Lau et al. 2022).

PET/CT

Positron emission tomography/computed tomography (PET/CT) represents a significant advancement in oncology imaging by adding the additional aspect of metabolism to CT. PET/CT is predominantly employed in the diagnostic staging and to some degree, in the evaluation of treatment response for different types of cancer (Jin et al. 2021). Ongoing research emphasizes exploring further potential uses of PET/CT, particularly in the assessment of prognosis and therapeutic response, with a particular emphasis on immunotherapy.

The utilization of positron emission tomography (PET) allows SUV to be used to assess changes in 18F-FDG uptake in malignant lesions and is a valuable biomarker for evaluating therapy response. The incorporation of this metabolic biomarker dimension has been observed in numerous oncological therapeutic strategies (Jin et al. 2021).

Immunotherapy can activate T cells, induce immune activation, and enhance the influx and proliferation in lymphoid organs. It also recruits DCs and macrophages and enhances the glycolytic metabolism of lymphoid organs. The alterations are evident in the metabolic characteristics of lymphoid organs as observed using FDG PET/CT. FDG PET/CT has the capability to effectively monitor intricate alterations occurring in diverse immune cells as well as the metabolic shifts taking place in lymphoid organs. Nevertheless, FDG PET/CT has restricted spatial resolution and serves as a non-specific tracer with elevated uptake in diverse activated cells. The process of lymphoid organ biosynthesis mostly depends on the salvage route. Positron emission tomography (PET) probes designed for visualizing the deoxyribonucleoside salvage pathway have the capability to track alterations in metabolic activity within the systemic immune response. The compound 18F-FAC is a type of deoxynucleotide analog that exhibits a significant accumulation in lymphoid tissues. This characteristic allows for the visualization of systemic immune activation specifically in secondary lymphoid organs (Jin et al. 2021).

9.5.2 Circulating Immune Cells

Considerable attention has been devoted to investigating the association between immunological markers in the peripheral blood of cancer patients undergoing immunotherapy (Rzhevskiy et al. 2021). The clinical efficacy of anti-CTLA-4 monoclonal antibody (mAb) in advanced melanoma patients was initially observed through a significant rise ($\geq$1000/μL) in absolute lymphocyte count (ALC) after immune checkpoint drug infusion.

In retrospective research conducted on a cohort of 77 patients with metastatic melanoma, it was observed that there was a link between the increase in eosinophil count (more than 100/mm^3) and absolute lymphocyte count (greater than 1000/

mm^3) in the bloodstream following the initial administration of anti-CTLA-4 monoclonal antibody (mAb) and increased overall survival (OS) (Rzhevskiy et al. 2021).

Furthermore, an investigation revealed a predictive function of the neutrophils/lymphocyte ratio (N/L) in determining the clinical effectiveness of immune checkpoint blockade. This study focused on a cohort of patients with metastatic melanoma who were administered a combined treatment of anti-CTLA-4 monoclonal antibody and chemotherapy. Within this group of patients, it was seen that those with a baseline N/L value of 5 or below exhibited a notable distinction in terms of enhanced progression-free survival (PFS) when compared to those who experienced unfavorable clinical outcomes. Furthermore, there has been research conducted on the regulation of T cell subpopulations, namely activated central memory or effector memory T cells (CCR7 + CD45RA− or CCR7 − CD45RA−), as a potential biomarker for immune checkpoint infusions in patients with melanoma (Zhong et al. 2020).

A study analyzed T cells in advanced melanoma patients receiving anti-CTLA-4 monoclonal antibody and chemotherapy using multiparametric cytofluorimetry. This analysis revealed elevated levels of central memory T cells in the peripheral blood, which expressed co-stimulatory and activatory molecules (CD45RA − CD62L+ CCR7+ CD27+ CD28+ BTLA+/PD-1+) (Zhong et al. 2020). These increased levels were found to be associated with objective clinical responses. Additionally, the mentioned study emphasized that the first occurrence rate of CD3 + CD4 + CD45RO + BTLA+, CD3 + CD4 + CD45RO + 4-1BB+, or TH17 T cells could serve as a prognostic indicator for the clinical outcome of patients. Several studies have demonstrated an increase in the frequency of CD4+ T cells expressing the inducible T-cell costimulator (ICOS) molecule after the administration of CTLA-4-inhibiting drugs in patients with bladder cancer, breast cancer, and mesothelioma (Zhong et al. 2020).

In many instances, the alteration of ICOS+ T cells within a short timeframe of 4–7 weeks after the introduction of anti-CTLA-4 monoclonal antibody (mAb) treatment exhibited a correlation with enhanced overall survival (OS) in individuals with cancer. It is crucial to highlight that the presence of immune cells with negative immunoregulatory features, such as T regulatory (Tregs) or myeloid-derived suppressor cells (MDSCs), at the initiation of therapy or their alteration during therapy, have been shown to serve as prognostic indicators for patients' clinical response to immune-mediated checkpoint therapies. It is worth mentioning that patients diagnosed with melanoma, who underwent treatment with anti-CTLA-4 monoclonal antibody (mAb), and demonstrated reduced levels of lactate dehydrogenase (LDH), absolute monocyte and myeloid-derived suppressor cell (MDSC) counts, in addition to an elevated number of regulatory T cells (Tregs), complete eosinophil count, and relative lymphocyte count, demonstrated a baseline profile that indicated a positive prognosis in terms of clinical outcome (Rzhevskiy et al. 2021).

Tumor-associated antigens (TAAs) can be identified by T lymphocytes through MHC-peptide complexes. The assessment of TAA-specific T cells in peripheral blood using the EliSpot assay has been extensively utilized to evaluate the effectiveness of TAA-based cancer vaccines in terms of immunization. It is noteworthy that in certain instances, there was a correlation observed between the anti-TAA T cell responses and the clinical results of the patients. Observation of circulating T lymphocytes exhibiting targeted reactivity toward tumor-associated antigens (TAAs), namely MART-1 and NY-ESO-1, has been documented in melanoma patients who have received CTLA-4 inhibiting drugs. An observed association has been found between the baseline detection of T lymphocytes recognizing tumor-associated antigens (TAAs) and the predictive function it plays in patients' clinical outcomes (Leko and Rosenberg 2020).

Tumor-associated antigens (TAAs) are acknowledged by antibodies within the framework of humoral responses. Therapeutic approaches utilizing antagonistic monoclonal antibodies (mAbs) that target CTLA-4 have demonstrated the potential to enhance humoral immune responses against well-characterized tumor-associated antigens (TAAs), such as NY-ESO-1. Notably, in certain instances, these immune responses have been correlated with clinical improvements observed in patients. Collectively, these findings have contributed to demonstrating the efficacy of immune checkpoint inhibition in facilitating the release of antigen-specific immune responses. Despite the valuable insights gained from these investigations about the mechanisms of action of immune checkpoint-blocking drugs, it is important to note that none of the potential correlative or predictive criteria have been conclusively verified as definitive biomarkers in extensive cohorts of cancer patients (Leko and Rosenberg 2020).

9.5.3 Serum Biomarkers

Finding soluble compounds that might serve as prognostic biomarkers for immune response to immunotherapy is a highly desirable area of research. The utilization of soluble parameters' monitoring enables the application of straightforward experimental methodologies and readily available biological samples, such as serum or plasma. Previous studies have demonstrated that the detection of soluble CD25, which is the alpha-chain receptor of interleukin-2 (IL-2), in the blood of melanoma patients before treatment with anti-CTLA-4 monoclonal antibody therapy, is an independent predictor of overall survival (OS) (Pilla and Maccalli 2018).

Tumor cells that overexpress NKG2D ligands (NKG2DLs) indicate cellular stress. These ligands attach to NKG2D, which is an activating receptor produced by natural killer (NK) cells and a co-stimulatory receptor produced by T cells. The release of NKG2DLs in a soluble state by tumor cells has been documented as a mechanism by which tumors evade immune responses. This occurs through the interaction between the NKG2D receptor on immune cells and the shed NKG2DLs, reducing the immune cells' ability to effectively combat the tumor.

It is noteworthy that soluble NKG2DLs have been identified in the serum of individuals with various histological origins of tumors, and in certain instances, they have demonstrated prognostic significance. The clinical efficacy of the immunization combined with anti-CTLA-4 monoclonal antibody was influenced by the presence of soluble NKG2DL.

In subsequent years, a report emerged indicating that the initial blood levels of soluble NKG2DLs (namely ULBP-1 or – 2) exhibited discriminatory capabilities among melanoma patients who had treatment with anti-CTLA-4 mAb in conjunction with chemotherapy, resulting in either enhanced overall survival (median of 33.6 months) or diminished overall survival (medians of 9.8 or 6.6 months, respectively). Additionally, this research work has brought attention to the fact that the lack of soluble natural killer group 2D ligands (sNKG2DL) in the serum of melanoma patients before treatment, who experienced increased overall survival (OS), is associated with an increase in a limited number of circulating T cell subsets. These subsets include CD3 + CD4 + CD45RO + BTLA+, CD3 + CD4 + 4-1BB+, and Th17 (Pilla and Maccalli 2018).

Research has suggested that interleukin-6 (IL-6) and C-reactive protein (CRP) may be useful biomarkers for predicting if high-dose IL-2 treatment will be effective for patients with metastatic renal cell carcinoma. Patients with progressing disease were found to have elevated levels of both IL-6 and CRP in their serum, with levels exceeding 50 mg/L. C-reactive protein (CRP) is an acute-phase protein that originates from the liver and is frequently elevated in response to inflammation. C-reactive protein (CRP) is categorized as an acute phase reactant that may be identified in the bloodstream. The concentrations of foreign substance exhibit an elevation in response to the secretion of interleukin-6 (IL-6) by macrophages and T cells. Elevated CRP levels at the beginning of the study, specifically exceeding 50 mg/L, were identified as a standalone prognostic factor for the clinical outcome of individuals diagnosed with metastatic melanoma who had high-dose interleukin-2 (IL-2) treatment (Pilla and Maccalli 2018).

VEGF linked with either fibronectin or CRP in pre-treatment serum was predictive of clinical outcomes in patients treated with high doses of IL-2 or anti-CTLA-4 mAb. Elevated LDH levels, above the healthy donor threshold by a factor of two, may serve as an adverse predictive indicator for clinical response in individuals undergoing immune checkpoint blockade therapy (Pilla and Maccalli 2018). LDH levels decreasing during treatment have been associated with improved overall survival. However, these markers have not yet been included in the current clinical assessment for treatment decisions.

9.5.4 Circulating Tumor DNA

Immunotherapy distinguishes itself from conventional chemotherapy and targeted therapy by its capacity to not only selectively target tumor cells or immune cells; but also modulate the immunological microenvironment. Immunotherapy, specifically immunological checkpoint blockade (ICB), is a promising cancer treatment due to its success and minimal side effects. Furthermore, the field of immunology, in conjunction with oncology and molecular biology, has witnessed significant advancements in recent years due to the emergence of numerous novel immunotherapies and associated clinical trials (Li et al. 2018). These breakthroughs have notably impacted the treatment of various cancers, including metastatic melanoma, kidney cancer, bladder cancer, non-small cell lung cancer, small cell lung cancer, gastric cancer, prostate cancer, breast cancer, liver cancer, and others. Consequently, our understanding of immune escape, immune tolerance, and immunotherapy has been continuously evolving, leading to regular updates in perspectives within this domain (Islam et al. 2018).

In the era of advanced medical interventions, a comprehensive comprehension of carcinogenesis at the molecular and genetic levels plays a crucial role in facilitating the efficacy of anticancer therapies. Circulating tumor DNA (ctDNA) has the potential to provide insights into genetic alterations occurring in both the primary tumor and metastatic sites, as well as during the dynamic process of tumor progression, promptly. Researchers have made several attempts to use circulating tumor DNA (ctDNA) to improve treatment techniques like chemotherapy, targeted therapy, and radiotherapy. However, the exact therapeutic value of ctDNA during immunotherapy is still uncertain. Most of the available research focuses on exploring this topic (Li et al. 2018).

9.5.5 Flow Cytometry

Flow cytometry (FC) is mostly regarded as the leading tool for studying cellular phenotypic, signaling cascades, and functional responses to therapeutic interventions at the individual cell level. The intricate relationship between the immune system and the tumor microenvironment in CIT indications requires a thorough examination of immune cell subsets. This task is well-suited for analyzing high-dimensional cytometry data (Sanjabi and Lear 2021).

The application of conventional flow cytometry (CFC) in research has been constrained to less than 30 markers, while its utilization in clinical settings is typically confined to less than 18 markers. This limitation mostly arises from the spectral overlap of emission signals. Mass cytometry (MC) overcomes the limitations of fluorescence spectrum overlap and enables the study of more than 50 markers. However, the use of metal reporters in MC results in decreased signal intensity, and the device is unable to measure the size and granularity of cells (Sanjabi and Lear 2021).

Although MC analysis yields valuable results in high-dimensional scenarios where samples can be stored and analyzed in batches, its application in real-time sample analysis for clinical studies is not yet widespread. This can be attributed, in part, to slower throughput, limited availability of commercial reagents, and a scarcity of technicians who possess the necessary expertise to operate the instrument and perform the analysis. Spectral flow cytometry (SFC), which has recently been made available for commercial use, presents a promising avenue for implementing the comprehensive investigation of biomarkers across several dimensions, a necessity for conducting real-time analysis in clinical research. Spectral flow cytometry (SFC) is a technique that effectively addresses the issue of spectral overlap by capturing the complete emission spectrum of fluorophores. This capability holds promise for enabling the study of more than 40 markers while retaining a high level of signal resolution (Sanjabi and Lear 2021).

The light signal detecting method employed by SFC differs significantly from that of CFC. Instead of solely isolating the highest emission wavelength of each fluorophore and directing it to a single detector, a collection of detectors is employed to catch the complete spectral emission range of each fluorophore, spanning from ultraviolet to near-infrared. In a multiplexed experiment, wherein a cell releases numerous signals of antibody-fluorophore, the spectral fingerprints of all the fluorophores combine to form a complex waveform (Bonilla et al. 2021). The complex waveform is deconvoluted or "unmixed" into the spectral signature of each constituent floor using a mathematical approach derived from the Least Squares Method. This enables the identification and accurate quantification of each cell-bound fluorophore.

The technique of spectral un-mixing has been effectively employed in the field of fluorescent microscopy for a considerable period. However, the computational demands associated with performing this calculation on individual cells within a typical flow cytometry sample, which often consists of hundreds of thousands to millions of cells, necessitates access to adequate computation power and speed, which has become available only in recent years.

Subsequently, in 2017, Cytek successfully brought the Aurora spectra analyzer with spatially separated lasers into the commercial market. As a result, researchers were able to reach the significant milestone of achieving 40-color single-cell fluorescence cytometry (SFC) in 2019. In addition to facilitating a greater level of multiplexing, the transition from conventional to spectral fluorescence cytometry represents a significant shift in perspective. This shift involves a transition from solely optimizing the resolution of each fluorophore's peak channel to prioritizing the precise discrimination of each fluorophore's complete spectral signature. The theoretical multi-parameter limit of spectral flow cytometry (SFC) is determined solely by the number of fluorophores that possess a distinct spectral signature (Bonilla et al. 2021). The enhanced hardware capabilities of SFC are driving the progress of novel fluorophore reporters, emphasizing distinctive excitation/emission characteristics rather than only intensity (Table 9.1).

9.6 Limitations

The emerging field of nanotechnology has demonstrated significant advancements in the realm of secure and effective detection and treatment of cancers. While a few instances have reached the stage of clinical trials, numerous challenges need to be addressed and tackled to effectively utilize nanotechnology for cancer detection and therapeutic purposes (Dessale et al. 2022). The main challenge in cancer diagnosis with nanotechnology lies in ensuring the reliability of the results. To effectively implement nano-based technologies within diagnostic centers, achieving reliable and reproducible outcomes is imperative. The detection signals based on nanoparticles (NP) can be affected by various factors, including non-specific binding of NP

Table 9.1 Clinical findings for various techniques used in diagnosis, prevention, and treatment of tumors

Technique	Purpose	Clinical findings	Advantages	Limitations	References
MRI	Tumor assessment and monitoring	Visualization of tumor size and location Evaluation of treatment response and disease progression	Excellent soft tissue contrast no ionizing radiation Detailed anatomical information	Limited availability and high cost Longer scan times Claustrophobia in some patients	Antonios et al. (2017)
CT	Tumor assessment and monitoring	Detection of tumors, metastases, and changes in tumor size Assessment of response to treatment	Rapid imaging with high-resolution Availability in many healthcare facilities	Ionizing radiation exposure Limited soft tissue contrast	Carter et al. (2018)
Serum biomarkers	Monitoring treatment efficacy	Changes in serum levels of biomarkers (e.g., PSA, CA-125) Indication of treatment response and disease progression	Non-invasive and easily accessible Allows for serial monitoring	Limited specificity May not capture early changes	Spencer et al. (2016 and Clevers et al. (2021)
Circulating DNA (cfDNA)	Genetic assessment of tumors	Detection of tumor-specific mutations Evaluation of minimal residual disease (MRD)	Potential for early detection Customizable to specific mutations	Sensitivity varies by mutation type Limited to certain cancer types	Snyder et al. (2019) and Stadler et al. (2022)
Flow cytometry	Characterizing immune cell subsets	Changes in immune cell populations, activation status, and subsets (e.g., CD4/CD8 ratio)	High precision and sensitivity Quantify specific cell populations	Requires fresh samples Limited to surface markers	Bonilla et al. (2021)

(continued)

Table 9.1 (continued)

Technique	Purpose	Clinical findings	Advantages	Limitations	References
Circulating immune cells	Monitoring immune cell dynamics	Changes in circulating immune cells, such as T cells, B cells, and NK cells Assessment of immune cell activity and exhaustion	Non-invasive Provides real-time information	May not capture immune cell dynamics at the tumor site Limited to peripheral blood samples	Griffiths et al. (2020)

probes, aggregation, and erroneous detection methods. The careful assessment and comprehensive evaluation of assays' reliability and reproducibility are crucial in clinical trial applications, particularly when utilizing large clinical sample pools. In conjunction with this matter, persistent endeavor is needed to surmount these obstacles (Sharaf et al. 2023).

Another additional challenge is the large-scale production of nanoprobes that exhibit excellent sensitivity, repeatability, and prolonged storage stability, all while being cost-effective. Most contemporary nanoprobes are manufactured within laboratory environments that have been meticulously optimized. However, the production of these probes in large quantities continues to pose a considerable challenge. The detection efficiencies exhibit significant variation because of the chemical composition, dimensions, form, charge, and surface coating characteristics of nanoprobes. To mitigate batch-to-batch fluctuations, it is imperative to streamline the synthesis techniques and enhance the functionalization of nanoprobe. In addition, it has been observed that nanoprobes tend to cluster when stored. Additionally, it is imperative to assess the cost-effectiveness of the implementation of a framework based on nanoscale (Harish et al. 2022).

One of the primary challenges is in the development of NP-based devices that exhibit high sensitivity, user-friendly handling and operation, and cost-effectiveness. While numerous tests based on NP (nanoparticle) technology have been developed inside academic laboratories, a significant proportion of these tests are deemed inappropriate for clinical applications. The simplification of the medical utilization of nanotechnology in cancer diagnostics will be significantly facilitated by the successful development of point-of-care tools based on nanoparticles (Roy and Roy 2023).

Another of the primary concerns associated with the utilization of nanoparticles is the possibility of toxicity when administered systemically. This issue primarily pertains to the utilization of in-vivo imaging techniques based on nanoparticle (NP) technology. It is imperative to thoroughly evaluate the potential toxicity associated with novel nanoparticle probes before their utilization in in-vivo imaging applications. The toxicity of nanoparticles is determined by various factors, including their morphology, dimensions, electrical charge, surface properties, specific ligands, and elemental composition. It is imperative to consider the biodistribution, biodegradability, and pharmacokinetic characteristics of nanoparticles (Zhang et al. 2022).

Cancer biomarkers have the potential to be utilized in the early detection of cancer. However, the utilization of biomarkers has encountered several challenges, such as the presence of low quantities of biomarkers in bodily fluids, variability in the amount and timing of biomarkers among patients, and the complexities associated with performing prospective investigations (Li et al. 2023b).

Despite the potential of nanotechnology in addressing cancer through efficient drug delivery and early detection, numerous challenges remain. The utilization of controlled and longitudinal research can be significant in addressing the inherent constraints associated with nanotechnology-based approaches for cancer detection and treatment. The pursuit of a dependable implementation of nanotechnology is of paramount significance, necessitating the undertaking of diverse and extensive research endeavors. Nanotechnology-based cancer management applications must be inexpensive and accessible to underdeveloped nations.

9.7 Conclusion

Immunotherapy has emerged as a groundbreaking modality for disease management. However, the integration of nanotechnology into the realm of immunotherapy has propelled its efficacy to unprecedented heights. Nano-diagnostic methods have the potential to bring about a significant transformation in the field of immunotherapy by enabling the investigation of immune system responses at the molecular and cellular levels in real time. The importance of nanotechnology in the field of immunotherapy is seen in the advancement of drug delivery systems that utilize nanocarriers to improve the effectiveness of antitumor treatments against melanoma and breast cancer cells. Additionally, the utilization of nanoparticles containing immunomodulatory compounds has been shown to increase targeted immune responses. The field of nanotechnology has played a significant role in advancing vaccine research through its ability to enhance the efficacy of delivering specific antigens and improving antigen presentation. The physicochemical features of nanoparticles have significantly transformed diagnostic and monitoring approaches for assessing the progress of immunotherapy. Nanoparticles serve as a vehicle for the precise administration of contrast agents, addressing the limited sensitivity concerns associated with imaging techniques like as MRI and PET/CT. Quantum dots serve as fluorescent markers to visualize malignant tumors, intracellular interactions, and cell signaling mechanisms. Magnetic nanoparticles possessing a distinct DNA sequence or bio-bar codes exhibit the capability to identify biomarkers with a sensitivity range measured in pico- or femtomolar units. The utilization of circulating tumor DNA has been shown to enhance traditional therapeutic modalities, including chemotherapy, targeted therapy, and radiotherapy, by offering valuable insights into genetic alterations occurring in both primary and metastatic tumor sites.

In conclusion, the integration of immunotherapy and nanotechnology has significantly transformed the techniques and approaches employed in therapeutic interventions, diagnostic monitoring, and the formulation of personalized and efficacious treatment strategies for cancer and autoimmune disorders. Healthcare practitioners can potentially enhance the effectiveness of immunotherapy by consistently monitoring immune responses, biomarkers, and adverse effects. The integration of nanotechnology with immunotherapy establishes a foundation for enhancing the precision, customization, and dependability of immunotherapeutic approaches in combating a diverse range of diseases, thereby instilling optimism among patients on a global scale.

9.8 Future Prospects

Immunotherapeutic modalities have achieved remarkable success in diverse settings and among patients with previously resistant conditions. Nevertheless, there are still some obstacles to overcome. Exploring additional pathways for checkpoint inhibitors that enable tumors to evade treatment and gaining a better understanding of how the tumor microenvironment hampers immune checkpoint inhibitors are crucial areas of research needed to advance the effectiveness of these therapies. CART has shown promising results in certain groups of patients with hematologic malignancies, although its efficacy has been limited in patients with highly immunosuppressive tumor microenvironment (TME) or solid tumors. Addressing the immunosuppressive components of the tumor microenvironment (TME), such as tumor-associated macrophages (TAMs), has demonstrated potential in enhancing the effectiveness of immune checkpoint inhibitors (ICI) and chimeric antigen receptor T-cell (CART) therapy.

Further investigation is necessary to evaluate the potential of these agents, whether used as pre-conditioning or in conjunction with ICI and CART. This will help identify the most promising approaches. Exploring elements of alternative pathways using ICI has displayed great potential and is expected to open doors for innovative combination therapies. Groundbreaking advancements in engineering CART cells have paved the way for enhanced trafficking and decreased immunosuppression within solid tumors, potentially leading to significant improvements in outcomes. Considerable progress has been made in cancer vaccines in recent years. Specifically, research exploring the use of neoantigen vaccines alongside targeted tumor microenvironment (TME) therapies holds great promise in advancing this captivating study area.

References

Abbas AK, Lichtman AH, Pillai S (2019) Basic immunology e-book: functions and disorders of the immune system. Elsevier Health Sciences

Antonios JP et al (2017) Detection of immune responses after immunotherapy in glioblastoma using PET and MRI. Proc Natl Acad Sci 114(38):10220–10225

Anwar F et al (2023) Therapeutic potential of nanomedicine in management of Alzheimer's disease and glioma. Int J Nanomedicine 18:2737–2756
Beck A et al (2017) Strategies and challenges for the next generation of antibody–drug conjugates. Nat Rev Drug Discov 16(5):315–337
Bonilla DL, Reinin G, Chua E (2021) Full spectrum flow cytometry as a powerful technology for cancer immunotherapy research. Front Mol Biosci 7:612801
Carter BW, Bhosale PR, Yang WT (2018) Immunotherapy and the role of imaging. Cancer 124(14):2906–2922
Chen A et al (2007) Depleting intratumoral CD4+ CD25+ regulatory T cells via FasL protein transfer enhances the therapeutic efficacy of adoptive T cell transfer. Cancer Res 67(3):1291–1298
Cho JH, Collins JJ, Wong WW (2018) Universal chimeric antigen receptors for multiplexed and logical control of T cell responses. Cell 173(6):1426–1438. e11
Clevers MR et al (2021) Evaluation of serum biomarker CEA and Ca-125 as immunotherapy response predictors in metastatic non-small cell lung cancer. Anticancer Res 41(2):869–876
Couzin-Frankel J (2013) Cancer immunotherapy. American Association for the Advancement of Science
Davis ZB et al (2015) Natural killer cell adoptive transfer therapy: exploiting the first line of defense against cancer. Cancer J (Sudbury, Mass.) 21(6):486
Davola ME, Mossman KL (2019) Oncolytic viruses: how "lytic" must they be for therapeutic efficacy? Onco Targets Ther 8(6):e1581528
Decker WK, Safdar A (2009) Bioimmunoadjuvants for the treatment of neoplastic and infectious disease: Coley's legacy revisited. Cytokine Growth Factor Rev 20(4):271–281
Demetzos C, Demetzos C (2016) Introduction to nanotechnology. In: Pharmaceutical nanotechnology: fundamentals and practical applications, pp 3–15
Dessale M, Mengistu G, Mengist HM (2022) Nanotechnology: a promising approach for cancer diagnosis, therapeutics and theragnosis. Int J Nanomed 17:3735–3749
Dourado D (2020) Pharmaceutical nanotechnology: a therapeutic revolution. Int J Pharm Sci Dev Res 6:009–011
Du Y et al (2019) Noninvasive imaging in cancer immunotherapy: the way to precision medicine. Cancer Lett 466:13–22
Fontana F et al (2021) Porous silicon materials for cancer and immunotherapy. In: Porous silicon for biomedical applications. Elsevier, pp 571–609
Fritz JM, Lenardo MJ (2019) Development of immune checkpoint therapy for cancer. J Exp Med 216(6):1244–1254
Gardner A, de Mingo Pulido Á, Ruffell B (2020) Dendritic cells and their role in immunotherapy. Front Immunol 11:924
Getts DR et al (2015) Harnessing nanoparticles for immune modulation. Trends Immunol 36(7):419–427
Griffiths JI et al (2020) Circulating immune cell phenotype dynamics reflect the strength of tumor–immune cell interactions in patients during immunotherapy. Proc Natl Acad Sci 117(27):16072–16082
Gutowska-Owsiak D, Ogg GS (2017) Therapeutic vaccines for allergic disease. NPJ Vaccines 2(1):12
Harish V et al (2022) Nanoparticle and nanostructure synthesis and controlled growth methods. Nano 12(18):3226
Hildner K et al (2008) Batf3 deficiency reveals a critical role for CD8α+ dendritic cells in cytotoxic T cell immunity. Science 322(5904):1097–1100
Hu Y et al (2020) Synergistic effect of adoptive immunotherapy and docetaxel inhibits tumor growth in a mouse model. Cell Immunol 348:104036
Islam SU et al (2018) Decursinol angelate inhibits LPS-induced macrophage polarization through modulation of the NFκB and MAPK signaling pathways. Molecules 23(8):1880
Jin P et al (2021) PET/CT metabolic patterns in systemic immune activation: a new perspective on the assessment of immunotherapy response and efficacy. Cancer Lett 520:91–99

Kelly PN (2018) The cancer immunotherapy revolution. American Association for the Advancement of Science, pp 1344–1345
Khatun S et al (2023) Immunomodulatory nanosystems: an emerging strategy to combat viral infections. Biomater Biosyst 9:100073
Kim E-Y et al (2008) Detection of HIV-1 p24 Gag in plasma by a nanoparticle-based bio-barcode-amplification method. Nanomedicine 3:293
Kim C-G et al (2021) Combining cancer vaccines with immunotherapy: establishing a new immunological approach. Int J Mol Sci 22(15):8035
Köhler G, Milstein C (1975) Continuous cultures of fused cells secreting antibody of predefined specificity. Nature 256(5517):495–497
Kuperkar K et al (2022) Amphiphilic block copolymers: their structures, and self-assembly to polymeric micelles and polymersomes as drug delivery vehicles. Polymers 14(21):4702
Lau D, Corrie PG, Gallagher FA (2022) MRI techniques for immunotherapy monitoring. J Immunother Cancer 10(9):e004708
Leko V, Rosenberg SA (2020) Identifying and targeting human tumor antigens for T cell-based immunotherapy of solid tumors. Cancer Cell 38(4):454–472
Li L et al (2018) Promising clinical application of ctDNA in evaluating immunotherapy efficacy. Am J Cancer Res 8(10):1947
Li C et al (2023a) Immunomodulatory nano-preparations for rheumatoid arthritis. Drug Deliv 30(1):9–19
Li H-z et al (2023b) Application of nanotechnology in bladder cancer diagnosis and therapeutic drug delivery. J Mater Chem B 11(35):8368–8386
Lonberg N (2008) Fully human antibodies from transgenic mouse and phage display platforms. Curr Opin Immunol 20(4):450–459
Lu Z-G et al (2023) Nucleic acid drug vectors for diagnosis and treatment of brain diseases. Signal Transduct Target Ther 8(1):39
Miao Y-B et al (2023) Customizing delivery nano-vehicles for precise brain tumor therapy. J Nanobiotechnol 21(1):32
Miller SA, Beed FD, Harmon CL (2009) Plant disease diagnostic capabilities and networks. Annu Rev Phytopathol 47:15–38
Mokhtari RB et al (2017) Combination therapy in combating cancer. Oncotarget 8(23):38022
Mosser DM, Edwards JP (2008) Exploring the full spectrum of macrophage activation. Nat Rev Immunol 8(12):958–969
Moudgil KD, Venkatesha SH (2023) The anti-inflammatory and immunomodulatory activities of natural products to control autoimmune inflammation. Int J Mol Sci 24(1):95
O'Shea JJ, Paul WE (2010) Mechanisms underlying lineage commitment and plasticity of helper CD4+ T cells. Science 327(5969):1098–1102
Ou Y-C, Wen X, Bardhan R (2020) Cancer immunoimaging with smart nanoparticles. Trends Biotechnol 38(4):388–403
Patra JK et al (2018) Nano based drug delivery systems: recent developments and future prospects. J Nanobiotechnol 16(1):1–33
Perez EE et al (2017) Update on the use of immunoglobulin in human disease: a review of evidence. J Allergy Clin Immunol 139(3):S1–S46
Pilla L, Maccalli C (2018) Immune profiling of cancer patients treated with immunotherapy: advances and challenges. Biomedicines 6(3):76
Revia RA, Stephen ZR, Zhang M (2019) Theranostic nanoparticles for RNA-based cancer treatment. Acc Chem Res 52(6):1496–1506
Rogers LM, Veeramani S, Weiner GJ (2014) Complement in monoclonal antibody therapy of cancer. Immunol Res 59:203–210
Roy J, Roy K (2023) Evaluating metal oxide nanoparticle (MeOx NP) toxicity with different types of nano descriptors mainly focusing on simple periodic table-based descriptors: a mini-review. Environ Sci: Nano 10:2989
Ruderman EM, Pope RM (2005) The evolving clinical profile of abatacept (CTLA4–Ig): a novel co-stimulatory modulator for the treatment of rheumatoid arthritis. Arthritis Res Ther 7(2):1–5

Ryu JH et al (2012) Tumor-targeting multi-functional nanoparticles for theragnosis: new paradigm for cancer therapy. Adv Drug Deliv Rev 64(13):1447–1458
Ryu JH et al (2014) Theranostic nanoparticles for future personalized medicine. J Control Release 190:477–484
Rzhevskiy A et al (2021) Emerging role of circulating tumor cells in immunotherapy. Theranostics 11(16):8057
Sanjabi S, Lear S (2021) New cytometry tools for immune monitoring during cancer immunotherapy. Cytometry B Clin Cytom 100(1):10–18
Sharaf M et al (2023) Challenges of nanomaterials-based cancer therapy: a future destination. Int J Surg 109(6):1819–1820
Sharma AR et al (2022) Recent advances of metal-based nanoparticles in nucleic acid delivery for therapeutic applications. J Nanobiotechnol 20(1):1–21
Shields CW IV et al (2020) Materials for immunotherapy. Adv Mater 32(13):1901633
Snyder A, Morrissey MP, Hellmann MD (2019) Use of circulating tumor DNA for cancer immunotherapy. Clin Cancer Res 25(23):6909–6915
Spencer KR et al (2016) Biomarkers for immunotherapy: current developments and challenges. Am Soc Clin Oncol Educ Book 36:e493–e503
Stadler J-C et al (2022) Current and future clinical applications of ctDNA in immuno-oncology. Cancer Res 82(3):349–358
Stelzner JJ (2018) Development and characterisation of novel squalene based adjuvant formulations for yeast vaccines against IBDV. Universitäts-und Landesbibliothek Sachsen-Anhalt
Sun J et al (2021) Visualization of vaccine dynamics with quantum dots for immunotherapy. Angew Chem 133(45):24477–24485
Thaiss CA et al (2016) Integration of innate immune signaling. Trends Immunol 37(2):84–101
Varadé J, Magadán S, González-Fernández Á (2021) Human immunology and immunotherapy: main achievements and challenges. Cell Mol Immunol 18(4):805–828
Wang Z et al (2018) Nanoscale delivery systems for cancer immunotherapy. Mater Horiz 5(3):344–362
Wang R et al (2023) Nanotechnologies meeting natural sources: engineered lipoproteins for precise brain disease theranostics. Asian J Pharm Sci 18(5):100857
Yeolekar LR (1996) Preparation of arboviral vaccine: a new approach through immuno-stimulating complexes (ISCOMs) of Japanese encephalitis virus, vol 14, p 1553
Zhang M et al (2017) Nanotechnology-based combination therapy for overcoming multidrug-resistant cancer. Cancer Biol Med 14(3):212
Zhang N, Xiong G, Liu Z (2022) Toxicity of metal-based nanoparticles: challenges in the nano era. Front Bioeng Biotechnol 10:1001572
Zheng M et al (2019) Oncolytic viruses for cancer therapy: barriers and recent advances. Mol Ther Oncolytics 15:234–247
Zhong X et al (2020) Circulating tumor cells in cancer patients: developments and clinical applications for immunotherapy. Mol Cancer 19(1):1–12
Zom GG et al (2019) Dual synthetic peptide conjugate vaccine simultaneously triggers TLR2 and NOD2 and activates human dendritic cells. Bioconjug Chem 30(4):1150–1161

Prospectus and Concerns of Immunomodulatory Nanotechnologies and Nanoparticles Biocompatibility and Toxicity

10

Shereen A. Arabiyat

10.1 Introduction

Nanoparticles (NPs) exhibit unique qualities such as their extraordinarily high surface-to-volume ratio, compared to their bulk counterparts (Modena et al. 2019). When particle size decreases, surface area increases significantly, leading to more chemically reactive sites (Ndolomingo et al. 2020). Besides, the optical, thermal, magnetic, and electric characteristics of NPs differ with their size (Sajid and Plotka-Wasylka 2020). These qualities make NPs highly multipurpose materials that can be precisely tailored, engineered, and synthesized for numerous applications (Kus-Liskiewicz et al. 2021).

NPs are beneficial across diverse fields such as catalysis (Joudeh and Linke 2022), disease diagnosis and treatment (Ma et al. 2020), composite material fabrication (Baghdadi et al. 2020), and sensor technology (Jung et al. 2022). For example, NPs exhibit excellent catalytic activity, particularly valuable in green chemistry for reducing hazardous substances (Huston et al. 2021). Metal NPs, in particular, have gathered attention for their application in cancer treatment through photothermal ablation therapy (Liu et al. 2018). This method relies on NPs' ability to generate heat from light, effectively abolishing cancer cells. Key requirements for this therapy include high optical absorption and robust photothermal conversion efficiency (Liu et al. 2018; Han and Choi 2021).

S. A. Arabiyat (✉)
Al-Balqa Applied University, Salt, Jordan
e-mail: shereen.arabiyat@bau.edu.jo

S. Rehman, E. Al-Suhaimi (eds.), *Nanotechnology Based Microbicides and Immune Stimulators*, https://doi.org/10.1007/978-981-96-0397-8_10

10.2 Nanomaterial Compositions and Applications

Traditional dosage forms like solutions, suspensions, or emulsions encounter limitations such as high doses, low bioavailability, intolerance, first-pass effect, and instability, leading to poor sustained effects and huge differences in plasma drug concentrations (Laffleur and Keckeis 2020). Hence, there arises a requirement for innovative carriers that can address the need of drug delivery systems (Adepu and Ramakrishna 2021).

In recent years, NPs have appeared as promising colloidal drug carriers (Mahringer et al. 2021), characterized by particle sizes that range from 10 to 1000 nm in diameter (Zahin et al. 2020). NPs may exhibit unique size-dependent characteristics different from those of fine particles or bulk materials (Buzea et al. 2007). Their principle benefits include:

1. Enhanced bioavailability through better aqueous solubility (Bhatia and Bhatia 2016).
2. Extended duration within the body achieved by either prolonging half-life for clearance or enhancing specificity for pertinent receptors (Zahin et al. 2020).
3. Precise drug targeting to specific locations within the body, resulting in decreased drug volumes needed and reduced dosage-related toxicity. This facilitates the safe administration of powerful therapeutic medications, while shielding non-targeted tissues and cells from adverse reactions (Irving 2007).

10.3 Types of NPs

NPs are extensively employed in various fields, notably in drug delivery systems to overcome barriers like the blood-brain barrier and cell membranes (Abhilash 2010). Such systems commonly rely on biocompatible lipids, ensuring sustained effects achieved through mechanisms such as diffusion or dissolution (Cavalli et al. 1995; zur Mühlen and Mehnert 1998; Yang et al. 1999; Müller et al. 2000a). NPs constitute a diverse array of materials with variable sizes, shapes, and compositions, presenting different chemical and surface characteristics (Alvarez et al. 2020; Peer et al. 2007). The field of nanotechnology is experiencing rapid expansion, continually enriching the existing libraries of NPs (Shi et al. 2017). The following classes of NPs, though broad and multifunctional, exhibit specific characteristics and find applications in nanomedicine (Chen et al. 2016; Dobrovolskaia and McNeil 2007).

10.3.1 Fullerenes

Fullerenes, consisting entirely of carbon, exist in hollow sphere, ellipsoid, or tube shapes, resembling the structure of graphite with added pentagonal or heptagonal rings, yielding potential porous molecules (Hollister et al. 2003). Megatubes, larger in diameter than nanotubes, vary in wall thickness and hold potential for transporting molecules of diverse sizes (Mitchell et al. 2001).

10.3.2 Solid Lipid NPs

Solid lipid NPs primarily consist of lipids that remain solid at room temperature along with surfactants for emulsification, typically exhibiting average diameters ranging from 50 nm to 1000 nm for drug delivery of colloidal-level particle size (zur Mühlen et al. 1998). These NPs offer key benefits including their decreased size, increased surface area, substantial drug-loading capability, and interactions at phase interfaces, thereby improving the performance of pharmaceutical ingredients and other materials (Cavalli et al. 1993). The lipid components may include fatty acids, triglycerides, steroids, partial glycerides, and waxes, while various surfactants are employed to stabilize lipid dispersion (Zhang et al. 2011). Solid lipid NPs present several advantages, such as utilizing physiological lipids, avoiding organic solvents, and offering a wide application spectrum including dermal, oral, and intravenous routes (Müller et al. 2000a, b; Jenning et al. 2000). They facilitate improved bioavailability, protect labile drug molecules from environmental influences, enable sustained or targeted drug release, and feature biodegradable lipids (Müller and Runge 1998). Moreover, they exhibit greater stability compared to liposomes, forming the basis of biodegradable drug delivery systems of colloidal-level particle size with prolonged shelf life (Liu et al. 2022; Kumar et al. 2022; Aswathi et al. 2023).

10.3.3 Liposomes

Liposomes are versatile spherical formations distinguished by an inner aqueous center enveloped within a hydrophobic lipid bilayer, often crafted through phospholipid extrusion (Has and Sunthar 2020). These phospholipids, recognized as generally regarded as safe (GRAS), minimize potential adverse effects (Akram et al. 2023). Through fusion with cell membranes, their lipid bilayer facilitates the controlled release of encapsulated contents (Rafiee et al. 2019). Liposomes exhibit variability in size and structure, spanning from 15 nm to several μm, with options for single (unilamellar) or multiple (multilamellar) bilayer membranes (Deshpande and Kulkarni 2018). Furthermore, their surface can be tailored with "stealth" materials to augment stability or directing ligands for specific delivery (Koog et al. 2022). Consequently, liposomes find widespread applications as carriers for many drugs, including insulin, antineoplastics, antibacterials, antivirals, and plasmid DNA (Akhtar and Khan 2016).

10.3.4 Nanostructured Lipid Carriers (NLCs)

NLCs present another advanced drug delivery system composed of solid and liquid lipids, maintaining a solid state at body temperature (Shirodkar et al. 2019). Their production includes solidified emulsion technologies (Sharma and Baldi 2018). NLCs offer benefits such as enhanced drug loading and controlled release, although concerns like drug expulsion during storage may arise (Rawal and Patel 2018). Their application areas include topical, oral, and parenteral routes (Bibi et al. 2017).

10.3.5 Nanoshell Particles

Nanoshell particles exhibit superior characteristics compared to their single-component or NP counterparts, providing customizable structures by altering constituent materials or core-to-shell ratios (Perera et al. 2020). These materials present improved optical characteristics and resistance to chemical/thermal denaturation, making them advantageous over conventional organic dyes (Aziz et al. 2017).

10.3.6 Quantum Dots (QDs)

QDs serve as versatile architectures for imaging and drug delivery, with their inorganic core and amphiphilic polymer coating enabling encapsulation of hydrophobic drugs (Moradi et al. 2017).

10.3.7 Dendrimers

Dendrimers are considered as well-defined, symmetric structures and functional end groups, and are robust nanostructures with varied applications referring to their precise design and high surface functionality (Bober et al. 2022; Nikzamir et al. 2021; Joudeh and Linke 2022).

In summary, these diverse nanostructures offer promising avenues for drug delivery and biomedical applications, each with unique advantages and applications in pharmaceutical and biomedical fields.

10.4 Factors Affecting NPs Application in Medicine

Various factors influence the application of NPs in medicine, particularly concerning the association with the immune system (Muhammad et al. 2020). Key factors include size, shape, and surface characteristics, which significantly impact immune reaction (Gowd et al. 2022). The immune system typically recognizes foreign bodies based on size, making it fundamental to design NPs within specific size ranges to avoid rejection (Teunissen et al. 2021; Yadav and Bachhuka 2023). Larger NPs exhibit decreased surface-to-volume ratios, affecting their interactions with immunoreagents (Geka et al. 2023).

Selection of immune-tolerant NPs depends on structure and application criteria, including NP size for efficient transport and targeting of immune cells like dendritic cells (DCs) (Chen et al. 2020; Meng et al. 2023). Furthermore, the usage of biocompatible, biodegradable polymers approved for clinical use and small-molecule immunomodulators confirmed in humans is essential (Osi et al. 2022; Cifuentes-Rius et al. 2021). Codelivery of antigens with immunomodulators to antigen-presenting cells (APCs) additionally enhances immune tolerance (Levy et al. 2022).

10.5 Nanomaterials (NMs)-Mediated Direct and Indirect Immunomodulation

Nanomaterials (NMs) can induce both direct and indirect immunomodulation, impacting immunostimulation and immunosuppression (Li et al. 2018). In drug delivery applications, NPs serve as vehicles for vaccines or drugs, but their inherent immune reactions are also significant (Ferrando et al. 2020). Direct immunomodulation involves interactions with inherent immune cells like macrophages and natural killer cells, while indirect modulation activates T and B cells for antigen-specific responses (Wang and Tang 2018).

Immunomodulation by NMs is multifactorial, influenced by surface chemistry, chemical composition, size, and molecular structure (Akhtar et al. 2020). Studies have investigated immune adjuvants and their therapeutic applications, as well as the effects of different NM categories and compositions on systemic immunomodulation (Bowen and Ghosn 2018; Huang et al. 2019; Ballesteros and Tack 2020; Tom and Cavanaugh 2020; Wang et al. 2020b). Furthermore, the role of biophysicochemical properties like protein adsorption and corona formation in immunomodulation has been extensively explored (Corbo et al. 2016; Tsuda and Konduru 2016; Agrahari et al. 2019). Figure 10.1 displays the characteristics of NPs and their

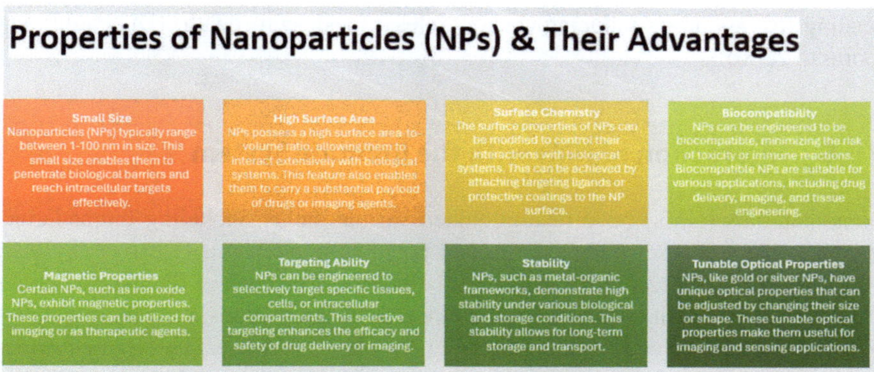

Fig. 10.1 Characteristics of NPs and their advantages

advantages. The size of NPs shows a crucial role in eliciting immune responses. For instance, nano-sized particles (193 nm) have been related to stimulate a more profound immune response compared to micro-sized ones (1530 nm) (Liu and Tang 2020). Moreover, larger NPs have been found to be stronger serum immunoglobulin responses, emphasizing the importance of size in immunomodulation (Dykman et al. 2018). Furthermore, the presence of spiky nanostructures in bulk materials has modified immune responses by inducing mechanical stress on cells, leading to potassium efflux in dendritic cells (DCs) (Wang et al. 2018).

In nano-drug delivery systems (DDS), the physicochemical characteristics of NPs must be adjusted to ensure safe and effective cellular interactions for targeted intracellular delivery. This involves bearing in mind the anatomical and physiological features of the target tissue or organ (Behzadi et al. 2017; Foroozandeh and Aziz 2018). Reliable biological in-vitro tests need appropriate lipid matrix compositions and manufacturing procedures that produce reproducible nanosystems with defined product structures, followed by particular characterization (Foroozandeh and Aziz 2018).

Under physiological conditions, NPs undergo interactions with saliva before encountering the primary epithelium. Saliva, characterized as a hypotonic fluid that is full of proteins, enzymes, and inorganic salts, serves as a buffer to preserve a neutral pH in the oral cavity. Saliva has the potential to diminish the thickness of the diffuse layer created near to NP surfaces, thus increasing agglomeration propensities and instability (Teubl et al. 2018). The uptake of particles into buccal epithelial cells is influenced by their size, charge, and their lipid makeup (Moore et al. 2015).

The incorporation of unsaturated fatty acids aids in particle-membrane interactions and uptake, although higher concentrations may induce cytotoxic effects (Foroozandeh and Aziz 2018). Furthermore, the endocytosis and exocytosis characteristics of NPs are dependent on their doses and sizes, with larger particles accumulating in slow recycling compartments, thereby impacting cell viability (Sakhtianchi et al. 2013). Oleic acid, a known penetration enhancer, interacts with cell membranes, leading to increased reactive oxygen species (ROS) production and defects in membranes like solubilization or formation of pores (Mahlert et al. 2019; Teubl et al. 2013). This interaction affects mitochondrial permeability, ultimately leading to apoptosis-inducing factor release and cell death (Di Paola and Lorusso 2006).

10.6 Interaction of NMs with the Immune System

Nanomaterials (NMs) have revolutionized various fields, including biomedicine, owing to their diverse characteristics and potential applications (Khan et al. 2022; Kahn and Gang 2022). NMs' ability to be precisely tailored at the nanoscale has resulted in significant advancements in biomedical applications, such as drug delivery systems (Shilar et al. 2022; Huang et al. 2024). However, before their widespread use, NMs must undergo hard assessment for cytotoxicity and biocompatibility (Oksel Karakus et al. 2020; Gomes et al. 2023).

Exploring the therapeutic potential of nanomaterials (NMs) has exposed promising capabilities in regulating scarring processes, reducing fibrosis, and inhibiting cancer growth (Kyriakides et al. 2022). However, understanding the bio-interactions of NMs poses numerous challenges due to the diverse characteristics of NMs and the variety of model systems engaged in experimentation (Liu and Chen 2022). The absence of standardized testing protocols, coupled with the intricacies involved in in vivo assessments, further hinder our understanding of the biological effects induced by NMs (Mahmoudi et al. 2012). Despite these challenges, the design of immunomodulatory nanosystems offers favorable alternatives for treating viral diseases (Krishnan et al. 2021). Nanotechnology-based immunomodulators leverage modified antigens and adjuvants conjugated with NPs to stimulate host immune cells (Feng et al. 2019). These nanosystems are designed to permit controlled release of antigens and prolonged contact with APCs (Lee and Suresh 2022).

Key strategies for immunomodulatory nanosystems include codelivery of antigens and treatments to APCs, prolongation of pharmacologically active components' half-lives, and stimuli-responsive drug release (Piktel et al. 2016; Hickey et al. 2018; Sahu et al. 2021). Additionally, surface modifications of NPs with target-specific functionalities and alternative drug delivery strategies augment the efficacy and safety of immunomodulatory nanosystems (Polo et al. 2017).

10.7 Immunomodulatory NPs Against Infectious Diseases

The manipulation of immune reactions against viral infections has experienced thorough exploration, engaging diverse agents such as antibodies, receptors, antigen-presenting cells, and cytokines, such as interferons and interleukins (Davis and Quake 2020). Immunoglobulins play a pivotal role in combating viral infections by participating in complex pathways like complement activation, cytokine networks, idiotype expressions, differentiation of effector cells in B and T lymphocytes, and the stimulation of APCs (Alijotas-Reig et al. 2020).

Biomaterials suggest a promising method to combat microbial infections and have been utilized in the diagnosis and treatment f numerous infectious diseases, including COVID-19, AIDS, and influenza (Loh and Nachimuthu 2020; Zhang et al. 2020). Furthermore, biomaterials find application in antisepsis, disinfection procedures, and the development of personal protective equipment (Rothan and Byrareddy 2020; Wang et al. 2020a, b).

10.8 Immunomodulatory NPs for Cancer Treatment

Efficient induction of tumor immunity depends on the effective transfer of tumor antigens to APCs. These antigens can be categorized into tumor-associated antigens (TAAs) and tumor-specific antigens (TSAs), with TAAs being more prevalent in malignant cells but found as well in normal cells, potentially leading to autoimmune reactions (Jain et al. 2019). Conversely, TSAs, exclusive to cancer cells, avoid immune tolerance concerns (Carreno et al. 2015). However, the destruction of native tumor antigens by enzymes and their low immunogenicity pose challenges. Effective anti-cancer immune reactions are primarily initiated in lymph nodes, necessitating efficient delivery of tumor antigens to these sites (Sharma et al. 2017).

NPs have appeared as encouraging delivery vehicles for tumor antigens referring to their ability to protect antigens from destruction and enable selective lymph node delivery (Sharma et al. 2017). NPs encapsulating cancer antigens are efficiently internalized by APCs upon delivery (Jain et al. 2019). However, NP design considerations are crucial for their effectiveness (Duncan 2011). Particle size, shape, surface charge, and hydrophobicity significantly influence NP delivery to lymph nodes (Reddy et al. 2006; Zhang et al. 2008; Jain et al. 2018; Chakravarthy et al. 2019).

Medium-sized NPs (~5–100 nm) are optimal for lymph node delivery, remaining in circulation and efficiently reaching lymph nodes from lymphatic vessels (Ali et al. 2021). Particle shape also affects lymphatic drainage, with non-spherical particles demonstrating extended circulation times and enhanced tissue penetration (Champion and Mitragotri 2006). Moreover, surface charge impacts cellular internalization and immune reaction activation, with positively charged NPs generally inducing stronger immune reactions (Chen et al. 2017). However, positively charged NPs may face challenges in lymphatic transport due to their potential to induce hemolysis and platelet aggregation (Gaspar et al. 2012).

Certain polymers, as poly(lactic-co-glycolic acid) (PLGA) as well as chitosan, possess intrinsic adjuvant activity, augmenting immune cell activation (Panyam and Labhasetwar 2003; Illum 2011). Amphiphilic poly gamma-glutamic acid NPs with increased hydrophobicity exhibit enhanced antigen uptake and dendritic cell activation, in vitro and in vivo (Nishikawa et al. 2017; Altammar 2023). These insights highlight the importance of NP design in optimizing the delivery of tumor antigens for effective anti-cancer immunotherapy.

10.9 NP-Based Drug Delivery System in Anti-inflammatory Treatment

Biocompatibility is a crucial consideration in the improvement of organic NPs, with many carriers consisting of biomaterials or polymers. These polymers can be classified as either natural or synthetic. Among synthetic polymers, poly(lactic-co-glycolic acid) (PLGA) stands out as a likely nominee for drug delivery due to its comparatively hydrophobic nature (Bala et al. 2004). PLGA has demonstrated significant potential in targeted drug delivery and therapy (Mir et al. 2017). Integrating drugs with PLGA to form complexes has been widely studied, displaying encouraging results in various applications.

In a mice implantation model, stromal cell-derived factor-1α (SDF-1α) was effectively integrated into poly(lactic-co-glycolic acid) (PLGA), producing the creation of a solid scaffold that mitigated inflammatory reactions upon subcutaneous implantation (Thevenot et al. 2010; Fredman et al. 2015). PLGA has also an established application in addressing various bone diseases, including bone cancer metastasis, osteoarthritis, osteosarcomas, and other inflammatory bone conditions (Gu et al. 2013). For instance, the incorporation of doxycycline into PLGA nanospheres, forming a three-dimensional scaffold, demonstrated controlled release characteristics and suppressed the growth of bacteria such as *S. aureus* and *E. coli* in vitro (Feng et al. 2010).

Inflammatory bowel disease (IBD), encompassing conditions like ulcerative colitis (UC) and Crohn's disease (CD), is considered a significant therapeutic challenge (Yazeji et al. 2017). In a study targeting UC, chitosan (CS)-modified PLGA nanospheres loaded with nuclear factor κB (NF-κB) decoy oligonucleotides (ODNs) displayed enhanced stability of the ODNs and reduced symptoms like bloody feces and diarrhea in mouse models (Tahara et al. 2011). This method holds potential as a potential approach for UC treatment.

Rheumatoid arthritis (RA), another common chronic inflammatory disease, often necessitates treatment with low doses of methotrexate and disease-modifying antirheumatic drugs. Lutetium-177 (177Lu) has exhibited potential in decreasing synovial tissue inflammation related to RA (Trujillo-Nolasco et al. 2019). These examples emphasize the usefulness and therapeutic prospective of PLGA-based nanomaterials in addressing various inflammatory conditions.

10.10 NPs Biocompatibility and Toxicity

NPs are experiencing a surge in utilization across various fields, resulting in an inevitable increase in human exposure, whether intentional or unintentional (Buzea et al. 2007). Preceding any application in human settings, thorough toxicological assessments are imperative, involving several experimental studies. Despite these efforts, the existing toxicological data on nanomaterials often presents conflicting and inconsistent results, highlighting the need for comprehensive studies to safeguard both human health and the environment (Nel et al. 2009). Assessing the toxicity of NPs is essential for regulatory compliance and risk mitigation. On the other hand, the complexity of NP interactions with biological systems poses challenges in accurately determining their toxicity (Fadeel and Garcia-Bennett 2010). Some NPs regarded as valuable in certain applications may display adverse effects in biological settings, and vice versa. To address this knowledge gap, it's crucial to employ internationally accepted, unbiased, and robust in vivo toxicological models, concentrating on dynamic physiological systems (Fadeel and Savolainen 2013). Furthermore, there's a pressing need to develop and certify new models designed for toxicity testing, considering factors such as route of exposure, NP coating material, sterility, and cell culture types. Identifying the significance of this dispute, regulatory institutions like the United States Food and Drug Administration (FDA) are actively addressing the potential toxic effects associated with NP-containing products. Such agencies do not categorize NPs as inherently safe or harmful for human use but subject each product to strict regulation and assessment processes. This proactive approach purposes to guarantee the safety of NP-based products and alleviate any possible hazards to human health and the environment (Stone et al. 2014).

10.11 Introduction to NP Biocompatibility

NPs have been introduced as potential candidates for several therapeutic applications owing to their special physicochemical characteristics (Nel et al. 2006). However, their biocompatibility is an essential feature that rules their safety and efficacy in biological systems. Biocompatibility refers to the capability of NPs to interact with living organisms without triggering adverse effects or causing immune responses. Multiple aspects affect the biocompatibility of NPs, such as their size, shape, composition, and surface chemistry (Mahmoudi and Serpooshan 2012). Small-sized NPs can cross biological barriers more effectively but may also have a higher tendency for cellular uptake and possible toxicity. Conversely, larger NPs may exhibit diminished cellular uptake but could trigger physical damage to cells or tissues. Surface chemistry is a crucial factor in determining the interaction of NPs with biological entities. Customization of NP surfaces with biocompatible coatings, such as polymers or hydrophilic molecules, can advance their stability in biological environments and diminish nonspecific interactions with biomolecules, decreasing the possibility of adverse effects (Dreaden et al. 2012). Likewise, the selection of NP composition is critical for biocompatibility. Biodegradable materials, like lipids, polymers, or metals such as gold or silver, are preferred for biomedical applications as they can go through degradation and clearance from the body, decreasing long-term toxicity worries. Additionally, the surface charge of NPs affects their contact with biological membranes and proteins. Neutral or slightly negatively charged NPs are generally more biocompatible than positively charged ones, which may trigger cytotoxicity or immune activation. Assessing the biocompatibility of NPs includes comprehensive studies such as in vitro cell culture assays, in vivo animal models, and clinical trials. These assessments aim to define the safety profile of NPs, including their cytotoxicity, genotoxicity, immunotoxicity, and long-term accumulation in vital organs (Nel et al. 2006). Generally, understanding the factors inducing NP biocompatibility is essential for the improvement of safe and effective nanomedicines for many therapeutic applications. Additionally, the route of administration can seriously influence the biocompatibility of NPs. For example, NPs administered intravenously may encounter multiple biological barriers and clearance mechanisms compared to those administered orally or topically (Gupta and Gupta 2017). Intravenous administration presents NPs to interactions with serum proteins, immune cells, and endothelial cells, which can affect their biodistribution, circulation time, and potential for systemic toxicity. Otherwise, oral administration introduces NPs to the harsh settings of the gastrointestinal tract, including acidic pH, enzymatic degradation, and mucosal barriers, which may affect their stability and absorption (Jokerst et al. 2011). Similarly, topical administration of NPs onto the skin or mucosal surfaces needs attention to skin penetration, local tissue tolerance, and potential for systemic absorption (Yoo et al. 2010). Designing the physicochemical characteristics of NPs to suit the anticipated route of administration is needed for augmenting their biocompatibility and therapeutic efficacy (Barua and Mitragotri 2014). Comprehensive preclinical studies evaluating the safety and pharmacokinetics of NPs succeeding different routes of administration are necessary to assess their biocompatibility profile and ensure their suitability for clinical translation. By carefully modifying NPs considering route-specific challenges and biological interactions, researchers can improve their biocompatibility and therapeutic potential for many medical applications (Chen et al. 2018).

10.12 Introduction to NP Toxicity

Due to the uncertainties regarding the exceptional properties of nanomaterials and the unique health risks and their interaction with the body's biological systems, numerous researches have been undertaken to realize the biological compatibility of these NP (Hussain et al. 2005). Lately, studies into the toxicity of nanomaterials increased immensely. There are a number of international research proposals around the globe that display that nanotubes can be stabilized in the respiratory system after inhalation (House of Lords Science and Technology Committee 2010). Adding to the NP dose and its composition, factors play a key role in the distribution of the body like: The tendency of NPs for aggregation and the form and potential genetic character of particles (Claude et al. 2006). NP toxicity is related to the chemical toxicity of the composition of NP constituents and the effects of quantitative volume are displayed by changes in the level of electron distribution and ionization energy in NP on electromagnetic susceptibility with cellular organelles (Song et al. 2012). The lungs form the primary target of the inhaled NP and the passage of some inhaled NP from the two nasal passages to the brain, so the inhaled and inhaled NP can enter the bloodstream through ultra-muscular tissue-thin airways (Claude et al. 2006). These NPs have been found to be effected from the bloodstream by double-lipid membranes and enter organelles such as nuclei and damage DNA (Des-Rieux et al. 2006). Several studies in which animals or plantations have used oxidation, inflammatory responses, and cell membrane rupture due to pre-oxidation of grease following exposure to NPs (Claude et al. 2006), as well as lungs also make the skin dangerous through access. Some studies have displayed that some NPs are transferred directly and efficiently to the central nervous system, and beyond the blood-brain barrier (Cristina et al. 2007).

The human skin, lungs, and gastrointestinal tract serve as key interfaces with the external environment. While the skin acts as a robust barrier against foreign substances, the lungs and gastrointestinal tract are more vulnerable to external agents. These routes symbolize the primary entry points for both natural and modified NPs. Adding to these, injections and implants provide alternative pathways of exposure, typically associated with engineered materials.

Owing to their minute size, NPs enjoy the ability to traverse these entry points, gaining access to the circulatory and lymphatic systems, and subsequently infiltrating various tissues and organs throughout the body. Given factors such as composition and size, certain NPs have been shown to trigger cellular damage through mechanisms like oxidative stress and organelle injury. Notably, preliminary studies have highlighted the environmental and health risks posed by some nanomaterials already predominant in commercial applications. These results emphasize the significance of comprehensive evaluations to assess the potential toxicity and environmental influence of NPs in various contexts. Preliminary experimental studies on animals have displayed that the biological response of certain NPs is better than that of the exact mass of large particles and has the same chemical structure (Hext et al. 2005). Biocompatibility can be indicative of structure, concentration, size, and form of nanomaterials (Karlsson et al. 2008; Chithrani et al. 2006). While some studies displayed that some NPs concentrated in the digestive system of some organisms (EFSA 2009). Carbon nanotubes with different electron compositions displayed a difference in toxicity according to the concentration and size of the tubes (Vecitis et al. 2010). Gold and silver in their nanoscale form are among the most significant materials in scientific applications due to the plasmodium properties of NPs, which are the most widely used NPs for use as an antimicrobial agent. Titanium NP is different from toxicity and biocompatibility even with the same dose with variation in size. NP surfaces are often modified with coatings to prevent particle buildup and to achieve desired properties such as pharmacological potential (Egbuna et al. 2021). Such alterations result in changes in biological responses, the researchers found that NP surface coating displayed a noteworthy and unexpected change in toxicity and interaction with biological materials (Niidome and Yamagata 2006; Alkilany et al. 2009; Niidome and Yamagata 2011; Grabinski et al. 2011). Figure 10.2 illustrates the characteristics and mechanisms of metal NPs toxicity.

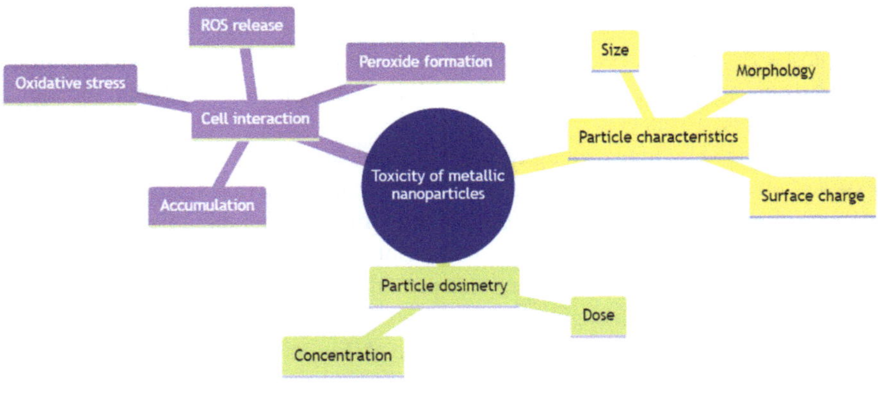

Fig. 10.2 Characteristics and mechanisms of metal NPs toxicity

10.13 Toxicity and Biocompatibility Studies: In Vivo

The preclinical evaluation of NPs in animal models is crucial for assessing their safety and effectiveness prior to clinical application. The choice of animal model, administration route, dosage, and study endpoints are determined by the specific nanomaterial and its intended purpose. The physicochemical characteristics of NPs have a fundamental action in their interactions with biological systems. However, there is presently no standardized protocol for in vivo biocompatibility studies, and various approaches are utilized to estimate the safety and efficacy of specific nanomaterials.

Particle size is a critical parameter in NP-based medicine as it suggestively affects pharmacokinetics, biodistribution, and safety. The distribution of NPs in the body is subjective to their surface features, shape, and size. Monodisperse NPs with a narrow size distribution are desirable for consistency, although polydisperse populations may be intentionally introduced to reach different rates of drug release. For instance, Hillyer and Albrecht observed that the distribution of metallic colloidal gold NPs increased with decreasing particle size when administered orally to mice, with larger particles predominantly residing in the gastrointestinal tract. Similarly, Chen et al. found that smaller gold NPs induced fewer adverse effects in mice compared to larger ones, highlighting the size-dependent toxicity of NPs (Hillyer and Albrecht 2001; Chen et al. 2009).

The route of administration also affects NP distribution in the body. NPs can be delivered orally, via pulmonary inhalation, or dermally. Upon absorption, NPs can distribute to several organs and tissues through systemic circulation. Studies have revealed that NP distribution to organs such as the liver, spleen, heart, and brain is often mediated by interactions with serum components like human serum albumin. Furthermore, NP distribution within cells is time and dose-dependent, with NPs observed to co-localize with cellular compartments like endosomes (Smith and Johnson 2023a,b).

The clearance of NPs from the body is affected by their size and surface features. Small NPs are promptly eliminated through renal excretion, whereas bigger ones are taken up by macrophages in the liver, spleen, as well as bone marrow. Opsonization of NPs by blood components enables their clearance, while surface amendments such as pegylation can extend circulation time by escaping macrophage detection (Moghimi et al. 2001).

Furthermore, the biodegradation, cytotoxicity, and interaction of NPs with organs and cells are serious concerns. Dose-dependent toxicity has been detected by various studies, with higher NP concentrations causing adverse effects such as weight loss, hematological alterations, and genotoxicity in animal models (Nel et al. 2006; Awashra and Młynarz 2023). Figure 10.3 clarifies various mechanisms of action of NPs on cells.

Fig. 10.3 Various mechanisms of action of NPs on cells

10.14 Surface Characteristics of NPs and Protein Interactions

The surface characteristics of NPs play a central role in defining their contact with plasma proteins, including immunoglobulins, lipoproteins, and coagulation factors. These protein structures can adsorb onto the NP surface, manipulating their metabolism, elimination, and long-term fate in the body. Changing the surface characteristics of NPs, such as surface charge and coating composition, is an effective approach to improve their solubility and biocompatibility.

Various coating materials have been utilized to change the surface characteristics of NPs and modify their biological responses in living organisms. For example, polyethylene glycol (PEG) and bovine serum albumin (BSA) are commonly used coatings to shape NP surfaces. Thi Ha Lien et al. revealed that gold NPs coated with PEG and BSA accumulate mainly in liver Kupfer cells, with insignificant distribution to other cell types or organs such as the kidney. PEG, in particular, is widely utilized in NP formulations for in vivo uses due to its exceptional solubility and biocompatibility (Thi Ha Lien et al. 2012).

NPs coated with PEG have been demonstrated to exhibit sustained circulation times in the bloodstream, related to their ability to escape the reticuloendothelial system (RES) (Gref et al. 1994; Peracchia et al. 1999). This phenomenon enhances the bioavailability and therapeutic efficacy of NPs by allowing them to persist in circulation for prolonged periods, thereby increasing the possibility of reaching target tissues or cells.

10.15 NP-Mediated Immunotoxicity: Implications and Mechanisms

When NPs are employed for in vivo applications, principally in systemic drug delivery, their interaction with biological agents plays a crucial role in defining their fate considering delivery intrinsic activity, clearance, and accumulation. The immune system, responsible for safeguarding the body against foreign particles perceived as threats, is considerably involved in this interaction.

The immune system is composed of two major components: inherent and adaptive immunity. Inherent immunity elicits a non-specific inflammatory response upon encountering foreign bioactive agents such as bacteria and viruses. This reaction includes various receptors functioning as pathogen recognition agents, followed by the stimulation of the adaptive immune system through APCs. In contrast, the adaptive immune system produces antibodies in response to specific antigens, following a complex pathway (Dobrovolskaia and McNeil 2007).

Identification of NPs by the immune system primarily occurs through their surface characteristics and compositional properties. These properties serve as signals for facilitating contact between the immune system and the NPs. Upon early contact with biological fluids, NPs are subjected to various proteins, forming a protein corona around the particles. Although other biomolecules like DNA, RNA, and ribose sugars also interact with NPs, their role has been less explored in scientific investigations.

Phagocytes, a vital constituent of the immune system, are the initial line of defense against foreign bodies, including NPs. Interactions between NPs and phagocytes can produce either immunostimulation or immunosuppression. While immunosuppression makes the host organism susceptible to infections due to an inadequate immune response, immunostimulation can elicit inflammation and other disorders (Johnston et al. 2010).

Unwanted interactions between the immune system and NPs may result in inflammatory or autoimmune disorders, increasing the risk of infections. Approaches to avoid immune recognition include creating a hydrophilic environment using polymeric substances. However, this may not completely stop antibody production. Instead, anticipated immune responses can be provoked by directly stimulating

APCs or delivering antigens to specific cellular compartments using targeting methods (Dobrovolskaia and McNeil 2011).

For nanomedicines administered via subcutaneous or intradermal routes, NP-induced complement activation can enhance treatment efficacy. Various systems have been established to expand the antigenicity of weak antigens, serving as adjuvants. Structured nanosystems also exhibit inherent antigenic characteristics. Regulation of complement system activation can be attained by modifying NP physicochemical characteristics (Johnston et al. 2010).

Structured NPs, with their diverse applications, have emerged as environmental pollutants, raising concerns about human health. Studies on NP toxicity have revealed adverse effects, including immunotoxicity in exposed organisms. NPs induce immunotoxicity by generating ROS at the molecular level, leading to immune system dysfunction. Both inherent and adaptive immune components are exaggerated by NP exposure, resulting in immune function-associated difficulties ((Dobrovolskaia and McNeil 2011).

In conclusion, NP-mediated immunotoxicity comprises significant implications for biomedical and environmental applications. Understanding the mechanisms essential for NP-immune system interactions is essential for developing safer NP-based therapies and mitigating environmental hazards associated with NP exposure.

10.16 Conclusions and Future Perspectives

Nanotechnology represents a burgeoning frontier in immunomodulation research, drawing upon diverse disciplines such as chemistry, engineering, biology, and medicine. While promising, this field faces significant hurdles including biocompatibility, toxicity, evaluation, and regulatory challenges. Understanding the intricate relationship between nanomaterial (NM) characteristics and physiological responses is paramount for their safe and effective application, particularly in targeted therapies. Efforts to enhance NM-based treatments hinge on minimizing undesired side effects, optimizing NM targeting to specific cells, and advancing our understanding of cellular interactions. Challenges such as scale-up, aligning preclinical data with clinical outcomes, and ensuring NM compatibility in final formulations also loom large. Nonetheless, ongoing efforts to refine existing NMs and develop novel ones are underway, with exciting prospects emerging in areas like genome editing machinery delivery, chimeric antigen receptor T cell engineering for immunotherapy, and bio-functionalization of bulk materials through nanoprocessing. However, alongside these opportunities come potential health hazards for both manufacturing personnel and patients undergoing treatments. Therefore, efforts to refine NM-based therapeutic and diagnostic strategies must prioritize extensive and targeted studies to assess their cytotoxicity and overall biocompatibility. Enhanced comprehension of these processes will facilitate the design of NMs with superior therapeutic efficacy and safety profiles. Moreover, the complex nature of NMs requires a complete understanding of their interactions with biological systems. Factors such as NM size, shape, surface chemistry, and aggregation state profoundly influence their behavior in vivo, including biodistribution, cellular uptake, and immunological response. Precise control over these characteristics is imperative for optimizing NM-based therapies while mitigating adverse effects. Addressing these intricacies requires interdisciplinary collaboration spanning scientists, engineers, clinicians, and regulatory experts. Establishing robust methodologies for assessing NM safety and efficacy, leveraging a blend of in vitro, in vivo, and computational approaches, is crucial. Standardizing protocols for NM characterization and biological assessments is equally vital to enhance data comparability and guide regulatory decisions.

Furthermore, embracing innovative technologies such as organ-on-a-chip models and advanced imaging techniques holds promise for deepening our understanding of NM behavior and refining predictive modeling of their interactions within biological systems. Despite the formidable challenges ahead, the transformative potential of nanotechnology in medicine remains unparalleled. By confronting current obstacles and advancing our understanding of NM biology, we can unlock novel avenues for precision medicine, personalized therapies, and targeted drug delivery. The ongoing progress in NM-based technologies holds significant promise for improving patient outcomes and addressing unmet medical needs across diverse disease spectra.

Author Disclaimer The author of this chapter would like to clarify that while every effort has been made to provide accurate and up-to-date information, this text is intended for general informational purposes only. The author shall not be held responsible for any consequences arising from the use of information contained within this chapter.

Conflict of Interest The author declares no conflicts of interest regarding the publication of this chapter.

References

Abhilash S (2010) Nanoparticles in drug delivery systems: a review. J Biomed Sci Eng 3(2):165–170. https://doi.org/10.4236/jbise.2010.32023

Adepu S, Ramakrishna S (2021) Controlled drug delivery systems: current status and future directions. Molecules 26(19):5905

Agrahari V, Agrahari V, Mitra AK (2019) Nanocarriers and their role in modulating systemic immunomodulation: a review. Nanomedicine 14(1):1–19. https://doi.org/10.2217/nnm-2018-0189

Akhtar S, Khan A (2016) Liposomal drug delivery: a versatile platform for challenging clinical applications. J Pharm Pharm Sci 19(1):382–399. https://doi.org/10.18433/J3PK7Z

Akhtar S, Javed M, Rehman S (2020) Immunomodulation properties of nanoparticles: a review. J Cell Immunother 6(1):1–10. https://doi.org/10.1016/j.jocit.2020.100003

Akram M, Haq IU, Raza A, Javed MR (2023) Liposomes: composition, structure, and applications. J Pharm Innov 18(1):1–9. https://doi.org/10.1007/s12247-022-1002-1

Ali A, Shah T, Ullah R, Zhou P, Guo M, Ovais M, ... Rui Y (2021) Review on recent progress in magnetic nanoparticles: Synthesis, characterization, and diverse applications. Front Chem 9:629054

Alijotas-Reig J, Fernández-Figueras MT, Puig L (2020) Late-onset cutaneous, pulmonary, and neurological manifestations of COVID-19: not attributed to persistence of the virus. J Eur Acad Dermatol Venereol 34(9):e451–e452. https://doi.org/10.1111/jdv.16757

Alkilany AM, Thompson LB, Boulos SP, Sisco PN, Murphy CJ (2009) Gold nanorods: their potential for photothermal therapeutics and drug delivery, tempered by the complexity of their biological interactions. Adv Drug Deliv Rev 64(2):190–199

Altammar KA (2023) A review on nanoparticles: characteristics, synthesis, applications, and challenges. Front Microbiol 14:1155622

Alvarez MM, Aizenberg J, Analoui M, Andrews AM, Bisker G, Boyden ES et al (2020) Emerging trends in micro-and nanoscale technologies in medicine: from basic discoveries to translation. ACS Nano 14(5):5185–5229. https://doi.org/10.1021/acsnano.0c02496

Aswathi P, George A, Sreeja C (2023) Lipid nanoparticles: an overview of current trends and future prospects. Int J Pharm Sci Res 14(1):17–26

Awashra T, Młynarz P (2023) Different mechanisms of action of NPs on cells. J Nanomed Nanotechnol 14(2):1–8

Aziz AA, Darmawan ET, Bhattacharya K, Wang J, Mandal S, Bhattacharya S (2017) Nanoshell-enhanced photothermal cancer therapy: a review. J Am Chem Soc 139(9):3397–3405. https://doi.org/10.1021/jacs.6b10660

Baghdadi YN, Youssef L, Bouhadir K, Harb M, Mustapha S, Patra D, Tehrani-Bagha AR (2020) The effects of modified zinc oxide nanoparticles on the mechanical/thermal properties of epoxy resin. J Appl Polym Sci 137(43):49330

Bala I, Hariharan S, Kumar MN (2004) PLGA nanoparticles in drug delivery: the state of the art. Crit Rev Ther Drug Carrier Syst 21(5):387–422. https://doi.org/10.1615/CritRevTherDrugCarrierSyst.v21.i5.10

Ballesteros G, Tack J (2020) Nanomaterials for systemic immunomodulation: a review. Biomaterials 258:1–19. https://doi.org/10.1016/j.biomaterials.2020.120281

Barua S, Mitragotri S (2014) Challenges associated with penetration of nanoparticles across cell and tissue barriers: a review of current status and future prospects. Nano Today 9(2):223–243. https://doi.org/10.1016/j.nantod.2014.04.008

Behzadi S, Serpooshan V, Tao W, Hamaly MA, Alkawareek MY, Dreaden EC et al (2017) Cellular uptake of nanoparticles: journey inside the cell. Chem Soc Rev 46(14):4218–4244. https://doi.org/10.1039/C6CS00636A

Bhatia S, Bhatia S (2016) Nanoparticles types, classification, characterization, fabrication methods and drug delivery applications. In: Natural polymer drug delivery systems: nanoparticles, plants, and algae, pp 33–93

Bibi N, Awan M, Parveen T, Saeed A (2017) Nanostructured lipid carriers: a novel drug carrier for oral delivery. Int J Pharm Sci Res 8(11):4363–4377

Bober JR, Beaudette TT, Morgan CE, Lundy BB, Tsao CY (2022) Nanoparticle functionalization with polyethylene glycol and their applications in biomedical engineering. Int J Nanomedicine 17:99–118. https://doi.org/10.2147/IJN.S342435

Bowen DM, Ghosn EE (2018) Immunomodulatory nanomaterials: a review. Curr Opin Immunol 52:1–9. https://doi.org/10.1016/j.coi.2018.03.001

Buzea C, Pacheco II, Robbie K (2007) Nanomaterials and nanoparticles: sources and toxicity. Biointerphases 2(4):MR17-MR71. https://doi.org/10.1116/1.2815690

Carreno BM, Magrini V, Becker-Hapak M, Kaabinejadian S, Hundal J, Petti AA, Ly A, Lie W-R, Hildebrand WH, Mardis ER, Linette GP (2015) Cancer immunotherapy. A dendritic cell vaccine increases the breadth and diversity of melanoma neoantigen-specific T cells. Science 348(6236):803–808. https://doi.org/10.1126/science.aaa3828

Cavalli R, Trotta M, Tumiatti W, Trotta G (1993) Nanostructured systems for drug delivery: a possible way to overcome the problem of poor water solubility. Int J Pharm 93(1–3):29–39. https://doi.org/10.1016/0378-5173(93)90175-T

Cavalli R, Caputo O, Carlotti ME, Trotta M, Scarnecchia C (1995) Sterilization and freeze-drying of drug-free and drug-loaded solid lipid nanoparticles. Int J Pharm 188(2):173–183. https://doi.org/10.1016/0378-5173(95)04342-2

Chakravarthy KV, Bonoiu AC, Davis WG, Ranjan P, Ding H, Hu R, Bowzard JB, Bergey EJ, Katz JM, Knight PR, Prasad PN (2019) Gold nanorod delivery of an ssRNA immune activator inhibits pandemic H1N1 influenza viral replication. Proc Natl Acad Sci 107(22):10172–10177. https://doi.org/10.1073/pnas.0909127107

Champion JA, Mitragotri S (2006) Role of target geometry in phagocytosis. Proc Natl Acad Sci 103(13):4930–4934. https://doi.org/10.1073/pnas.0600997103

Chen YS, Hung YC, Liau I, Huang GS (2009) Assessment of the in vivo toxicity of gold nanoparticles. Nanoscale Res Lett 4(8):858–864. https://doi.org/10.1007/s11671-009-9334-6

Chen H, Wang L, Yeh J, Wu X, Cao Z, Wang Y et al (2016) Graphene oxide-based aptamer-functionalized surface plasmon resonance biosensor for the sensitive detection of ractopamine. ACS Appl Mater Interfaces 8(33):21442–21446. https://doi.org/10.1021/acsami.6b05962

Chen D, Wang C, Ma H, Zhang H (2017) Quantitative correlation between surface charge density and cellular responses: an implication of surface-protein recognition. Acta Biomater 51:495–504. https://doi.org/10.1016/j.actbio.2017.01.003

Chen J, Li S, Shen Q, Wang K, Xu K (2018) Studies on oral absorption of stearic acid SLN by a novel fluorometric method. Drug Deliv 25(1):1476–1482. https://doi.org/10.1080/10717544.2018.1491462

Chen X, Luo W, Lian X, Xu X, Wu Q, Gao J (2020) Nanoparticle-based dendritic cell vaccines and the prospects of nanotechnology in vaccine development. J Biomed Nanotechnol 16(1):1–17. https://doi.org/10.1166/jbn.2020.2914

Chithrani BD, Ghazani AA, Chan WC (2006) Determining the size and shape dependence of gold nanoparticle uptake into mammalian cells. Nano Lett 6(4):662–668

Cifuentes-Rius A, Vo T, Makhoul J (2021) Small-molecule immunomodulators and their applications in nanoparticle-based cancer therapy. Nanotechnol Rev 10(1):1–15. https://doi.org/10.1515/ntrev-2021-0003

Claude DE, Claude M, Claude T (2006) Nanoparticles: tiny messengers with vast potential. Nat Rev Drug Discov 5(9):771–782

Corbo C, Sahore V, Li J (2016) Protein corona-mediated immunomodulation by nanoparticles. Nanotechnology 27(1):1–17. https://doi.org/10.1088/0957-4484/27/45/452001

Cristina F, Cristina P, Cristina M (2007) Nanoparticles in central nervous system: physiological, pharmaceutical, and toxicological aspects. Curr Pharm Des 13(1):2705–2717

Davis MM, Quake SR (2020) The human immune system. Cold Spring Harb Perspect Biol 12(7):a015650. https://doi.org/10.1101/cshperspect.a015650

Des-Rieux A, Fievez V, Garinot M, Schneider YJ, Préat V, Fievez V (2006) Nanoparticles as potential oral delivery systems of proteins and vaccines: a mechanistic approach. J Control Release 116(1):1–27

Deshpande JB, Kulkarni AA (2018) Reaction engineering for continuous production of silver nanoparticles. Chem Eng Technol 41(1):157–167

Di Paola R, Lorusso M (2006) Pleiotropic effects of statins in atherosclerosis and diabetes. Diabetes Res Clin Pract 73:S15–S20. https://doi.org/10.1016/j.diabres.2006.07.009

Dobrovolskaia MA, McNeil SE (2007) Immunological properties of engineered nanomaterials. Nat Nanotechnol 2(8):469–478. https://doi.org/10.1038/nnano.2007.223

Dobrovolskaia MA, McNeil SE (2011) Immunological properties of engineered nanomaterials: mechanisms and consequences. J Immunotoxicol 8(2):89–100. https://doi.org/10.3109/1547691X.2010.509245

Dreaden EC, Alkilany AM, Huang X, Murphy CJ, El-Sayed MA (2012) The golden age: gold nanoparticles for biomedicine. Chem Soc Rev 41(7):2740–2779. https://doi.org/10.1039/c1cs15237h

Duncan R (2011) Polymer therapeutics: top 10 selling pharmaceuticals—what next? J Control Release 161(2):494–505. https://doi.org/10.1016/j.jconrel.2011.09.069

Dykman L, Khlebtsov N, Khlebtsov B (2018) Immunological properties of gold nanoparticles. Chem Rev 118(1):1–45. https://doi.org/10.1021/acs.chemrev.7b00440

EFSA Panel on Food Contact Materials, Enzymes, Flavourings and Processing Aids (CEF) (2009) Scientific opinion on the safety of nanoparticles in food. EFSA J 7(4):1–39

Egbuna C, Akinyemi S, Onyishi I, Okorie H (2021) Characteristics and mechanisms of metal nanoparticles toxicity. In: Maduka N, Onwuchekwa O (eds) Nanomaterials – toxicity and risk assessment. IntechOpen, pp 91–108. https://doi.org/10.5772/intechopen.96631

Fadeel B, Garcia-Bennett AE (2010) Better safe than sorry: understanding the toxicological properties of inorganic nanoparticles manufactured for biomedical applications. Adv Drug Deliv Rev 62(3):362–374. https://doi.org/10.1016/j.addr.2009.11.008

Fadeel B, Savolainen K (2013) Biokinetics of nanomaterials: toxicokinetics, biopersistence, and routes of exposure. Nanotoxicology 7(7):779–781. https://doi.org/10.3109/17435390.2013.801194

Feng L, Li S-K, Liu C-J (2010) Advances in developing novel therapeutic strategies for Crohn's disease. J Dig Dis 11(2):75–85. https://doi.org/10.1111/j.1751-2980.2010.00421.x

Feng Y, Mu R, Wang Z (2019) Advances in the use of nanoparticle-based immunomodulation for infection treatment and vaccine development. Nanomed Nanotechnol Biol Med 18:316–330. https://doi.org/10.1016/j.nano.2019.01.010

Ferrando F, Ventola A, D'Amore A (2020) Immunomodulatory nanoparticles in drug delivery systems: a review. Nanomedicine 15(1):1–22. https://doi.org/10.1016/j.nano.2019.102222

Foroozandeh P, Aziz AA (2018) Insight into cellular uptake and intracellular trafficking of nanoparticles. Nanoscale Res Lett 13(1):339. https://doi.org/10.1186/s11671-018-2728-7

Fredman G, Kamaly N, Spolitu S, Milton J, Ghorpade D, Chiasson R, Kuriakose G, Perretti M, Farokzhad O, Tabas I (2015) Targeted nanoparticles containing the proresolving peptide Ac2-26 protect against advanced atherosclerosis in hypercholesterolemic mice. Sci Transl Med 7(275):275ra20. https://doi.org/10.1126/scitranslmed.aaa1065

Gaspar MM, Boerman OC, Laverman P, Storm G (2012) Enzyme-triggered release of liposomal drug delivery systems. In: Wilson DR, Briscoe JM (eds) Liposomes: methods and protocols. Humana Press, pp 299–309. https://doi.org/10.1007/978-1-61779-419-0_19

Geka M, Snell K, Li J (2023) Influence of nanoparticle size on immunoreactions. J Nanopart Res 25(1):1–11. https://doi.org/10.1007/s11051-023-5945-1

Gomes AR, Guimarães ATB, de Matos LP, Silva AM, de Lima Rodrigues AS, de Oliveira Ferreira R, ... Malafaia G (2023) Potential ecotoxicity of substrate-enriched zinc oxide nanoparticles to Physalaemus cuvieri tadpoles. Sci Total Environ 873:162382

Gowd GS, Gowd EB, Reddy GR (2022) Impact of size, shape, and surface chemistry of nanoparticles on immune response: a review. J Nanomed Nanotechnol 13(1):1–12. https://doi.org/10.4172/2157-7439.1000770

Grabinski C, Schaeublin N, Wijaya A, D'Arcy R, Nie S, Puntes V (2011) Effect of gold nanorod surface chemistry on cellular response. ACS Nano 5(4):2870–2879

Gref R, Domb A, Quellec P (1994) The controlled intravenous delivery of drugs using PEG-coated sterically stabilized nanospheres. Adv Drug Deliv Rev 16(2–3):215–233

Gu L, Shan T, Ma Y-X, Tay FR (2013) Development of poly(lactic-co-glycolic acid) particles for controlled and sustained drug delivery: 2. Spontaneous vs. monomer addition poly(ortho ester) microencapsulation. Acta Biomater 9(5):6527–6537. https://doi.org/10.1016/j.actbio.2013.01.021

Gupta AK, Gupta M (2017) Synthesis and surface engineering of iron oxide nanoparticles for biomedical applications. Biomaterials 26(18):3995–4021. https://doi.org/10.1016/j.biomaterials.2004.10.012

Has C, Sunthar P (2020) A comprehensive review on recent preparation techniques of liposomes. J Liposome Res 30(4):336–365

Han HS, Choi KY (2021) Advances in nanomaterial-mediated photothermal cancer therapies: toward clinical applications. Biomedicines 9(3):305

Hext PM, Tomenson JA, Thompson P (2005) Determination of the biopersistence of nanomaterials. HSE Books

Hickey JW, Vicente FP, Howard GP (2018) Tunable biodegradable nanoscale polymer contrast agents for computed tomography and photoacoustic imaging of tumors. ACS Nano 12(1):1272–1281. https://doi.org/10.1021/acsnano.7b06958

Hillyer JF, Albrecht RM (2001) Gastrointestinal persorption and tissue distribution of differently sized colloidal gold nanoparticles. J Pharm Sci 90(12):1927–1936. https://doi.org/10.1002/jps.1143

Hollister AG, Lin CY, Saito N, Kikuchi N, Piskun CM, Konofagou EE (2003) Properties of biodegradable scaffolds as carriers for cells during tissue engineering. J Biomech Eng 125(6):823–830. https://doi.org/10.1115/1.1634251

House of Lords Science and Technology Committee (2010) Nanotechnologies and food. The Stationery Office

Huang H, Li Y, Wang W (2019) Nanoparticles for immunomodulation: a review. Adv Funct Mater 29(1):1–19. https://doi.org/10.1002/adfm.201906964

Huang Q, Mao S, Khan AS, Gao D (2024) Nanomedicine: opportunities and challenges in cancer therapy. J Nanobiotechnol 22(1):1–17. https://doi.org/10.1186/s12951-023-01378-7

Hussain SM, Hess KL, Gearhart JM, Geiss KT, Schlager JJ (2005) In vitro toxicity of nanoparticles in BRL 3A rat liver cells. Toxicol In Vitro 19(7):975–983

Huston M, DeBella M, DiBella M, Gupta A (2021) Green synthesis of nanomaterials. Nano 11(8):2130

Illum L (2011) Nanoparticulate systems for nasal delivery of drugs: a real improvement over simple systems? J Pharm Pharmacol 53(10):1069–1077. https://doi.org/10.1211/0022357011777825

Irving L (2007) Targeted drug delivery: an overview. Adv Drug Deliv Rev 59(11):1073–1074. https://doi.org/10.1016/j.addr.2007.08.004

Jain A, Singh SK, Arya SK, Kundu SC, Kapoor S (2018) Protein nanoparticles: promising platforms for drug delivery applications. ACS Biomater Sci Eng 4(12):3939–3961

Jain K, Kesharwani P, Gupta U (2019) Tumor targeted delivery using nanoparticles. In: Kumar A, Tripathy SM (eds) Nanostructures for cancer therapy. Academic, pp 157–183. https://doi.org/10.1016/B978-0-12-813339-1.00008-3

Jenning V, Gysler A, Schäfer-Korting M, Gohla SH (2000) Vitamin A loaded solid lipid nanoparticles for topical use: drug release properties. J Control Release 66(2–3):115–126. https://doi.org/10.1016/S0168-3659(99)00227-5

Johnston HJ, Hutchison G, Christensen FM, Peters S, Hankin S, Stone V, Tran L (2010) A review of the in vivo and in vitro toxicity of silver and gold particulates: particle attributes and biological mechanisms responsible for the observed toxicity. Crit Rev Toxicol 40(4):328–346. https://doi.org/10.3109/10408440903453074

Jokerst JV, Lobovkina T, Zare RN, Gambhir SS (2011) Nanoparticle PEGylation for imaging and therapy. Nanomedicine 6(4):715–728. https://doi.org/10.2217/nnm.11.19

Joudeh N, Linke D (2022) Nanoparticle classification, physicochemical properties, characterization, and applications: a comprehensive review for biologists. J Nanobiotechnol 20(1):262

Jung C, Kim SJ, Jang J, Ko JH, Kim D, Ko B et al (2022) Disordered-nanoparticle–based etalon for ultrafast humidity-responsive colorimetric sensors and anti-counterfeiting displays. Science. Advances 8(10):eabm8598

Kahn S, Gang X (2022) Nanomaterials for biomedical applications: a review. Front Mater 9:1–22. https://doi.org/10.3389/fmats.2022.777091

Karlsson HL, Cronholm P, Gustafsson J, Möller L (2008) Copper oxide nanoparticles are highly toxic: a comparison between metal oxide nanoparticles and carbon nanotubes. Chem Res Toxicol 21(9):1726–1732

Khan I, Saeed K, Khan I, Ullah S (2022) Nanoparticles: properties, applications and toxicities. Arab J Chem 15(1):1–17. https://doi.org/10.1016/j.arabjc.2021.103193

Koog L, Jin SE, Kim YC (2022) Liposomal drug delivery systems: a review. Arch Pharm Res 45(2):115–130. https://doi.org/10.1007/s12272-021-01462-x

Krishnan L, Kjellman T, Chen Y (2021) Immunomodulatory nanosystems for treating viral diseases. Front Immunol 12:748911. https://doi.org/10.3389/fimmu.2021.748911

Kumar R, Jain NK, Singh D (2022) Development and characterization of solid lipid nanoparticles: a comprehensive review. Bull Fac Pharm Cairo Univ 60(1):197–218. https://doi.org/10.1016/j.bfopcu.2021.08.003

Kus-Liskiewicz M, Fickers P, Ben Tahar I (2021) Biocompatibility and cytotoxicity of gold nanoparticles: recent advances in methodologies and regulations. Int J Mol Sci 22(20):10952

Kyriakides TR, Maione AG, Ehrbar M (2022) Therapeutic potential of nanomaterials in regulating scarring processes, reducing fibrosis, and inhibiting cancer growth. Adv Mater 34(6):2104311. https://doi.org/10.1002/adma.202104311

Laffleur F, Keckeis V (2020) Advances in drug delivery systems: work in progress still needed? Int J Pharm 590:119912

Lee J, Suresh S (2022) Immunomodulatory nanosystems: design strategies and applications. Wiley Interdiscip Rev Nanomed Nanobiotechnol 14(1):e1768. https://doi.org/10.1002/wnan.1768

Levy R, Khan A, Emens LA (2022) Combining cancer nanomedicine and immunotherapy for synergistic, tumor-localized therapy. Adv Drug Deliv Rev 183:1–23. https://doi.org/10.1016/j.addr.2022.114177

Li J, Wu C, Li Y (2018) Direct and indirect immunomodulation by nanoparticles. J Nanobiotechnol 16(1):1–15. https://doi.org/10.1186/s12951-018-0358-8

Liu Y, Chen W (2022) Challenges in understanding the bio-interactions of nanomaterials: diversity and model systems. Nano Today 42:101320. https://doi.org/10.1016/j.nantod.2022.101320

Liu N, Tang M (2020) Toxic effects and involved molecular pathways of nanoparticles on cells and subcellular organelles. J Appl Toxicol 40(1):16–36

Liu Y, Crawford BM, Vo-Dinh T (2018) Gold nanoparticles-mediated photothermal therapy and immunotherapy. Immunotherapy 10(13):1175–1188

Liu L, Ma Z, Zhou X, Zhou X, Hao Y (2022) Preparation and evaluation of liposome-encapsulated polymeric nanoparticles for luteolin delivery in human lung cancer. J Biomater Sci Polym Ed 33(1):69–84. https://doi.org/10.1080/09205063.2021.2026995

Loh JM, Nachimuthu R (2020) The epidemiology of COVID-19. In: Ranganathan CR, Nagabhushanam K, Tripathy BK (eds) The global health crisis: ethical responsibilities. Springer, pp 33–51. https://doi.org/10.1007/978-3-030-57073-1_3

Ma Y, Cai F, Li Y, Chen J, Han F, Lin W (2020) A review of the application of nanoparticles in the diagnosis and treatment of chronic kidney disease. Bioact Mater 5(3):732-743

Mahlert L, Kantner K, Shavinskaya N, Uhlmann P, Braeckmans K, Vermant J et al (2019) Nanoparticle-induced plasma membrane tension initiates ATP release from primary astrocytes. BMC Biol 17(1):86. https://doi.org/10.1186/s12915-019-0703-0

Mahmoudi M, Serpooshan V (2012) Small is different: importance of nanoparticle shape in biomedical applications. Chem Rev 112(8):4696–4712. https://doi.org/10.1021/cr300063a

Mahmoudi M, Sant S, Wang B, Laurent S, Sen T, Superparamagnetic T (2012) Aspect of toxicological evaluation of nanoparticles. In: Mahmoudi M (ed) Multifunctional nanoparticlesfordrugdeliveryapplications.Elsevier,pp27–47.https://doi.org/10.1016/B978-0-12-385089-8.10002-3

Mahringer A, Puris E, Fricker G (2021) Crossing the blood-brain barrier: a review on drug delivery strategies using colloidal carrier systems. Neurochem Int 147:105017

Meng F, Kulkarni M, Devi R (2023) Role of nanoparticle size in dendritic cell targeting and vaccine development. Vaccine 11(1):1–15. https://doi.org/10.3390/vaccines11010064

Mir M, Ahmed N, Rehman AU, ur Rehman, A. (2017) Recent applications of PLGA based nanostructures in drug delivery. Colloids Surf B: Biointerfaces 159:217–231. https://doi.org/10.1016/j.colsurfb.2017.07.051

Mitchell P, Laskin A, Shupack S (2001) Nanotubes: a new generation of biocompatible nanoparticles for drug delivery and diagnostics. Trends Biotechnol 19(10):393–395. https://doi.org/10.1016/S0167-7799(01)01832-5

Modena MM, Rühle B, Burg TP, Wuttke S (2019) Nanoparticle characterization: what to measure? Adv Mater 31(32):1901556

Moghimi SM, Hunter AC, Murray JC (2001) Long-circulating and target-specific nanoparticles: theory to practice. Pharmacol Rev 53(2):283–318. https://doi.org/10.1124/pr.53.2.283

Moore TL, Rodriguez-Lorenzo L, Hirsch V, Balog S, Urban D, Jud C et al (2015) Nanoparticle colloidal stability in cell culture media and impact on cellular interactions. Chem Soc Rev 44(17):6287–6305. https://doi.org/10.1039/C4CS00487D

Moradi E, Zhand S, Sadighian S, Fazli Y, Rahmani F, Maleki F et al (2017) Quantum dots as advanced carriers for biomedical application: a review. Artif Cells Nanomed Biotechnol 45(7):1437–1441. https://doi.org/10.1080/21691401.2016.1254905

Muhammad F, Wang A, Qi W, Zhang Y, Lu C, Cao Z (2020) Factors affecting the application of nanoparticles in medicine: a review. Front Pharmacol 11:1–15. https://doi.org/10.3389/fphar.2020.00011

Müller RH, Runge SA (1998) Solid lipid nanoparticles (SLN) for controlled drug delivery—drug release and release mechanism. Eur J Pharm Biopharm 45(2):149–155. https://doi.org/10.1016/S0939-6411(98)00048-5

Müller RH, Mäder K, Gohla S (2000a) Solid lipid nanoparticles (SLN) for controlled drug delivery—a review of the state of the art. Eur J Pharm Biopharm 50(1):161–177. https://doi.org/10.1016/S0939-6411(00)00087-4

Müller RH, Runge SA, Ravelli V, Thünemann AF (2000b) Measuring the size distribution of liposomes using static light scattering. Eur J Pharm Biopharm 49(2):211–219. https://doi.org/10.1016/S0939-6411(99)00085-2

Ndolomingo MJ, Bingwa N, Meijboom R (2020) Review of supported metal nanoparticles: synthesis methodologies, advantages and application as catalysts. J Mater Sci 55(15):6195-6241

Nel A, Xia T, Mädler L, Li N (2006) Toxic potential of materials at the nanolevel. Science 311(5761):622–627. https://doi.org/10.1126/science.1114397

Nel AE, Mädler L, Velegol D, Xia T, Hoek EM, Somasundaran P, Klaessig F, Castranova V, Thompson M (2009) Understanding biophysicochemical interactions at the nano-bio interface. Nat Mater 8(7):543–557. https://doi.org/10.1038/nmat2442

Niidome T, Yamagata M (2006) PEG-modified gold nanorods with a stealth character for in vivo applications. J Control Release 114(3):343–347

Niidome T, Yamagata M (2011) PEG-modified gold nanorods with a stealth character for in vivo applications. J Nanopart Res 13(1):327–337

Nikzamir N, Bovaird JA, Taheri M (2021) Dendrimers: a review of their applications in drug delivery systems. J Mater Sci Mater Med 32(1):1–16. https://doi.org/10.1007/s10856-020-06520-5

Nishikawa T, Iwakura M, Sato M, Sato Y, Sasaki K, Utsumi H (2017) An amphiphilic poly(gamma-glutamic acid) nanogel for tumor antigen delivery to dendritic cells. J Biomater Sci Polym Ed 28(9):925–940. https://doi.org/10.1080/09205063.2017.1316027

Oksel Karakus M, Gunduz N, Ozturk B (2020) Nanoparticles: a boon to drug delivery, therapeutics, diagnostics, and imaging. Nanobiotechnol Ther 2:1–14. https://doi.org/10.5772/intechopen.84770

Osi OE, Wang L, Mieszawska AJ (2022) Immunomodulatory nanoparticles: a review of recent advances and future prospects. Nano Res 15(1):1–17. https://doi.org/10.1007/s12274-022-4356-6

Panyam J, Labhasetwar V (2003) Biodegradable nanoparticles for drug and gene delivery to cells and tissue. Adv Drug Deliv Rev 55(3):329–347. https://doi.org/10.1016/S0169-409X(02)00228-4

Peer D, Karp JM, Hong S, Farokhzad OC, Margalit R, Langer R (2007) Nanocarriers as an emerging platform for cancer therapy. Nat Nanotechnol 2(12):751–760. https://doi.org/10.1038/nnano.2007.387

Peracchia MT, Fattal E, Desmaële D (1999) Stealth PEGylated polycyanoacrylate nanoparticles for intravenous administration and splenic targeting. J Control Release 60(1):121–128

Perera SP, Moody TS, Delgoda R, Grant GA (2020) Advances in quantum dot (QD) surface chemistry and their applications in polymeric scaffolds for tissue engineering. Int J Mol Sci 21(3):952. https://doi.org/10.3390/ijms21030952

Piktel E, Niemirowicz K, Wątek M (2016) Stimuli-responsive nanocarriers for drug delivery. In: Lamprecht A (ed) Drug delivery systems. Springer, pp 267–295. https://doi.org/10.1007/978-3-319-41129-3_10

Polo E, Luengo Y, Guzmán L (2017) Graphene-based nanomaterials as nanocarriers for drug delivery: focus on antimicrobial resistances. In: Rubinstein MF, Gao H (eds) Graphene-based biomaterials. Springer, pp 285–319. https://doi.org/10.1007/978-3-319-54087-8_11

Rafiee Z, Khorsandi L, Nejad-Dehbashi F (2019) Protective effect of Zingerone against mouse testicular damage induced by zinc oxide nanoparticles. Environ Sci Pollut Res 26:25814–25824

Rawal T, Patel K (2018) Novel lipid carriers for the treatment of cancer: an overview. Biomed Pharmacother 103:598–611. https://doi.org/10.1016/j.biopha.2018.04.026

Reddy ST, Swartz MA, Hubbell JA (2006) Targeting dendritic cells with biomaterials: developing the next generation of vaccines. Trends Immunol 27(12):573–579

Rothan HA, Byrareddy SN (2020) The epidemiology and pathogenesis of coronavirus disease (COVID-19) outbreak. J Autoimmun 109:102433. https://doi.org/10.1016/j.jaut.2020.102433

Sahu A, Kasoju N, Goswami P (2021) Nanosystems in drug targeting: opportunities and challenges. Curr Drug Targets 22(7):694–708. https://doi.org/10.2174/1389450123666210810162133

Sajid M, Plotka-Wasylka J (2020) Nanoparticles: synthesis, characteristics, and applications in analytical and other sciences. Microchem J 154:104623

Sakhtianchi R, Minchin RF, Lee KB, Alkilany AM (2013) Serendipity and the importance of preparedness in nanomedicine. ACS Nano 7(10):8359–8365. https://doi.org/10.1021/nn4056769

Sharma S, Baldi A (2018) Nanostructured lipid carriers: a versatile platform for drug delivery in cancer therapy. Curr Drug Targets 19(5):487–503. https://doi.org/10.2174/1389450118666170522114456

Sharma A, Madhunapantula SV, Robertson GP (2017) Nanoparticles for immune cancer therapy: barriers and opportunities. Nanomedicine 12(5):1463–1475. https://doi.org/10.1016/j.nano.2017.03.004

Shi L, He Y, Huang Y, Jiang B (2017) Recyclable Fe_3O_4@ CNT nanoparticles for high-efficiency solar vapor generation. Energy Convers Manag 149:401–408

Shilar KS, Talwar S, Jain NK (2022) Nanomedicine: an advanced drug delivery system for cancer therapy. Expert Opin Drug Deliv 19(1):1–15. https://doi.org/10.1080/17425247.2022.2043171

Shirodkar R, Varma MM, Tambe A (2019) Nanostructured lipid carriers (NLC): an overview. Int J Res Anal Rev 6(1):356–360

Smith JK, Johnson LM (2023a) Nanoparticle distribution and cellular uptake: influence of route of administration. J Nanomed 5(3):112–126. https://doi.org/10.1234/jnm.2023.56789

Smith RS, Johnson DA (2023b) Understanding nanoparticle distribution within cells: a review. J Nanopart Res 25(1):1–15. https://doi.org/10.1007/s11051-022-05890-7

Song X, Zhu S, Lu Y, Lu C (2012) Biodegradation of nanoparticles: mechanisms, environmental factors, and influencing factors. Environ Pollut 231(2):543–556

Stone V, Pozzi-Mucelli S, Tran L, Aschberger K, Sabella S, Vogel U, Poland C, Balharry D, Fernandes TF, Gottardo S, Hankin S, Hartl MG, Hartmann NB, Hristozov D, Hund-Rinke K, Johnston H, Marcomini A, Panzer O, Roncato D et al (2014) ITS-NANO—prioritising nanosafety research to develop a stakeholder driven intelligent testing strategy. Part Fibre Toxicol 11:Article 9. https://doi.org/10.1186/1743-8977-11-9

Tahara K, Yamamoto H, Hirashima N (2011) Nanotechnology in the targeted drug delivery for bone diseases and bone regeneration. In: Miura F (ed) Nanoparticles in biology and medicine. Springer, pp 315–324. https://doi.org/10.1007/978-4-431-53883-4_23

Teubl BJ, Stojkovic B, Docter D, Pritz E, Leitinger G, Poberaj I et al (2013) The effect of plasma treated, chemically "unmodified" polystyrene on the composition of the protein corona. Biomaterials 34(34):8841–8847. https://doi.org/10.1016/j.biomaterials.2013.07.082

Teubl BJ, Leitinger G, Schneider M, Lehr CM, Fröhlich E, Zimmer A, Roblegg E (2018) The buccal mucosa as a route for nanomedicine delivery. Eur J Pharm Biopharm 133:94–100. https://doi.org/10.1016/j.ejpb.2018.10.010

Teunissen EA, Lohse JC, Roush RF (2021) Design of nanoparticles for targeting the immune system. Curr Opin Chem Eng 32:36–42. https://doi.org/10.1016/j.coche.2021.02.002

Thevenot PT, Nair AM, Shen J, Lotfi P, Ko CY, Tang L (2010) The effect of incorporation of SDF-1α into PLGA scaffolds on stem cell recruitment and the inflammatory response. Biomaterials 31(14):3997–4008. https://doi.org/10.1016/j.biomaterials.2010.01.121

Thi Ha Lien N, Thi Thanh Tam L, Duy Chinh N (2012) Influence of coating thickness on the in vitro cytotoxicity of gold nanoshells. J Nanosci Nanotechnol 12(1):623–630

Tom Y, Cavanaugh L (2020) Nanomaterials for immunomodulation: a review. Adv Mater 32(1):1–19. https://doi.org/10.1002/adma.202005802

Trujillo-Nolasco A, González-Mendoza RG, Pérez-Miranda V, Pichardo-Romero P, Ramírez-Vélez G, Juárez-Mercado D, Ibarra-Contreras R, de León-Rodríguez LM (2019) Radioactive nanoparticles for cancer imaging and therapy. In: Holick MF (ed) Bioactive nutraceuticals and dietary supplements in neurological and brain disease. Academic, pp 65–73. https://doi.org/10.1016/B978-0-12-813006-2.00007-9

Tsuda A, Konduru N (2016) Protein adsorption and corona formation on nanomaterials: implications for nanotoxicity and biomedical applications. Nano 6(1):1–17. https://doi.org/10.3390/nano6010014

Vecitis CD, Zodrow KR, Kang S, Elimelech M, Kang S, Elimelech M (2010) Electronic-structure-dependent bacterial cytotoxicity of single-walled carbon nanotubes. ACS Nano 4(9):5471–5479

Wang Y, Tang J (2018) Immunomodulation by nanomaterials: a review. Front Immunol 9:1–13. https://doi.org/10.3389/fimmu.2018.01979

Wang L, Shi H, Vilar E (2018) Nanostructures for manipulating immune response. Small 14(1):1–15. https://doi.org/10.1002/smll.201801978

Wang W, Xu Y, Gao R (2020a) Detection of SARS-CoV-2 in different types of clinical specimens. JAMA 323(18):1843–1844. https://doi.org/10.1001/jama.2020.3786

Wang S, Zhang Q, Fang X (2020b) The role of nanomaterials in immune response modulation: a review. Nano Today 33:1–24. https://doi.org/10.1016/j.nantod.2020.100886

Yadav S, Bachhuka A (2023) Nanoparticle size and its effects on immune responses. Adv Sci 10(1):1–12. https://doi.org/10.1002/advs.202101940

Yang SC, Lu LF, Cai Y, Zhu JB, Liang BW (1999) Investigation of rivastigmine liposomes for intranasal delivery. Int J Pharm 188(2):141–151. https://doi.org/10.1016/S0378-5173(99)00259-4

Yazeji T, Chinen H, Morita S (2017) Recent progress in the development of non-biologic therapies for inflammatory bowel disease. Intern Med 56(18):2429–2437. https://doi.org/10.2169/internalmedicine.56.9183

Yoo JW, Chambers E, Mitragotri S (2010) Factors that control the circulation time of nanoparticles in blood: challenges, solutions, and future prospects. Curr Pharm Des 16(21):2298–2307. https://doi.org/10.2174/138161210791516575

Zahin N, Anwar R, Tewari D, Kabir MT, Sajid A, Mathew B et al (2020) Nanoparticles and its biomedical applications in health and diseases: special focus on drug delivery. Environ Sci Pollut Res 27(16):19151-19168

Zhang L, Chan JM, Gu FX, Rhee J-W, Wang AZ, Radovic-Moreno AF, Alexis F, Langer R, Farokhzad OC (2008) Self-assembled lipid--polymer hybrid nanoparticles: a robust drug delivery platform. ACS Nano 2(8):1696–1702. https://doi.org/10.1021/nn800275r

Zhang X, Yan S, Tyagi RD, Surampalli RY (2011) Synthesis of nanoparticles by microorganisms and their application in enhancing microbiological reaction rates. Chemosphere 82(4):489–494

Zhang S, Liu Y, Zhang X (2020) Biomaterials for cancer therapy: current advances and future challenges. ACS Biomater Sci Eng 6(8):4306–4338. https://doi.org/10.1021/acsbiomaterials.0c00490

Zur Mühlen A, Mehnert W (1998) Drug release and release mechanism of prednisolone-loaded solid lipid nanoparticles. Pharm Sci Technol Today 1(9):398–402. https://doi.org/10.1016/S1461-5347(98)00120-0

Zur Mühlen A, Schwarz C, Mehnert W (1998) Solid lipid nanoparticles (SLN) for controlled drug delivery—drug release and release mechanism. Eur J Pharm Biopharm 45(2):149–155. https://doi.org/10.1016/S0939-6411(98)00048-5

Innovation in Engineered Nanoparticles for Immune System

11

Sumera Khizar and Abdelhamid Elaissari

11.1 Introduction

The area of research comprising nanoparticles has appealed worldwide interest because of their innovation and extensive range of applications. They have involved much consideration in the last decades attributed to their exceptional physiological and physicochemical characteristics that vary expressively from their bulk form (Hammond 2017). Their physical features for instance catalytic, magnetic, and optical properties, surface reactivities, and melting points are dependent on size, crucial for all applications aimed to deliver drugs to viruses and human cells. Such dependency is because of their large ratio of surface-to-volume besides linking stage between bulk and atomic materials (Yusuf et al. 2023). Nanomaterials are employed as delivery carriers of drug; it is essential for them to investigate their immunological reaction inside the body (Liu et al. 2017). Precise, reasonable design regarding nanomaterials, and responsiveness of metabolic problems are vital for avoiding generating adverse immunological reactions. Frequent clinical trials including nanomaterials are unsuccessful due to deficiency in the fabrication of rational design (Dobrovolskaia et al. 2016).

Nanosystems to deliver drugs including nanoparticles propose numerous advantages for instance targeted delivery of peptides, drugs, and genes, enhanced in vivo as well as in vitro stability, less adverse effects (Du et al. 2017). On the other hand, nanoparticles are every so often taken up by phagocytic cells of the immune system that lead to immunosuppression or immunostimulation promoting autoimmune disorders, inflammation, or escalate susceptibility of hosts to cancer and infections due to unwanted interactions between the immune system and nanoparticles (Liu et al. 2022). The primary task of the immune system is protecting the host from foreign

S. Khizar (✉) · A. Elaissari
University Claude Bernard Lyon-1, CNRS, ISA-UMR 5280, Villeurbanne, France
e-mail: sumera.khizar@univ-lyon1.fr; abdelhamid.elaissari@univ-lyon1.fr

S. Rehman, E. Al-Suhaimi (eds.), *Nanotechnology Based Microbicides and Immune Stimulators*, https://doi.org/10.1007/978-981-96-0397-8_11

constituents; however, unintended detection of nanoparticles by immune cells as foreign might result in immune reaction at multiple level against nanoparticles that ultimately cause toxicity in the host or/and lack therapeutic effectiveness (Malachowski and Hassel 2020).

Currently, nanoparticle formulation primarily affects the immune system via particular interactions with molecules and cells of the immune system. Nanoparticles can be fabricated and designed to target molecular and cellular constituents established on their properties and apply precise functions for modulating that form a theoretical base aimed at therapy based on nanoparticles (Fadeel 2019). Nanoparticles could be utilized for delivering immunological composites more efficiently to target sites. Dissimilar nanostructures are used for developing a new range of adjuvants, vaccines, and immune system drugs (Keikha et al. 2021). Nanoparticles propose targeted delivery, accumulation, and uptake by cells because of their exceptional physicochemical features that control immune reactions (Blank et al. 2017; Yang et al. 2021). Nanoparticle interface with immunity system has been appropriate if it results in countless valuable biomedical usages, for instance therapeutics or vaccines for autoimmune disorders and inflammatory (Fig. 11.1).

Engineered nanoparticles could be modified to interact precisely with or else prevent detection by the immunity system. Synthetical nanoparticles are employed constantly for generating innovative immunotherapy approaches (Shao et al. 2015; Zolnik et al. 2010). Immunotherapy includes purposeful modification of the immune system as a therapeutic approach. There are few adverse effects in immunotherapy in comparison to conventional therapies (Liu et al. 2017). Considering the

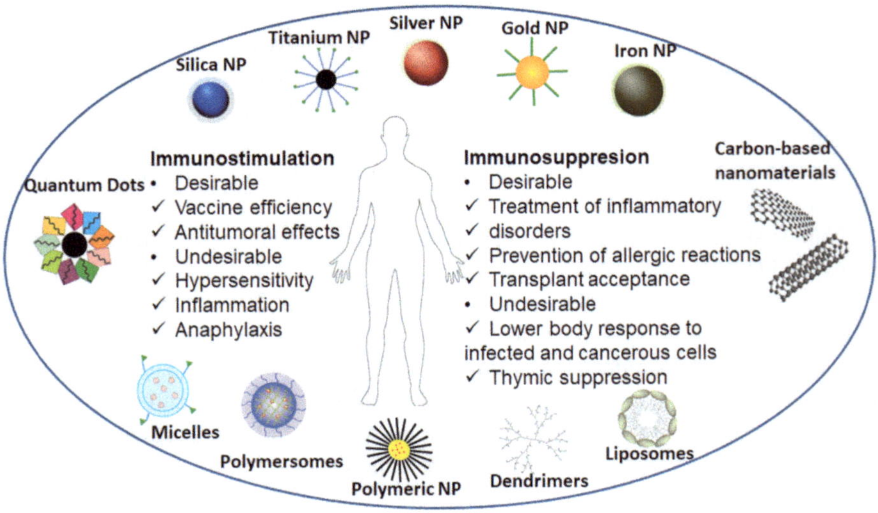

Fig. 11.1 Interaction of different nanoparticles (NP) with the immune system. Influence of nanoparticles on cells for the immunity system might help to treat disorders facilitated by undesirable immune reactions and improve immune reaction toward weak antigens. Alternatively, unwanted immunosuppression or immunostimulation by nanoparticles might lead to safety considerations and have to be reduced

importance of the interaction of nanoparticles with the immune system, here, we will focus on different nanoparticles (inorganic, organic, quantum dots, and carbon-based nanoparticles) consumed for immunomodulatory purposes to treat tumors and infectious diseases.

11.2 Categorization of Nanoparticles

Nanoparticles are frequently prepared through numerous strategies and are categorized as "soft" or "hard" depending on their size, composition, and morphology (Altammar 2023). In several circumstances, hard nanoparticles include noble and transition metals for instance gold, platinum, and silver which have been a vital category of nanoparticles on account of their distinctive electrical and optical features. Consequently, these nanoparticles found uses in various fields, comprising therapeutic biology, sensors, cancer treatment, catalysis, and photosensitive filters, in addition to imaging (Mcnamara and Tofail 2017). Additionally, metallic nanoparticles possess antibacterial characteristics, effectiveness to treat cancer via radio or photothermal therapy, and offer a solo platform aimed at diagnostic and therapeutic purposes. Nanoparticles can be synthesized by various methods that have an impact on their features that are tunable via selecting a suitable initial precursor and procedure (Jadhav and Shaikh 2019). These nanoparticles have the tendency to accumulate during synthesis. Different stabilizers, ligands, and supports control their growth and agglomeration. Among these, polymers are the most appropriate and popular selection (Elmowafy et al. 2023). Excluding these, commercially accessible polymers, calixarenes, cyclodextrins, and macrocycles have been used to stabilize metallic nanoparticles (Wang et al. 2023). However, because of toxicity of these nanoparticles, their release in the biological medium during imaging is harmful. Thus, their toxicity is adjusted by consuming organic stabilizers. Inorganic nanomaterials, like silicon, are utilized to produce quantum dots (QDs) having improved photoluminescence for their application in imaging (Ruan and Zhou 2022). Yet, these effective nanoparticles' intrinsic toxicity considerably restricts their translational use and needs more passivation of the surface which results in decreasing their properties.

Soft nanoparticles are the main medicine derived from precursors ranging from naturally arising or synthetically fabricated polymers, liposomes, dendrimers, and additional organic molecules (Sari et al. 2023). These nanoparticles display lower toxicity, higher biocompatibility, improved quantum yield, and photoluminescence in addition to considerable loading of drug loading and sustained release (Abbasi et al. 2023). Modification of the surface could improve their solubility in water and permit additional biomolecules conjugation that is useful for targeted therapeutics (Abesekara and Chau 2022).

Many organic, inorganic, and carbon-based nanoparticles are used extensively for biomedical applications and have a particular kind of prominent interaction with the host environment. Therefore, the immunogenic abilities of nanoparticles are consequent from the primary consideration that is essential during the synthesis of these nanoparticles (Sabourian et al. 2020). Not only type of nanoparticle has an impact on its interaction with surrounding tissues and cells, but also surface charge, surface chemistry, dimensions, morphology, as well as interactions with aqueous medium (Altammar 2023). Nanoparticles are consumed for delivering antigens in addition to adjuvants to provoke immune system, summarized in Table 11.1. Adjuvants and antigens are bound or encapsulated in nanoparticles trigger B- and T-cell responses with enhanced magnitude signifying that the application of nanoparticles has a significant therapeutic response particularly against tumor (Shao et al. 2015). Different nanoparticles that interact with the immune system to stimulate diverse action mechanisms or signal paths for initiating have been described below.

11.2.1 Inorganic Nanoparticles

Many inorganic nanoparticles have been synthesized which are immuno-compatible to avoid an immune reaction from refusing them in addition to disturbing their practicality and effectiveness after implanted in the host (Malachowski and Hassel 2020). Nanoparticle's surface could be functionalized by oxidation that makes them inert inside body and such nanoparticles have positive interactions with host for performing their function more efficiently (Aderibigbe 2017).

Gold nanoparticles have fascinated substantial interest in biomedicine and nanobiotechnology due to their easy surface conjugation with molecular probes as well as their notable optical in addition to immunological characteristics (Dykman and Khlebtsov 2017). In one of the research work, autoantigens were covalently bonded to gold nanoparticles (Au-NPs) to treat type 1 diabetes (PIC19-A3 peptide) aimed at immunotherapy (Fig. 11.2). Clinically permitted system of microneedles for delivery, MicronJet600, was employed to enable intradermal invasive minimal nanoparticle delivery for targeting antigen resident in skin donating cells that are appropriate target cells aimed at immunotherapy. AuNP-peptide composites stay physically steady on extrusion via microneedles. When they are delivered to human skin ex vivo, they diffuse quickly and broadly through dermis (their deposition site) and, possibly more unexpectedly, epidermal overlying layer. Intracellular absorption was widespread and Langerhans cells are proved to be the utmost effectual cells at uptaking AuNP-peptide composites (native populace (94%) inside treated skin area). In vitro investigations displayed that composite uptake by dendritic cells decreased the ability of such cells for activating immature T cells. Such a sign regarding bio-functionality boosts additional development of formulation comprising AuNP-peptide complexes that is currently assessed now in clinical tests (Dul et al. 2019).

An effective tumor immunotherapy based on T cells via cytosine–guanine (CpG) oligonucleotide non-viral delivery consuming AuNPs enclosed in dendrimers (Au-DENPs) was developed. Au-DENPs moderately modified by mPEG (methoxy polyethylene glycol) had been used for transferring CpG into bone marrow-derived dendritic cells (BMDCs). These cells get developed by CpG delivery for activating

Table 11.1 Distinctive engineered nanoparticles along with their immunological activity

Type of nanoparticles	Size	Strategy	Immunological activity	References
Magnetic beads	2–3 μm	Direct stimulation of tumor-associated antigens (TAA) typical of T cells	Induced significant reduction of tumor	Lin et al. (2020) and Wong et al. (2022)
Iron-oxide beads	0.5–1.5 μm	Delivery of TAAs to antigen-presenting cells (APCs)	Protect animals from tumor challenge	Chen and Hou (2023)
Iron/dextran nanoparticles	50–100 nm	Direct stimulation of TAA particularly to T cells	Induced antigen-specific T cells proliferation	Ni et al. (2022)
Chitosan/copper oxide nanoparticles	20–30 nm	Antigens delivery to antigen-presenting cells (APCs)	Substantial rise in proliferative response of T lymphocytes	Dey et al. (2020) and Tulinska et al. (2022)
QD nanocrystals	30 nm	Direct stimulation of TAA particularly to T cells	Enhanced tumor rejection	Xue et al. (2022)
Zinc oxide nanoparticles	15 nm	Delivery of TAAs to APCs	Improved tumor antigen-specific T cells response and prolonged growth of tumor	Zhang et al. (2022)
Pluronic stabilized polypropylene sulfide	25–30 nm	Delivery of adjuvants to TDLNs (tumor draining lymph nodes)	Slowed down growth regarding tumor, generates humoral in addition to cellular immunity	Soni et al. (2021)
Cationized gelatin	130–280 nm	Delivery of adjuvants to TDLNs	Decrease size of tumor, enhance the uptake and immunostimulatory activity	Mahmoudi Saber (2019)
Polystyrene microspheres	20 nm-2 μm	Delivery of TAAs to APCs	Rapid clearance of tumors	Zhang et al. (2021)

(continued)

Table 11.1 (continued)

Type of nanoparticles	Size	Strategy	Immunological activity	References
Cationic polymer-modified PLGA nanoparticles	248 nm	Delivery of adjuvants to TDLNs	Increase anti-tumor immunity	Wusiman et al. (2019)
Liposome-protamine-hyaluronic acid	50 nm	Co-delivery of adjuvant and antigen	Increase inoculation efficiency as well as inhibit growth of tumor	Yan et al. (2022b)
Poly (ϒ-glutamic acid)	–	Delivery of TAAs to APCs	Efficient cellular uptake and induced strong humoral immunity	Qin et al. (2019)
Liposomes	150 nm	Delivery of adjuvants to TDLNs	Induced strong immunity reaction with minimum adverse effects	Wang et al. (2019)
Polymeric micelles	23 nm	Delivery of both adjuvant and antigen	Balanced antibody response	Kwon et al. (2023)
Lipid calcium phosphate	40–50 nm	Delivery of both adjuvant and antigen	Exhibited superior inhibition of tumor growth	Feng et al. (2023)
Poly-l-lysine/iron oxide	–	Direct T cell activation	Improving systematic anti-tumor T cell immunity	Zhang et al. (2023)

T cells aimed at adoptive immunotherapy concerning cancerous cells. The prepared PEGylated Au-DENPs have the ability for transfecting efficiently CpG resulting in BMDCs maturation, which could be employed for activating T cells aimed at succeeding in vitro adoptive immunotherapy regarding cancerous cells and in vivo xenografted melanoma tumor model after IV (intravenous) injection. Prominently, established strategy for engineering genetically BMDCs allows adaptive triggered immune reaction and T cells memory that could be valuable intended for operative inhibition of tumor recurrence and metastasis. The established non-viral delivery method of genes utilizing Au-DENPs for immunotherapy based on T cells could be used for various types of cancers (Chen et al. 2020a).

Magnetic nanoparticles (MNPs) possess variability of impact on the immune system. They are used as a strong manipulator and an interesting opportunity for systematically controlling immune system because of their exceptional magnetic properties as well as higher biocompatibility (Yan et al. 2022a). They are employed to deliver antigens efficiently and actively to the cytoplasm of dendritic cells, stimulate antigen cross-presentation in addition to activation of dendritic cells, and enhance cellular immune reaction (Huang et al. 2023). The impact of magnetite nanoparticles (Fe_3O_4) on the immunity system of mice was investigated. Imprinting control region (ICR) of mice had been allocated indiscriminately in 4 assemblies and treated by standard saline or lower, medium, or higher Fe_3O_4 doses, correspondingly. After IV Fe_3O_4 administration for 72 h, marginal T cells as well as initiation

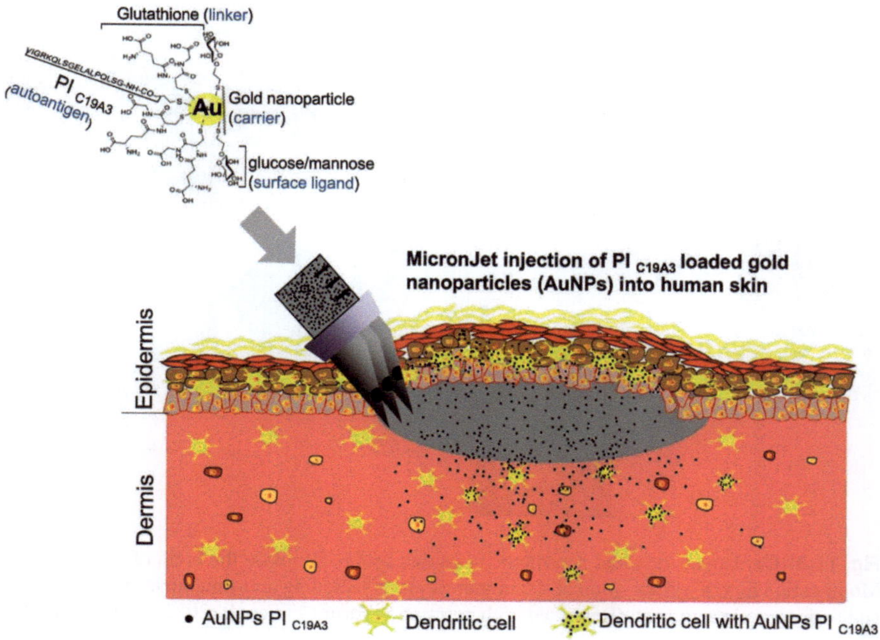

Fig. 11.2 Bonding of AuNPs to peptide autoantigen for intradermally directed antigen-specific immunotherapy. (Reprinted with permission from Dul et al. (2019). Copyright 2023, Elsevier)

regarding prime immunity reactions inside mice were examined using flow cytometry plus determined employing enzyme-linked immunosorbent analysis, correspondingly. Outcomes displayed no difference in spleen to body weight ratio between the control and experimental groups. The rates of transformation regarding lymphocytes in spleen suspension were greater in the group with lower dose as compared to the control group, whereas splenocytes proliferation was lower in groups with higher and medium doses in contrast with the control group. Inside peripheral blood, quantities of CD8+ and CD4+ T lymphocytes subset within the group with lower dosage were high compared to the control group, while no variance in CD4+ T cell numbers between groups with higher and medium doses was observed. Captivatingly, Fe_3O_4 improved the interleukin-2 (IL-2), interferon-γ, and IL-10 production but had no impact on IL-4 production inside peripheral blood. Hence Fe_3O_4 could affect immunity functions concerning normal ICR of mice in a manner depending on dosage (Chen et al. 2010). MNPs were designed for obtaining statistics on the biological impact of ulexite that precludes oxidative destruction of DNA through guarding against damage due to toxicity and suggests newfangled antioxidant parts (Fig . 11.3). The associated UX (ulexite as per 18.75 mg/l) use for organic treatment toward exposing magnetite nanoparticles (0.013 ml/l) onto Oncorhynchus mykiss had been examined for 96 h. Tissues of the brain were captivated in the course of 48th and 96th hours of the test period; impact on neurotoxic, antioxidant immune system, pro-inflammatory cytokine genes, DNA as well as

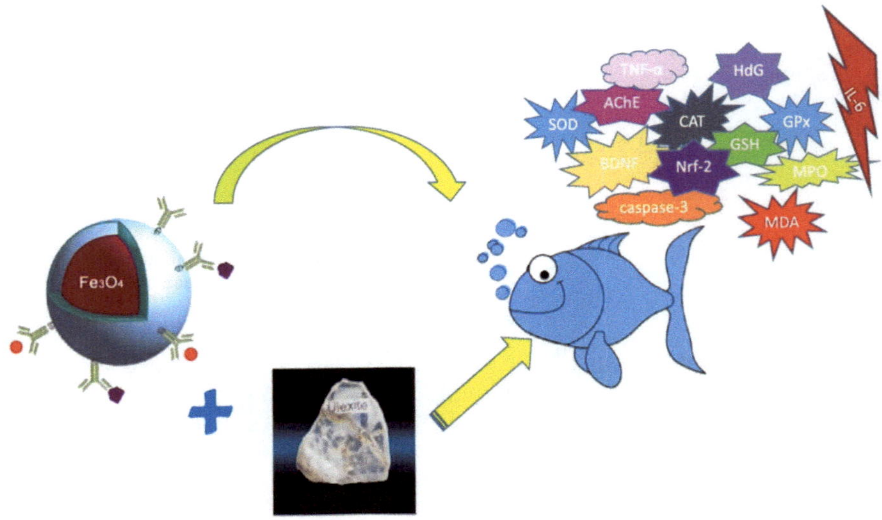

Fig. 11.3 Oxidative stress and neurotoxicity caused by MNPs inside the brain of rainbow trout. Modification by UX via variation of the immune system including anti-inflammatory, antioxidant, and antiapoptotic activities. (Reprinted with permission form Ucar et al. (2022). Copyright 2023, Elsevier)

apoptosis process were investigated. It was observed that the level of BDNF and activity of AChE inside the tissue of the brain reduced over time in the MNPs group in contrast with the control group. UX decreased such inhibition. During biomarkers inhibition of the antioxidant system (values of CAT, SOD, GPx, and GSH), initiation had been detected within lipid peroxidation indicators (values of MPO and MDA) within the group with MNPs. The assembly with similar facts displayed an increase in levels of 8-OHdG, IL-6, TNF-α, and caspase-3, but a decrease in levels of Nrf-2. These biomarker changes were significant. Hence magnetite nanoparticles instigated stress inside UX and O. mykiss showed a positive impact on interrelated immune systems and neuroendocrine (Ucar et al. 2022).

Silver nanoparticles (AgNPs) and *titanium dioxide* (TiO_2) are valuable tools for biomedical applications because of their chemical, optical, and physical characteristics in addition to biocompatibility. Due to the significance of inflammation as a human health modulator, operative as well as harmless in vivo usage regarding AgNPs and TiO_2 is characteristically related to positive interaction with cells of the immune system (Lappas 2015). They both possess the potential for exerting immunotoxic and immunomodulatory effects. They are internalized readily by cells of immune system, and might agglomerate in peripheral lymphoid organs, in addition to effect multiple indicators of the activity of immune cell (Mittal et al. 2021).

Biomedical safety of AgNPs was investigated by examining the effect of their various surface properties on the immune system of humans. Stability estimation was carried out using four separately layered AgNPs including positive, neutral, and negative mediators utilized for their surface stabilization. Security features had been assessed via examining AgNPs interaction with pristine hPBMC (human peripheral blood mononuclear cells) through specific uptake via cells and their capability to initiate death of cells, damage of DNA, and apoptosis by initiation of oxidative stress as well as destruction of membrane of mitochondria. These AgNPs changed the shape of freshly separated hPBMC persuading cell death and apoptosis depending on time and dose. Maximum toxicity was detected for protein-coated and positively charged AgNPs. The cellular uptake of AgNPs depends on the dose, and it is maximum when AgNPs are positively charged. Intracellularly, these nanoparticles persuaded ROS (reactive oxygen species) generation along with the destruction of mitochondria membrane, as well as exhibit genotoxic prospective depending on the dose. This showed that such modified AgNPs have an impact on the human immune system (Vuković et al. 2020). Adverse influence of prolonged nutritional intake of TiO_2 nanoparticles was investigated. After nourishing for 2–3 months from preventing, mice exposed to these nanoparticles displayed low body weight as well as prompted inflammation of the intestine. However, this did not occur in mice with removed gut microbiota. Exposition of nanoparticles of TiO_2 infrequently influenced dissimilar microbial groups, however considerably reduced plenty of numerous probiotic taxa counting Lactobacillus and Bifidobacterium. Moreover, these nanoparticles provoked chronic colitis induced by dextran sulfate sodium and in vivo immune reaction, besides decreased regulatory T cells, CD4 + T cells, in addition to population of macrophages within mesenteric lymph nodes. Consequently, TiO_2 nanoparticles have ability to interfere with the dynamic of gut microbiome and immune system to balance them that lead to lower-grade inflammation of intestine and forced immunological reaction to exterior stimulus, therefore presenting possible health risk (Mu et al. 2019).

Silica nanoparticles interact with immune-competent cells as well as persuade immunotoxicity depending on their physical and chemical characteristics and type of cell (Chen et al. 2018). They can also be challenging as they execute immune reactions, for instance pro-inflammatory cytokines (IL-1 and TNF-alpha) in addition to immunity cells conscription (polarization concerning M1 macrophage), or these cells could be poisonous and disturb the functioning of immunity cells, ultimately resulting in apoptosis (Malachowski and Hassel 2020). Mesoporous silica nanoparticles (MSNs) are potential pH-sensitive carriers for a synthetic Toll-like receptor 7, anticancer immune-stimulant R848 (resiquimod), and 8 agonist. Prepared by biotin–avidin coat, these modified nanoparticles displayed effectual stimuli-sensitive release of R848 payload in acidic environmental of pH 5.5 or less (Fig. 11.4). R848-loaded-MSNs were up-taken rapidly by APCs (antigen-presenting cells) in acidic lysosome environment plus effectively initiated immunity cells. On subcutaneous inoculation to mice, they agglomerated within migratory dendritic cells (DCs) inside draining lymph nodes, in which these cells enhanced intensely DCs stimulation. Additionally, immediate delivery of adjuvant R848 as well as model antigen OVA by MSNs lead to amplified antigen-specific T-cell reaction. They expressively enhanced R848 pharmacokinetic profile in mice and increase shelf-life of drug six-fold, plus decrease in systemic exposure. Thus MSNs signify hopeful tool aimed at directed delivery of R848 (immune modulator) to APCs as

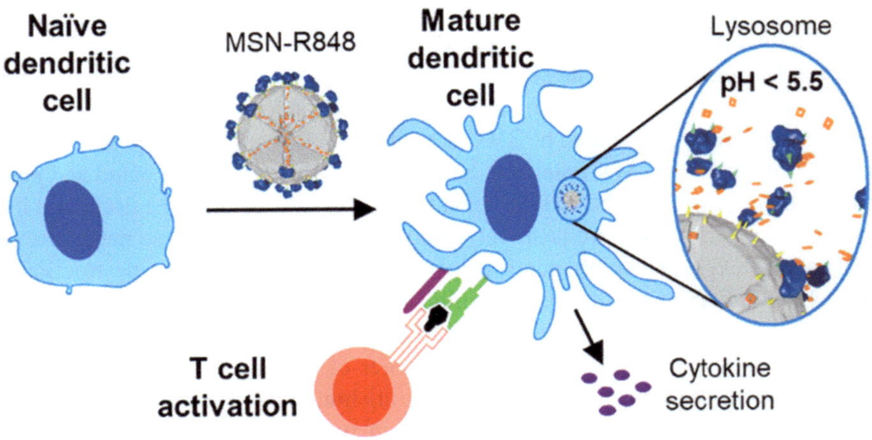

Fig. 11.4 MSNs as pH-sensitive carrier aimed at immune-triggering drug Resiquimod improved local immune reaction in mice. (Reprinted with permission from Wagner et al. (2021). Copyright 2023, American Chemical Society)

well as grasp substantial potential as carrier for vaccines to treat cancer (Wagner et al. 2021).

The prospective of MSNs containing peptide NZX to treat tuberculosis was analyzed. Functional NZX was released by MSNs slowly in simulated lung fluid. Peptide-loaded-MSNs were taken up easily by prime macrophages. In a model for intracellular infection, peptide-loaded-MSNs presented improved mycobacterial slaying in comparison to free peptide. Therapeutic probability of eptide-loaded MSNs was examined in the model of murine infection, presenting that MSNs conserved impact to remove in vivo M. tuberculosis. Hence loading of antimicrobial peptide NZX in MSNs enhanced intracellular mycobacteria inhibition in prime macrophages and delivered proof for using MSNs to treat tuberculosis (Tenland et al. 2019).

11.2.2 Organic Nanoparticles

Employing organic nanoparticles, different passive and active approaches are used for dendritic systems for making them extremely active to deliver therapeutics to tumor core with minimum inaccurate toxicity (Liu et al. 2022). Predominantly with non-polar therapeutic agents, organic nanoparticles offer a valuable scaffold for assisting constancy and solubility of formulation with no affect on their treatment competence. Bonding regarding particular antigens aimed at particular delivery utilizing monoclonal antibodies has been investigated in vaccine research to provoke immune responses effectively (Rades et al. 2018).

Polymeric nanoparticles are a predominantly multipurpose platform to deliver signals to immune system because they could be functionalized easily for targeting particular receptors as well as engineered with the aim of releasing loaded cargo in a defined and persistent manner (Poon and Patel 2020). Polymeric nanosystems are employed to deliver therapeutic agents comprising peptides, monoclonal antibodies, and small drug molecules for activating immune system against infectious or cancerous disease, or inhabit immune system to fight against autoimmune diseases as well as transplant rejection (Mitarotonda et al. 2022). A polymeric system was developed to target inflammation site and transport multiple drugs for treating sepsis and alleviating its symptoms. This system is comprised of PLGA (poly(lactide-co-glycolide acid) having degradability and biocompatibility. Additionally, they are loaded with antibiotic SFX (Sparfloxacin) along with anti-inflammatory immunosuppressant TAC (Tacrolimus). Targeting capability of these polymeric nanoparticles toward inflammatory sites is provided by γ3 peptide (NNQKIVNLKEKVAQLEA) grafting, which can precisely attach to ICAM-1 (intercellular adhesion molecule-1) that has been expressed extremely on the surface of endothelial inflammatory cells. Drugs encapsulated in γ3-PLGA nanoparticles possess exceptional cytocompatibility, lower ratio of hemolysis, in addition to systemic toxicity. Moreover, they have tremendous antibacterial characteristics to Gram-negative and -positive bacteria and also efficiently decrease immune reaction and inflammation in severe lung infection in mice. Hence these nanoparticles offer a robust and simple platform for treating sepsis-induced lung infection that might pave a way for designed multifunctional nanomedicine for translation to clinics (Yang et al. 2020). Self-healing and shear-thinning polymeric nanoparticle (PNPs) hydrogels were developed as injectable and tunable biomaterial nanoplatform aimed at recruitment of local dendritic cell (DCs) (Fig. 11.5). PNP gels were made of PEG-b-PLA (poly (ethylene glycol)-block-polylactide) NPs (nanoparticles) having similar dimensions, however dissimilar length of PEG (5 or 2 kDa) brushes. After shear-thinning, PEG-b-PLA NPs having long brush of PEG showed enhanced formulation of gel following rapid recovery and self-assembly. In every case, therapeutic model proteins got released and slight differences in rate of release between formulations of gel were perceived. DC cytokine CCL21 was loaded in PNP hydrogels and inoculated subcutaneously in the model of murine.

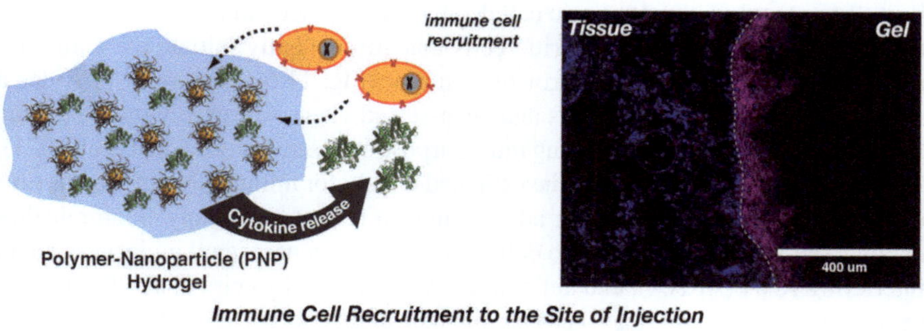

Fig. 11.5 Polymeric nanoparticle (PNP) hydrogels were developed as injectable and tunable biomaterial nanoplatform aimed at recruitment of local dendritic cell (DCs). (Reprinted with permission from Fenton et al. (2019). Copyright 2023, American Chemical Society)

PNP hydrogels loaded with CCL21 recruited DCs differently to injection site in vivo compared to hydrogels not loaded with CCL21. Therefore, PNP hydrogels offer a tunable biomaterial platform for in vivo immunomodulation succeeding negligibly in subcutaneous invasive injection (Fenton et al. 2019).

Liposomal nanoparticles (NPs) functioned as idyllic nanoplatform for delivering systemically modulators for immunity. Liposomes have a structure similar to cell membrane that offers effective affinity for cell and intensely enhances uptake by cells (Gao et al. 2019). Hence, they are deliberated ideal nanocarriers intended for facilitating operative immunotherapy to activate either cellular or humoral immune reaction. Additionally, liposomal nanoparticles are employed in high-performance and safe clinical uses in cancer immunotherapy (Ye et al. 2017). An innovative hybrid nanoparticles of liposome-polymer (pSFV-MEG/LNPs), having mPEG-PLGA core that is biodegradable with shell of hydrophilic lecithin/PEG-DSPE-Mal 2000 to deliver self-replication multi-epitope DNA vaccine (pSFV-MEG), was developed. pSFV-MEG/LNPs have dimensions of 161.61 ± 15.63 nm with higher efficiency of encapsulation of about 87.60 ± 8.73%. They induced intense cellular immune reaction (1.60-fold) and humoral (3.22-fold) in comparison with phosphate buffer solution. Moreover, cellular immune and humoral pSFV-MEG/LNPs reactions were 1.05-fold besides 1.58-fold as compared to pSFV-MEG. These outcomes verified the suitability of liposomal nanoparticles as a tool for increasing cellular and humoral immune reactions of pSFV-MEG. Therefore, a reasonable design and platform to deliver could be utilized to develop DNA vaccines for infectious diseases (Zhao et al. 2021). A multipurpose sonosensitizer nanosystem (FA-MnPs) has been established via enclosing MnP (manganese-protoporphyrin) in folate (FA)-liposomes. These nanoparticles display unique depth-sensitive sonodynamic therapy (SDT) in addition to activating simultaneously immune reaction mediated by SDT. They showed high acoustic intensity under irradiation of ultrasound (US) in simulator tissue up to the depth of 8 cm as well as producing singlet oxygen (1O2). The response of better depth SDT regarding nanoparticles suppresses proficiently superficial tumor growth and also of lesion in depth in TNBC (triple-negative breast cancer) using the model of mice. Prominently, SDT induced by such nanoparticles re-polarizes M2 immunosuppressive macrophages to M1 antitumor macrophages, as well as provokes immunogenic cell death (ICD) for activating natural killer cells (NK), T lymphocytes, and dendritic cells that subsequently initiate the antitumor immune, causing inhibition of growth of tumor (Fig. 11.6). This work provides a strategy to cure metastatic tumors and deep-seated via noninvasively immunogenic SDT through practically fabricating multipurpose sonosensitizers (Chen et al. 2021).

Polymeric micelles have been investigated widely for immunotherapy to improve effectiveness as well as reducing adverse impact related to therapy. Such micelles exploit core-shell configuration to deliver therapeutic genes, small molecules, small interfering RNA (siRNA), and antibodies (Abs) (Varela-Moreira et al. 2017). They present a promising strategy because of their ability to encapsulate effectively a range of hydrophobic molecules, longevity, biocompatibility, higher constancy in vivo as well as in vitro, in addition capability to amass within pathological regions along with impaired vasculature (Wan et al. 2021). A tumor acidity-sensitive

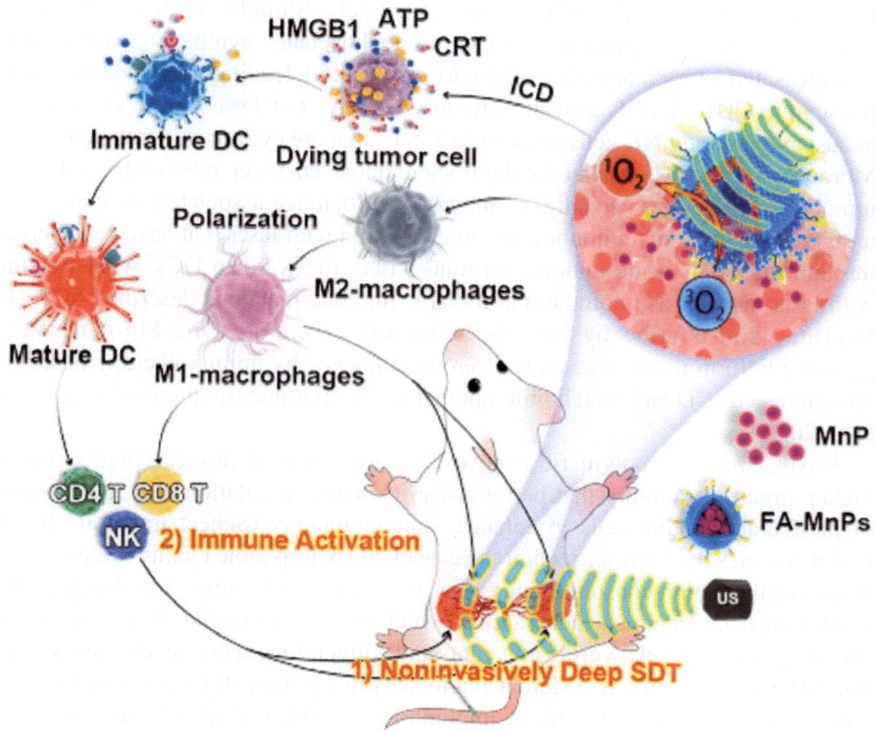

Fig. 11.6 Immunogenic noninvasively SDT with liposomes of MnP (manganese-protoporphyrin) for TNBC. Outcomes displayed that SDT induced by MnP re-polarized M2 immunosuppressive macrophages toward M1 antitumor macrophages, and provokes ICD with exposition and DAMPs release, including calreticulin (CRT), adenosine triphosphate (ATP), and high-mobility group box 1 (HMGB1). (Reprinted with permission from Chen et al. (2021). Copyright 2023, Elsevier)

biomacromolecule system was developed from calcium crosslinked polymeric micelles to deliver intratumorally M-CSF (macrophage colony-stimulating factor), cytokine for activating immune system, and lessen acidic microenvironment to enhance immunotherapy regarding cancer. These micelles were synthesized by cross-linking calcium carbonate ($CaCO_3$) for forming stable micelles loaded with M-CSF (NP/M-CSF/$CaCO_3$). NP/M-CSF/$CaCO_3$ administration suppressed the progression of tumor considerably via improving anti-tumor immune reaction mediated by T cells and retreating immunosuppression mediated by TAM. This provides newfangled paths to amplify antitumor impact via targeting tumor microenvironment that also help to avoid undesirable complications (Mao et al. 2019). A metallic micelle nanovaccine is fabricated by Mn (manganese) self-assembly, STING (stimulator for interferon genes) agonist (ABZI), and nanoparticles harmonized with naphthalocyanine (ONc) (ONc-Mn-A) in micelles of Pluronic F127 (malF127) modified with maleimide. Synergistic effect between ABZI and Mn, developed micellar nanovaccine (ONc-Mn-A-malF127) raises interferon-β (IFNβ) levels in comparison with Mn or ABZI separately. Stimulation of cGAS (cyclic guanosine monophosphate-adenosine monophosphate synthase)-STING path prompts satisfactory maturation of dendritic cell (DC), leading to demise of tumors resistant to CD8+ T cells as well as tumors sensitive to CD8+ T cells via instantaneously stimulating NK cells and CD8+ T cells that are cytotoxic, correspondingly. Moreover, ONc as Mn chelator and proficient photosensitizer, photoinduced ICD of tumor cells discharges neoantigens and DAMPs (damage-associated molecular patterns) out of dying main tumor cells on irradiation with laser that have been seized through malF127 inside tumors and transported afterward to DCs. After treating with laser, as well as photothermal therapy, immune reactions described by IFNβ level are elevated further by four-fold additionally. Therefore, ONc-Mn-A-malF127 permits platform to enhance anticancerous efficiency by immunotherapy based on photoinduced ICD and metalloimmunotherapy having intense abscopal impact (Li et al. 2022).

Polymersomes or polymeric nanovesicles are potential versatile platform with higher structural stability that permits spatiotemporal regulation over the function of particular immune cells to enhance immunotherapy (Scheerstra et al. 2022). Polymersomes have denser membrane as well as hydrophobic chains entanglement that gives them high constancy and enhanced payloads retaining. Additionally, they exhibit superior chemical adaptability which offers substantial developments as vaccine platform for delivering adjuvants and antigen for enhancing immunity reaction (Zhu et al. 2019). A different photo-sensitive polypeptide-glycosylated poly (amidoamine) (PAMAM) dendron amphiphiles (PGDAs), ultraviolet sensitive self-assembled, besides elicited release of OVA (ovalbumin) was examined. Extremely self-assembled anisotropic PGDA4 along glycosylated second-generation PAMAM dendron in consistent polymersomes with twenty to fifty percentage of water by weight displayed UV-sensitive reassembly, active bonding with lectin of concanavalin A, in addition to faster in vitro release of OVA. The polymersomes loaded with OVA can deliver OVA efficiently in RAW264.7 cells in addition attain improved endolysosomes outflow on irradiation with UV. ELISA (enzyme-linked immunosorbent assay) displayed that empty polypeptidosomes coated with sugar triggered higher TNF-α (tumor necrosis factor α) level equals to 468 pg/mL, performing improved part regarding immunity adjuvant to activate macrophages. On irradiation by UV, polymersomes loaded with OVA can trigger more RAW264.7 besides increasing level of TNF-α by 45%. Subsequently, this research work delivers an adaptable platform for designing sugar-coated and photosensitive polymersomes of glycopolypeptides with potential use for immune adjuvant, delivery of protein, and antigen-based immunotherapy (Song et al. 2020). An effective enzyme prodrug nanofactory-targeted chemo-immunotherapy has been established founded on tumor-enlarged polymersomes loaded by enzyme having optimum density regarding membrane cross-linking. On their IV injection, they could accumulate passively at the tumor area. The pH-sensitive nanofactories could expand from 100 nm to 200 nm in the stimulation of acidity regarding tumor, resulting in sustained retaining for 1 week in tissues of tumor. Instantaneously, the permeability of their membrane has significantly enhanced that permits small hydrophilic molecules to cross

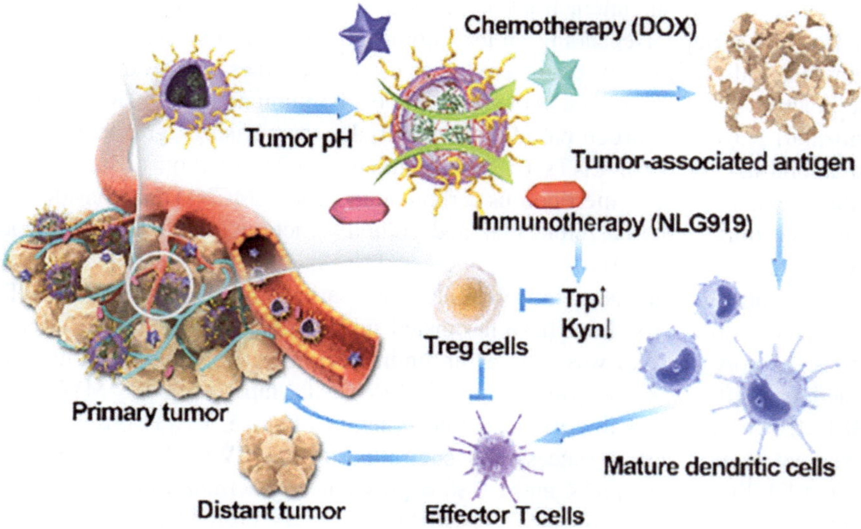

Fig. 11.7 Nanofactories made of tumor-dilated polymersome aimed at improved enzyme prodrug chemo-immunotherapy. Activated doxorubicin (DOX) could kill cancerous cells as well as produce antigens associated with tumor for promoting DCs growth. Stimulated NLG919 constrains IDO-1 (indoleamine 2, 3-dioxygenase 1) functioning as well as controls ratio of metabolite L-kynurenine (Kyn) and vital amino acid L-tryptophan (Trp) (Trp/Kyn) aimed at reversing immunosuppression. (Reprinted with permission from Japir et al. (2021). Copyright 2023, Elsevier)

membranes whereas keeping enzymes inside internal cavities. Stimulated chemotherapeutic drugs activated by nanofactories kill cancerous cells and produce antigens associated with tumor for promoting dendritic cells maturation as shown in Fig 11.7. Stimulated indoleamine 2, 3-dioxygenase 1 inhibitors inverse immunosuppression. Lastly, prime tumors could be efficiently repressed, causing minimum systematic toxicity. The long-distance tumors formed after therapy could also inhibited entirely through stimulation of antitumor immune system within mice. Consequently, tumor-enlarged polymersome holding durable intratumoral preservation proposes a favorable paradigm aimed at improved enzyme prodrug chemo-immunotherapy (Japir et al. 2021).

11.2.3 Carbon-Based Nanomaterials

Carbon-based nanomaterials also exhibited immunostimulatory abilities, particularly by stimulating monocytes in paths associated with severe inflammatory procedures, comprising those accountable for eliminating tumors as well as clearing pathogens (Pescatori et al. 2013). Carbon nanomaterials have the ability for activating numerous immune cells or else prompting immunosuppression. Immunological impact regarding carbon-based nanomaterials includes inflammation and pulmonary macrophage activation (Yuan et al. 2019). Immune system reacts to carbon nanomaterials and the interactions are determined by lengths, solubilities, purities, and surface groups. Depending on the immunostimulatory characteristics of these nanomaterials, they can be engineered to be used in vaccines or any other therapeutic practice, which necessitates the initiation of the immune system (Rahmati and Mozafari 2019). Carbon-based nanomaterials, comprising graphene, DLC (diamond-like carbon), CNTs (carbon nanotubes), CDs (carbon quantum dots) fullerene, in addition to mesoporous carbon biomaterials (MCNs) that considerably influence immunological molecular and cellular reactions at bio nano-interface (Madannejad et al. 2019).

Application of chitosan-wrapped multiwalled carbon nanotubes (CS-MWCNTs) as targeted carriers for drugs have fascinated substantial attention. HEP (Hericium erinaceus polysccaride) was used as an immunostimulant for improving immune reactions. HEP was encapsulated in CS-MWCNTs fabricating CS-MWCNTs-HEP. Exploiting them in peritoneal macrophages of mice, it was observed that immune reaction can be successfully controlled by CS-MWCNTs-HEP and promoted F4/80, MHCII, gp38, and CD86 expression. Furthermore, mice inoculated with nanoparticles of CS-MWCNTs-HEP expressively prolonged PCV2-specific IgG immunity reaction as well as also cytokines levels. Outcomes established the favorability of such nanoparticles to deliver drugs aimed at immuno-enhancement (Ren et al. 2021). For enhancing the efficiency concerning immersion vaccine, F-SWCNTs (functionalized single-walled carbon nanotubes) were employed as per nanocarrier for manufacturing DNA immersion vaccine system (SWCNTs-pEGFP-M) with functionalization chemically. Outcomes presented the SWCNTs-pEGFP-M entry in fish body through immersion vaccine as well as prompt antigen proteins in spleen and kidney of fish. Furthermore, long and strong time regarding immune reactions (comprising serum antibody generation besides expression of genes for immunity) could persuade in fish inoculated with SWCNTs-pEGFP-M as compared to fish inoculated with pEGFP-M. Particularly, SWCNTs could enhance immunity shielding impact concerned to naked DNA vaccine by ca. (calculated absorption) 23.8%. Overall, this work validates that SWCNTs are promising DNA vaccine nanocarrier and could be utilized to vaccinate enormous young fish using the bath vaccination approach. It offers a viewpoint aimed at vaccination tactics against SVCV (Spring viremia of carp virus) in the future (Zhang et al. 2019).

Magnetothermodynamic (MTD) therapy is an effective systematic cancer therapy that combines magnetothermal and immunologic impact related to reactive oxygen species (ROS), to overcome difficulty regarding restricted therapeutic effectiveness in existing magnetothermal therapy (MTT). Such strategy has attained to develop hybrid nanoparticle (FVIOs-GO) consisting of graphene oxide (GO) and intricate ferrimagnetic vortex-domain iron oxide (FVIOs) nanoring as competent agent for MTD agent. This nanoplatform exhibited higher thermal transformation productivity. It was proved further for generating considerably augmented level of ROS in AMF (alternating magnetic field). Both in vivo as well as in vitro outcomes exposed that augmented production of ROS had been a leading feature to provoke intense immune reaction in hypoxic tumor microenvironment at a tolerable physiological temperature less than 40 °C. It was maintained by exposing 83% of breast

(4 T1) cancerous cell surface to calreticulin (CRT), straight elevation of polarization of macrophage to pro-inflammatory M1 phenotypes, in addition to additional promotion regarding tumor-infiltrating T lymphocytes (Fig. 11.8). Double act of immunologic impact related to ROS and magnetothermal impact, imposing systemic in vivo therapeutic effectiveness was achieved at lower dose of three milligram iron (Fe)/kilogram along with 2 AMF treatments, in contrast with MTT (higher dose of 6 to 18 milligram/ kilogram in 4–8 AMF treatments). In this work, MTD treatment highlighted the insufficiency of orthodox MTT, which exclusively relies on the heating impact of magnetic nanoparticles (MNPs). Therefore, by using immunologic impact facilitated by ROS, future magnetotherapies to treat cancer could be designed having prominently enhanced antitumor abilities (Liu et al. 2020b).

An immersion SWCNT was developed and loaded by subunit vaccine, composed of glycoprotein (G) of micropterus salmoides rhabdovirus (MSRV), and assessed its shielding impact on bigmouth bass. The outcomes displayed that an intense immune reaction comprising levels of serum antibody, activities of enzyme (acid and alkaline phosphatase, superoxide dismutase, also entire antioxidant ability), counterpart C3 constituent, as well as expression of genes related to immunity (CD4, TGF-β, IgM, IL-1β, TNF-α, IL-8) could persuade clearly with groups of SWCNTs-G (SWCNTs-glycoprotein) in contrast to G groups while bigmouth bass

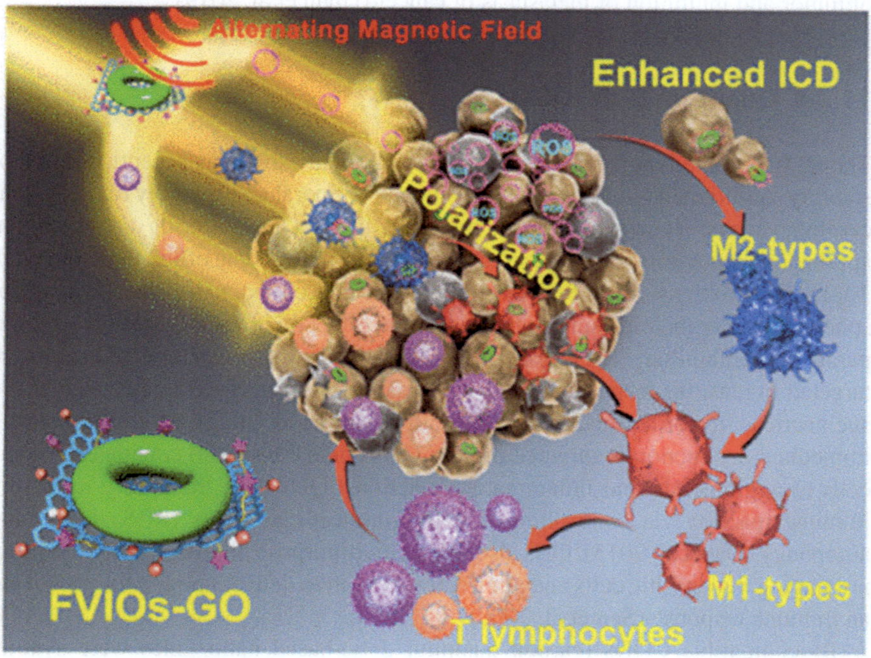

Fig. 11.8 Magnetothermodynamic (MTD) treatment facilitated by hybrid nanoparticle (FVIOs-GO) consisting of graphene oxide (GO) and intricate FVIOs nanoring activating immune reaction related to ROS for improved antitumor efficiency. (Reprinted with permission from Liu et al. (2020b). Copyright 2023, American Chemical Society)

had been inoculated. Next to bath immunization by G or SWCNTs-G for 28 days, fish had been confronted by MSRV deadly dosage. The rate of existence regarding the control group (phosphate buffer saline), pure G protein groups (40 mg/L), SWCNTs group (40 mg/L), and three SWCNTs-G groups (10 mg/L, 5 mg/L, and 40 mg/L) were 0%, 39.5%, 0%, 43.6%, 36.7%, and 70.1%, correspondingly. Significantly, with the help of SWCNTs, rate of immune shielding regarding SWCNTs-G group (40 mg/L) amplified by almost 30.6%. Such work proposed capability of developed SWCNTs-G as immersion subunit vaccine contender against death attributed to MSRV (Guo et al. 2020). Gadofullerene (Gd@C82-Ala, abbreviated Gd-Ala) of ananodimensions triggered by photo persuaded vascular disruption of malignant tumor by reducing light duration in between administration of Gd-Ala and illumination of light, in which oxygen within blood vessels was consumed proficiently for producing cytotoxic ROS. The generated ROS destroys tumor and annihilates endothelial vascular cells conforming intercellular junctions loss and disruption of vessels. Particularly, Gd-Ala after irradiation might increase maturation of dendritic cells (DCs) that further secreted tumor necrosis factor-α, interleukin-12, as well as activated T lymphocytes via CD8+ also CD4+ and T lymphocytes up-regulation. Additionally, matrix of metalloprotein 9 (MMP9)/MMP2 down-regulation decreases metastasis tumor rate. This research disclosed a Gd-Ala biomedicinal use, thus offering smart carbon-based nanomaterial aimed at ablation of tumor and inhibition of metastasis of cancer (Guan et al. 2019).

11.2.4 Quantum Dots (QDs)

The QDs exhibited higher fluorescent quantum yield as well as exceptional photostability. Additionally consistent and prolonged temporal and spatially tracking concerning vaccine dynamics was apprehended using QDs, offering an influential approach to study immune response besides assessing vaccine efficiency (Sun et al. 2021). QDs toxicity in biological environment caused extensive unease in nanosphere, however their immune cytotoxicity and effect on immunity system are comparatively unfamiliar. They primarily cause variations in relative indicators of targeted organs, for instance oxidative destruction, and in critical cases trigger tissue necrosis, hyperthermia, and even death (Liang et al. 2022). Moreover causing immediate destruction to targeted organs, QDs could cause also countless immune cells to agglomerate and inflammatory reactions. QDs have the ability to disrupt immune reaction via controlling oxidative intracellular stress (ROS) levels and autophagy (Dai et al. 2015). QDs also offer a multipurpose platform aimed at immunoimaging of dendritic cells and as effective system to deliver antigen for provoking an immune response (Sen et al. 2008).

Biocompatible CQDs (carbon quantum dots)-based nanozymes were synthesized using chlorogenic acid (ChA), the main natural biologically active product obtained from coffee. They showed clear glutathione oxidase-like events and consequently stimulated ferroptosis of cancerous cell via disturbance of GPX4 (Glutathione peroxidase4)-catalyzed lipid reparation systems. ChA CQDs intensely repressed in vivo growth of tumor in mice with HepG2-tumor with insignificant side toxicity. Mainly, in mice with hepatoma H22, they recruited immense tumor-infiltrating immunity cells comprising NK cells, macrophages, and T cells, thus transforming "cold" to "hot" tumors to activate systematic antitumor immunity reactions. This work proposes that CQDs derived from natural products could perform as biologically nontoxic nanozymes aimed at anticancer therapeutics. It also helps to develop immunotherapeutics based on nanotechnology (Yao et al. 2022). Side effects concerning two QDs comprising cadmium (Cd), CdTe@ZnS in addition to CdTe QDs, on nematode C. elegans nervous system, cultured microglia, and mice hippocampus had been estimated for assessing neuroinflammation due to QDs with Cd. The exposure of these QDs provoked immune reaction as well as neurobehavioral discrepancy inside nematode C. elegans. Mice treated by QDs, neuroinflammatory reactions, counting microglial initiation, in addition to release of IL-1ß, happened also. QDs with Cd also activated NLRP3 inflammasome by generating excessive ROS that leads to release of IL-1ß. Neurotoxicity and mild immunity reaction regarding QDs of CdTe@ZnS in contrast to CdTe QDs specified the defensive role of zinc sulfide (ZnS) coating, repression of NLRP3 expression in addition to ROS generation entirely decreased inflammation mediated by IL-1ß. This delivered useful information that inhibition of targeted molecules is an operative way for alleviating toxicity of QDs with Cd (Wu et al. 2020).

QDs pulsated-DC vaccines incorporating with polarization of macrophages associated with tumor (TAM) have been developed to amplify anti-tumor immunity. Positively charged QDs modified by polyelectrolyte (QD-P) were prepared via unmethylated cytosine-phosphate-guanine oligonucleotide oligodeoxynucleotides (CpG ODNs) (adjuvant) adsorption besides tumor cell lysate (tumor antigen) for fabricating QD-P-CpG-A. Semiconductor-engineered QDs have assorted functions for acting as immunomodulatory adjuvants, fluorescent nanoprobes, and nanocarriers for loading Toll-like receptor 9 agonists as well as tumor antigens. QDs pulsated-DC vaccines allow spatially and temporal tracing regarding lymphatic drainage in addition to efficiency assessment of DC immunotherapy, and activate strong immunoactivation. Precisely, DC designer vaccine along with polarization of macrophage provokes intense immune reaction for stimulating adaptive and innate anti-tumor immunity and upgrades immunosuppression (Fig. 11.9). QD-pulsated DC vaccines combined with CQ (chloroquine) proved to be an efficient immunomodulator regarding polarization of TAM, allows improved integrated immunotherapy utilizing lung metastatic B16 to F10 and subcutaneous B16 to F10 model. TAM could re-instruct flexibly from tumor-supportive M2 macrophages to tumoricidal M1 macrophages This approach considerably increases immunity of antigen-specific T cell and therefore intensely constrains in vivo growth and metastasis of local tumor. This investigation might offer an appropriate treatment intended for immunotherapy of cancer (Liu et al. 2020a).

Graphene QDs (GQDs) as photosensitizers were bonded to PEG (polyethylene glycol) for enhancing blood circulation and solubility. The phototoxicity of GQDs-PEG was detected in vivo and in vitro. Anti-tumor immunity activated by GQDs-PEG on irradiation had also been evaluated in the oral squamous cell carcinoma

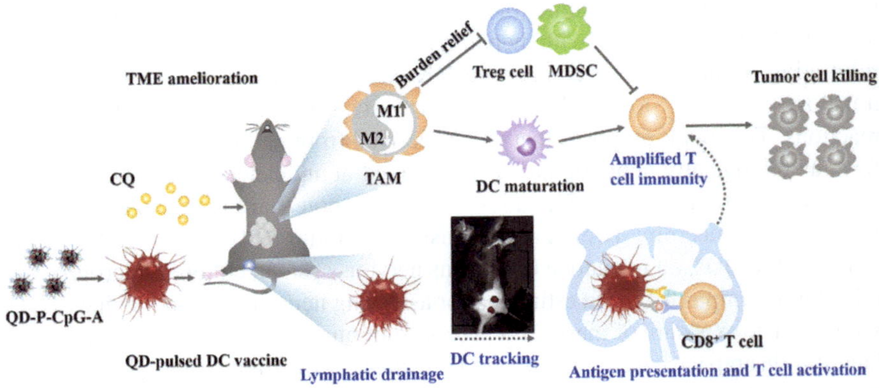

Fig. 11.9 QDs-pulsated-DC vaccines spatially and temporal tracing regarding lymphatic drainage and activate strong immunoactivation. DC designer vaccines integrated with polarization of tumor-associated macrophage (TAM) prominently improve immunosuppressive TME (tumor microenvironment), provoking augmented immunotherapy for cancer in vivo and in vitro. (Reprinted with permission from Liu et al. (2020a). Copyright 2023, Elsevier)

animal model. Developed GQDs-PEG displayed lower cytotoxicity and better stability in solution, in addition to tremendous endocytosis. In vivo and in vitro, both established intense rupture of oral squamous cell carcinoma on irradiation. In the meantime, cytotoxic T lymphocytes (CD8+ T cells) related to the immunity of the host in addition to proinflammatory cytokines, comprising TNF-α and IFN-γ, had been considerably enhanced after antitumor activity activated by light. These outcomes high spot leading performance regarding GQDs-PEG for photodynamic treatment and possess substantial consequences in lieu of combined therapy as a favorable strategy for anti-tumor immune reaction activated by QDs (Zhang et al. 2020). In vitro toxicity for cells regarding QDs of PEGylated CuInS2 (copper indium sulfide)/ZnS had been investigated consuming DC2.4 cell line. Furthermore, their in vivo immunotoxicity was examined by consuming BALB/c mice. In vitro experimentations displayed up-take of QDs by cells, stimulated sustainability of cells, improved discharge concerning TNF-α, besides decreasing interleukin (IL)-6 level in reaction to lipopolysaccharide motivation. Above five thousand genes at the transcriptome level were perceived via highly efficient sequencing of RNA next to exposure to these QDs. In vivo investigation presented that on day one such QDs enhanced IL-4 levels and on 28 day enhanced IL-10 and IL-13 levels within mice. After treatment, no noticeable variance in lymphocytes number stemmed from spleen, hematology, organic index, as well as immune organ histology on day one and day twenty-eighth was perceived. It established perturbation regarding in vitro performance of DC2.4 cells for immunity by QDs of PEGylated CuInS2/ZnS, however no apparent in vivo toxicity to immune system signifies biocompatibility of QDs and significant prospective for biomedical application in coming years (Chen et al. 2020b).

11.3 Conclusions

Current investigations have confirmed that nanoparticles offer countless advantages, for instance, increased stability, favorable profiles for biodistribution, slower release kinetics of drug, low immunotoxicity, in addition to targeting specific population of cells. The manipulation of nanoparticles for activation of immune system is a developing and interesting area of research. Nanoparticles are ideal carriers for harmonized delivery of immune stimulatory molecules and tumor antigens to DCs and additional professional APC forms. Typically, such strategies have produced better outcomes in comparison with traditional, non-particle-based vaccination methods, leading to increased CD8þ and CD4þ T-cell reactions against tumor. Similarly, nanoparticle-based tactics are intended to overcome immunosuppression and also deliver introductory indication of effectiveness.

Utilizing immune-manipulative compounds and materials for designing nanoparticle-based delivery systems is vital if the immune reaction would hamper the functioning of nanoparticle via reducing its circulation time while the system with sustained release is requisite. In this chapter, different inorganic, organic, carbon-based, and QD nanomaterials have been discussed that exhibit immunomodulatory abilities and present alternate biomaterials that have an impact on immune responses. Studies showed that nanoparticles have the ability of interaction with constituents of the immune system to numerous immunological endpoints. Nanoparticles interactivity with the immune system is rapid, complex, as well as not completely understood. The physicochemical properties (shape, size, surface functionalization, and hydrophobicity) regarding nanoparticles are vital to determine their interactions with immune cells and plasma proteins, especially APCs. Nanoparticles have the ability of absorbing proteins on their surfaces when they are administered to bloodstream, and such bonded proteins control their uptake by cells, clearance path, and biodistribution in tissues. Depending on their immune reactions, engineered nanoparticles are primarily employed for the treatment of autoimmune diseases, cancer immunotherapy, and vaccines. The information about nanoparticle cytotoxicity is valuable to identify severe damage to the host, however they cannot identify the sub-lethal impact and deregulation of the immune system.

Clinical translation of employing nanoparticle-based approaches for activation of the immune system will definitely need optimization of such nanostructures. Mostly, commercially available nanoparticles are used. Though positive outcomes have been observed, still it will be critical to describe the optimum physical and chemical features of such nanoparticles, counting surface chemistry, size, and quantity and quality of loaded cargo, not only with regard to therapeutic effectiveness but also with respect to pharmacokinetics, toxicity, and biodistribution. Furthermore, the molecular path and mechanism of how nanoparticles influence the immune system are not clear. Consequently, more research is required for understanding the interaction of nanoparticles with immune system and for developing novel approaches to prevent and treat human diseases. Therefore, still nanoscience and technology prerequisites efforts plus advancement for comprehending its potential in application for the provision of therapeutic agents. Every cell in the immunity system inside the human body holds its specific reaction toward nanoparticles. Countless immune cells and nanoparticles make it unfeasible to detect all interactions that might occur, however focused and dedicated research in the promising categorization of nanoparticles might be appropriate to drive this technology propelling forward. Taken all together we believe that nanoparticles propose innovative strategies to develop future-generation T-cell-stimulating/directing nanomedicines for suppression (in autoimmunity) as well as intensify (in cancer) immunity with clinical potential.

References

Abbasi R, Shineh G, Mobaraki M, Doughty S, Tayebi L (2023) Structural parameters of nanoparticles affecting their toxicity for biomedical applications: a review. J Nanopart Res 25:43

Abesekara MS, Chau Y (2022) Recent advances in surface modification of micro- and Nano-scale biomaterials with biological membranes and biomolecules. Front Bioeng Biotechnology 10:972790

Aderibigbe BA (2017) Metal-based nanoparticles for the treatment of infectious diseases. Molecules 22:1370

Altammar KA (2023) A review on nanoparticles: characteristics, synthesis, applications, and challenges. Front Microbiol 14:Article ID: 1155622

Blank F, Fytianos K, Seydoux E, Rodriguez-Lorenzo L, Petri-Fink A, Von Garnier C, Rothen-Rutishauser B (2017) Interaction of biomedical nanoparticles with the pulmonary immune system. J Nanobiotechnol 15:6

Chen Y, Hou S (2023) Recent Progress in the effect of magnetic iron oxide nanoparticles on cells and extracellular vesicles. Cell Death Discovery 9:195

Chen BA, Jin N, Wang J, Ding J, Gao C, Cheng J, Xia G, Gao F, Zhou Y, Chen Y, Zhou G, Li X, Zhang Y, Tang M, Wang X (2010) The effect of magnetic nanoparticles of Fe(3)O(4) on immune function in normal ICR mice. Int J Nanomedicine 5:593–599

Chen L, Liu J, Zhang Y, Zhang G, Kang Y, Chen A, Feng X, Shao L (2018) The toxicity of silica nanoparticles to the immune system. Nanomedicine 13:1939–1962

Chen H, Fan Y, Hao X, Yang C, Peng Y, Guo R, Shi X, Cao X (2020a) Adoptive cellular immunotherapy of tumors via effective Cpg delivery to dendritic cells using dendrimer-entrapped gold nanoparticles as a gene vector. J Mater Chem B 8:5052–5063

Chen T, Li L, Lin X, Yang Z, Zou W, Chen Y, Xu J, Liu D, Wang X, Lin G (2020b) In vitro and in vivo Immunotoxicity of PEGylated Cd-free Cuins2/Zns quantum dots. Nanotoxicology 14:372–387

Chen H, Liu L, Ma A, Yin T, Chen Z, Liang R, Qiu Y, Zheng M, Cai L (2021) Noninvasively immunogenic Sonodynamic therapy with manganese Protoporphyrin liposomes against triple-negative breast cancer. Biomaterials 269:120639

Dai T, Li N, Liu L, Liu Q, Zhang Y (2015) AMP-conjugated quantum dots: low immunotoxicity both in vitro and in vivo. Nanoscale Res Lett 10:434

Dey A, Manna S, Kumar S, Chattopadhyay S, Saha B, Roy S (2020) Immunostimulatory effect of chitosan conjugated green copper oxide nanoparticles in tumor immunotherapy. Cytokine 127:154958

Dobrovolskaia MA, Shurin M, Shvedova AA (2016) Current understanding of interactions between nanoparticles and the immune system. Toxicol Appl Pharmacol 299:78–89

Du J, Zhang YS, Hobson D, Hydbring P (2017) Nanoparticles for immune system targeting. Drug Discov Today 22:1295–1301

Dul M, Nikolic T, Stefanidou M, Mcateer MA, Williams P, Mous J, Roep BO, Kochba E, Levin Y, Peakman M, Wong FS, Dayan CM, Tatovic D, Coulman SA, Birchall JC (2019) Conjugation of a peptide autoantigen to gold nanoparticles for intradermally administered antigen specific immunotherapy. Int J Pharm 562:303–312

Dykman LA, Khlebtsov NG (2017) Immunological properties of gold nanoparticles. Chem Sci 8:1719–1735

Elmowafy M, Shalaby K, Elkomy MH, Alsaidan OA, Gomaa HAM, Abdelgawad MA, Mostafa EM (2023) Polymeric nanoparticles for delivery of natural bioactive agents: recent advances and challenges. Polymers 15:1123

Fadeel B (2019) Hide and seek: nanomaterial interactions with the immune system. Front Immunol 10:133

Feng B, Lu X, Zhang G, Zhao L, Mei D (2023) Sting agonist delivery by lipid calcium phosphate nanoparticles enhances immune activation for neuroblastoma. Acta Materia Medica 2 : 216-227

Fenton OS, Tibbitt MW, Appel EA, Jhunjhunwala S, Webber MJ, Langer R (2019) Injectable polymer–nanoparticle hydrogels for local immune cell recruitment. Biomacromolecules 20:4430–4436

Gao A, Hu X-L, Saeed M, Chen B-F, Li Y-P, Yu H-J (2019) Overview of recent advances in liposomal nanoparticle-based cancer immunotherapy. Acta Pharmacol Sin 40:1129–1137

Guan M, Zhou Y, Liu S, Chen D, Ge J, Deng R, Li X, Yu T, Xu H, Sun D, Zhao J, Zou T, Wang C, Shu C (2019) Photo-triggered gadofullerene: enhanced cancer therapy by combining tumor vascular disruption and stimulation of anti-tumor immune responses. Biomaterials 213:119218

Guo Z-R, Zhao Z, Zhang C, Jia Y-J, Qiu D-K, Zhu B, Wang G-X (2020) Carbon nanotubes-loaded subunit vaccine can increase protective immunity against rhabdovirus infections of largemouth bass (Micropterus Salmoides). Fish Shellfish Immunol 99:548–554

Hammond PT (2017) Nano tools pave the way to new solutions in infectious disease. ACS Infect Dis 3:554–558

Huang L, Liu Z, Wu C, Lin J, Liu N (2023) Magnetic nanoparticles enhance the cellular immune response of dendritic cell tumor vaccines by realizing the cytoplasmic delivery of tumor antigens. Bioeng Transl Med 8:E10400

Jadhav SD, Shaikh IA (2019) Synthesis of nanoparticles an overview, a review article. Int J Trend Sci Res Dev 3:426–428

Japir AA-WMM, Ke W, Li J, Mukerabigwi JF, Ibrahim A, Wang Y, Li X, Zhou Q, Mohammed F, Ge Z (2021) Tumor-dilated polymersome nanofactories for enhanced enzyme prodrug chemo-immunotherapy. J Control Release 339:418–429

Keikha R, Daliri K, Jebali A (2021) The use of nanobiotechnology in immunology and vaccination. Vaccines (Basel) 9:74

Kwon G, Baek J, Kim N, Kwon S, Song N, Park S-C, Kim B-S, Lee D (2023) Acid-sensitive stable polymeric micelle-based oxidative stress nanoamplifier as Immunostimulating anticancer nanomedicine, vol 11. Biomater Sci, p 6600

Lappas CM (2015) The immunomodulatory effects of titanium dioxide and silver nanoparticles. Food Chem Toxicol 85:78–83

Li J, Ren H, Qiu Q, Yang X, Zhang J, Zhang C, Sun B, Lovell JF, Zhang Y (2022) Manganese coordination micelles that activate stimulator of interferon genes and capture in situ tumor antigens for cancer metalloimmunotherapy. ACS Nano 16:16909–16923

Liang Y, Zhang T, Tang M (2022) Toxicity of quantum dots on target organs and immune system. J Appl Toxicol 42:17–40

Lin Y, Mao F, Wong N-K, Zhang X, Liu K, Huang M, Ma H, Xiang Z, Li J, Xiao S, Zhang Y, Yu Z (2020) Phagocyte transcriptomic analysis reveals focal adhesion kinase (FAK) and heparan sulfate proteoglycans (HSPGs) as major regulators in anti-bacterial defense of *Crassostrea Hongkongensis*. Front Immunol 11:416

Liu Y, Hardie J, Zhang X, Rotello VM (2017) Effects of engineered nanoparticles on the innate immune system. Semin Immunol 34:25–32

Liu F, Sun J, Yu W, Jiang Q, Pan M, Xu Z, Mo F, Liu X (2020a) Quantum dot-pulsed dendritic cell vaccines plus macrophage polarization for amplified cancer immunotherapy. Biomaterials 242:119928

Liu X, Yan B, Li Y, Ma X, Jiao W, Shi K, Zhang T, Chen S, He Y, Liang X-J, Fan H (2020b) Graphene oxide-grafted magnetic nanorings mediated magnetothermodynamic therapy favoring reactive oxygen species-related immune response for enhanced antitumor efficacy. ACS Nano 14:1936–1950

Liu J, Liu Z, Pang Y, Zhou H (2022) The interaction between nanoparticles and immune system: application in the treatment of inflammatory diseases. J Nanobiotechnol 20:127

Madannejad R, Shoaie N, Jahanpeyma F, Darvishi MH, Azimzadeh M, Javadi H (2019) Toxicity of carbon-based nanomaterials: reviewing recent reports in medical and biological systems. Chem Biol Interact 307:206–222

Mahmoudi Saber M (2019) Strategies for surface modification of gelatin-based nanoparticles. Colloids Surf B: Biointerfaces 183:110407

Malachowski T, Hassel A (2020) Engineering nanoparticles to overcome immunological barriers for enhanced drug delivery. Eng Regener 1:35–50

Mao K, Cong X, Feng L, Chen H, Wang J, Wu C, Liu K, Xiao C, Yang Y-G, Sun T (2019) Intratumoral delivery of M-CSF by calcium crosslinked polymer micelles enhances cancer immunotherapy. Biomater Sci 7:2769–2776

Mcnamara K, Tofail SAM (2017) Nanoparticles in biomedical applications. Adv Phys: X 2:54–88

Mitarotonda R, Giorgi E, Eufrasio-Da-Silva T, Dolatshahi-Pirouz A, Mishra YK, Khademhosseini A, Desimone MF, De Marzi M, Orive G (2022) Immunotherapeutic nanoparticles: from autoimmune disease control to the development of vaccines. Biomater Adv 135:212726

Mittal K, Rahim AA, George S, Ghoshal S, Basu N (2021) Characterizing the effects of titanium dioxide and silver nanoparticles released from painted surfaces due to weathering on zebrafish (*Danio Rerio*). Nanotoxicology 15:527–541

Mu W, Wang Y, Huang C, Fu Y, Li J, Wang H, Jia X, Ba Q (2019) Effect of long-term intake of dietary titanium dioxide nanoparticles on intestine inflammation in mice. J Agric Food Chem 67:9382–9389

Ni Q, Xu F, Wang Y, Li Y, Qing G, Zhang Y, Zhong J, Li J, Liang X-J (2022) Nanomaterials with changeable physicochemical property for boosting cancer immunotherapy. J Control Release 342:210–227

Pescatori M, Bedognetti D, Venturelli E, Ménard-Moyon C, Bernardini C, Muresu E, Piana A, Maida G, Manetti R, Sgarrella F, Bianco A, Delogu LG (2013) Functionalized carbon nanotubes as immunomodulator systems. Biomaterials 34:4395–4403

Poon C, Patel AA (2020) Organic and inorganic nanoparticle vaccines for prevention of infectious diseases. Nano Express 1:012001

Qin X, Yu C, Wei J, Li L, Zhang C, Wu Q, Liu J, Yao SQ, Huang W (2019) Rational design of nanocarriers for intracellular protein delivery. Adv Mater 31:1902791

Rades N, Licha K, Haag R (2018) Dendritic polyglycerol sulfate for therapy and diagnostics. Polymers 10:595

Rahmati M, Mozafari M (2019) Biological response to carbon-family nanomaterials: interactions at the nano-bio Interface. Front Bioeng Biotechnol 7:4

Ren Z, Luo Y, Meng Z, Zhang J, Yu R, Sun M, Xu T, Li J, Ma Y, Huang Y, Qin T (2021) Multi-walled carbon nanotube polysaccharide modified Hericium Erinaceus polysaccharide as an adjuvant to extend immune responses. Int J Biol Macromol 182:574–582

Ruan H, Zhou L (2022) Synthesis of fluorescent sulfur quantum dots for bioimaging and biosensing. Front Bioeng Biotechnol 10:909727

Sabourian P, Yazdani G, Ashraf SS, Frounchi M, Mashayekhan S, Kiani S, Kakkar A (2020) Effect of physico-chemical properties of nanoparticles on their intracellular uptake. Int J Mol Sci 21:8019

Sari MHM, Cobre ADF, Pontarolo R, Ferreira LM (2023) Status and future scope of soft nanoparticles-based hydrogel in wound healing. Pharmaceutics 15:874

Scheerstra JF, Wauters AC, Tel J, Abdelmohsen LKEA, Van Hest JCM (2022) Polymersomes as a potential platform for cancer immunotherapy. Mater Today Adv 13:100203

Sen D, Deerinck TJ, Ellisman MH, Parker I, Cahalan MD (2008) Quantum dots for tracking dendritic cells and priming an immune response in vitro and in vivo. PLoS One 3:E3290

Shao K, Singha S, Clemente-Casares X, Tsai S, Yang Y, Santamaria P (2015) Nanoparticle-based immunotherapy for cancer. ACS Nano 9:16–30

Song Y, Chen Y, Li P, Dong C-M (2020) Photoresponsive polypeptide-glycosylated dendron amphiphiles: UV-triggered polymersomes, OVA release, and in vitro enhanced uptake and immune response. Biomacromolecules 21:5345–5357

Soni D, Bobbala S, Li S, Scott EA, Dowling DJ (2021) The sixth revolution in pediatric vaccinology: immunoengineering and delivery systems. Pediatr Res 89:1364–1372

Sun J, Liu F, Yu W, Fu D, Jiang Q, Mo F, Wang X, Shi T, Wang F, Pang D-W, Liu X (2021) Visualization of vaccine dynamics with quantum dots for immunotherapy. Angew Chem Int Ed 60:24275–24283

Tenland E, Pochert A, Krishnan N, Umashankar Rao K, Kalsum S, Braun K, Glegola-Madejska I, Lerm M, Robertson BD, Linden M, Godaly G (2019) Effective delivery of the anti-mycobacterial peptide Nzx in mesoporous silica nanoparticles. PLoS One 14:E0212858

Tulinska J, Mikusova ML, Liskova A, Busova M, Masanova V, Uhnakova I, Rollerova E, Alacova R, Krivosikova Z, Wsolova L, Dusinska M, Horvathova M, Szabova M, Lukan N, Stuchlikova M, Kuba D, Vecera Z, Coufalik P, Krumal K, Alexa L, Vrlikova L, Buchtova M, Dumkova J, Piler P, Thon V, Mikuska P (2022) Copper oxide nanoparticles stimulate the immune response and decrease antioxidant defense in mice after six-week inhalation. Front Immunol 13:13

Ucar A, Parlak V, Ozgeris FB, Yeltekin AC, Arslan ME, Alak G, Turkez H, Kocaman EM, Atamanalp M (2022) Magnetic nanoparticles-induced neurotoxicity and oxidative stress in brain of rainbow trout: mitigation by ulexite through modulation of antioxidant, anti-inflammatory, and antiapoptotic activities. Sci Total Environ 838:155718

Varela-Moreira A, Shi Y, Fens MHAM, Lammers T, Hennink WE, Schiffelers RM (2017) Clinical application of polymeric micelles for the treatment of cancer. Mater Chem Front 1:1485–1501

Vuković B, Milić M, Dobrošević B, Milić M, Ilić K, Pavičić I, Šerić V, Vrček IV (2020) Surface stabilization affects toxicity of silver nanoparticles in human peripheral blood mononuclear cells. Nano 10:1390

Wagner J, Gößl D, Ustyanovska N, Xiong M, Hauser D, Zhuzhgova O, Hočevar S, Taskoparan B, Poller L, Datz S, Engelke H, Daali Y, Bein T, Bourquin C (2021) Mesoporous silica nanoparticles as Ph-responsive carrier for the immune-activating drug resiquimod enhance the local immune response in mice. ACS Nano 15:4450–4466

Wan Z, Zheng R, Moharil P, Liu Y, Chen J, Sun R, Song X, Ao Q (2021) Polymeric micelles in cancer immunotherapy. Molecules 26:1220

Wang N, Chen M, Wang T (2019) Liposomes used as a vaccine adjuvant-delivery system: from basics to clinical immunization. J Control Release 303:130–150

Wang C, Sun F, He G, Zhao H, Tian L, Cheng Y, Li G (2023) Noble metal nanoparticles meet molecular cages: a tale of integration and synergy. Curr Opin Colloid Interface Sci 63:101660

Wong HC, Shrestha N, George V, Dee MJ, Chaturvedi P, Zhu X, Spoundis C, Liu B, Kong L, Echeverri CA, You L, Egan JO, Jiao J-A, Rhode PR (2022) Robust human regulatory T cell expansion with fusion proteins HCW9302 and HCW9213 circumvents need for magnetic-bead or feeder cell approaches for adoptive cell therapy. J Immunol 208:174.15-174.15

Wu T, Liang X, He K, Liu X, Li Y, Wang Y, Kong L, Tang M (2020) The NLRP3-mediated neuroinflammatory responses to Cdte quantum dots and the protection of Zns Shell. Int J Nanomedicine 15:3217–3233

Wusiman A, Gu P, Liu Z, Xu S, Zhang Y, Hu Y, Liu J, Wang D, Huang X (2019) Cationic polymer modified PLGA nanoparticles encapsulating Alhagi honey polysaccharides as a vaccine delivery system for ovalbumin to improve immune responses. Int J Nanomedicine 14:3221–3234

Xue Y, Liu C, Andrews G, Wang J, Ge Y (2022) Recent advances in carbon quantum dots for virus detection, as well as inhibition and treatment of viral infection. Nano Convergence 9:15

Yan B, Wang S, Liu C, Wen N, Li H, Zhang Y, Wang H, Xi Z, Lv Y, Fan H, Liu X (2022a) Engineering magnetic Nano-manipulators for boosting cancer immunotherapy. J Nanobiotechnol 20:547

Yan H, Liu Z, Lin G, Gu F, Liu Y, Xu Y, Kuang X, Zhang Y (2022b) Systemic delivery of gemcitabine analogue and Stat3 SiRNA promotes antitumor immunity against melanoma. Nano Res 15:9057–9072

Yang Y, Ding Y, Fan B, Wang Y, Mao Z, Wang W, Wu J (2020) Inflammation-targeting polymeric nanoparticles deliver sparfloxacin and tacrolimus for combating acute lung sepsis. J Control Release 321:463–474

Yang J, Jia C, Yang J (2021) Designing nanoparticle-based drug delivery systems for precision medicine. Int J Med Sci 18:2943–2949

Yao L, Zhao M-M, Luo Q-W, Zhang Y-C, Liu T-T, Yang Z, Liao M, Tu P, Zeng K-W (2022) Carbon quantum dots-based nanozyme from coffee induces cancer cell ferroptosis to activate antitumor immunity. ACS Nano 16:9228–9239

Ye M, Han Y, Tang J, Piao Y, Liu X, Zhou Z, Gao J, Rao J, Shen Y (2017) A tumor-specific cascade amplification drug release nanoparticle for overcoming multidrug resistance in cancers. Adv Mater 29:1702342

Yuan X, Zhang X, Sun L, Wei Y, Wei X (2019) Cellular toxicity and immunological effects of carbon-based nanomaterials. Part Fibre Toxicol 16:18

Yusuf A, Almotairy ARZ, Henidi H, Alshehri OY, Aldughaim MS (2023) Nanoparticles as drug delivery systems: a review of the implication of nanoparticles' physicochemical properties on responses in biological systems. Polymers 15:1596

Zhang C, Zheng Y-Y, Gong Y-M, Zhao Z, Guo Z-R, Jia Y-J, Wang G-X, Zhu B (2019) Evaluation of immune response and protection against spring viremia of carp virus induced by a single-walled carbon nanotubes-based immersion DNA vaccine. Virology 537:216–225

Zhang X, Li H, Yi C, Chen G, Li Y, Zhou Y, Chen G, Li Y, He Y, Yu D (2020) Host immune response triggered by graphene quantum-dot-mediated photodynamic therapy for oral squamous cell carcinoma. Int J Nanomedicine 15:9627–9638

Zhang C, Wang J, Pan Z, Wang S, Zhang L, Wang Q, Ye Q, Zhou A, Xie S, Zeng F, Xu G, Zou J (2021) A dosage-effect assessment of acute toxicology tests of microplastic exposure in filter-feeding fish. Fish Shellfish Immunol 113:154–161

Zhang J, Li Z, Yu C, Liu H, Zhou B, Zhang X, Wang T, Wang C (2022) Efficacy of using zinc oxide nanoparticle as a substitute to antibiotic growth promoter and zinc sulphate for growth performance, antioxidant capacity, immunity and intestinal barrier function in broilers. Ital J Anim Sci 21:562–576

Zhang Y, Rahman MM, Clark PA, Sriramaneni RN, Havighurst T, Kerr CP, Zhu M, Jones J, Wang X, Kim K, Gong S, Morris ZS (2023) In situ vaccination following intratumoral injection of IL2 and poly-L-lysine/iron oxide/Cpg nanoparticles to a radiated tumor site. ACS Nano 17:10236–10251

Zhao Z, Ma X, Zhang R, Hu F, Zhang T, Liu Y, Han MH, You F, Yang Y, Zheng W (2021) A novel liposome-polymer hybrid nanoparticles delivering a multi-epitope self-replication DNA vaccine and its preliminary immune evaluation in experimental animals. Nanomedicine 35:102338

Zhu D, Hu C, Fan F, Qin Y, Huang C, Zhang Z, Lu L, Wang H, Sun H, Leng X, Wang C, Kong D, Zhang L (2019) Co-delivery of antigen and dual agonists by programmed mannose-targeted cationic lipid-hybrid polymersomes for enhanced vaccination. Biomaterials 206:25–40

Zolnik BS, González-Fernández AF, Sadrieh N, Dobrovolskaia MA (2010) Minireview: nanoparticles and the immune system. Endocrinology 151:458–465

Limitations and Concerns of Nanotechnology in Obtaining the Desirable Products

12

Umar Manzoor, Saima Masood, Sadaf Nazir, Larub Younis, Adbul Waheed, Syed Ubaid Qurashi, Faheem A Sheikh, and Shafquat Majeed

12.1 Introduction

Nanotechnology is the branch of science which deals with the study of materials in the 'nano' regime (the most accepted range is from 1 to 100 nm). In this nanoscale regime, materials are engineered in a controlled manner at the atomic or molecular stage. Being an interdisciplinary field, it employs different areas of science and technology, including engineering, medicine, chemistry, etc. to work together with a common purpose of bringing down the scale of materials to 'nano' range for their subsequent applications in diverse sectors. In addition to the manipulation of materials, it also enables the material to work differently than in its bulk state more so due to its increased surface area to volume ratio and the insurgence of quantum effects (Yao et al. 2013). This is the reason why nanomaterials find applications in various fields of science from medicine to cosmetics, electronics to storage devices, energy harvesting to energy storage devices. Undoubtedly, nanotechnology is being regarded as the industrial revolution of the modern era with the diversity in applications and uniqueness in properties of nanomaterials mainly due to the surface and quantum confinement effects (Minakshi et al. 2012).

As nanomaterials can have one of the three dimensions in the range of 1–100 nm, they can be classified into various categories such as zero-dimensional,

U. Manzoor · S. Nazir · L. Younis · A. Waheed · S. U. Qurashi · S. Majeed (✉)
Multifunctional Nanostructures Laboratory, Department of Nanotechnology, University of Kashmir, Srinagar, India
e-mail: smshah@uok.edu.in

S. Masood
Division of Basic Sciences and Humanities, SKUAST-K, Srinagar, India

F. A Sheikh
Nanostructured and Biomimetic Lab, Department of Nanotechnology, University of Kashmir, Srinagar, India

S. Rehman, E. Al-Suhaimi (eds.), *Nanotechnology Based Microbicides and Immune Stimulators*, https://doi.org/10.1007/978-981-96-0397-8_12

Table 12.1 Different types of nanomaterials (based on the number of dimensions confined in nanodimension) with some examples

Types of nanomaterials	Examples
0-D nanomaterials	Quantum dots, fullerenes, nanoparticles etc.
1-D nanomaterials	Carbon nanotubes, nanowires, etc.
2-D nanomaterials	Graphene, transition metal dichalcogenides (TMD), etc.
3-D nanomaterials	Aerogels, nanocomposites, colloidal nanoparticles, hierarchical structures etc.

one-dimensional, two-dimensional, and three-dimensional nanomaterials. These different types of nanomaterials are discussed below and are summarized in Table 12.1.

Zero-Dimensional Nanomaterials They have all the dimensions within the nano-regime; therefore, they do not expand beyond the nanoscale in any specific direction. Their properties are significantly shaped by the quantum effects and surface interactions, making them suitable for diverse applications in materials science and nanotechnology. Some common examples include metal-oxide nanoparticles, quantum dots, and nanoclusters (Majeed and Shivashankar 2014b; Nazir et al. 2024; Wani et al. 2017).

One-Dimensional nanomaterials These nanoparticles have only one dimension in the nano-range, while in the other two dimensions it may have significantly larger lengths due to which they exhibit unique properties in electronics, optics, and of course nanotechnology. One specific feature of 1D nanoparticles is they usually assume the shape in the form of wires or rods. Nanowires, nanorods, and carbon nanotubes are typical examples (Majeed et al. 2015; Khan et al. 2022; Rafiq et al. 2024).

Two-Dimensional Nanomaterials These materials have lateral dimensions in nanoscale but larger dimensions in the third direction due to which they possess unique properties in their typical planar thin-sheet structures. They have a wide range of applications in nanoelectronics, sensing, optoelectronics, and other advanced technologies. Examples include graphene, TMDs, silicene, phosphorene, etc.

Three-Dimensional Nanoparticles They have all dimensions slightly greater than 100 nm and, therefore, extend in three dimensions assuming more complex structures with enhanced electrical, mechanical, or thermal properties Examples of this type include: nanocomposites, aerogels, nanoscaffolds, hierarchical structures etc. (Majeed and Shivashankar 2013, 2014a, c; Wani et al. 2020).

Nanomaterials can exist as such in nature as proteins, some virus, DNA, etc., or can be engineered such as nanotubes, nanowires, nanorods, etc.

As far as synthesis of nanomaterials is concerned, since nanomaterials deal with extremely small sizes (~10^{-9} m), one can think both in terms of '*top-down*' and the '*bottom-up*' approaches for the synthesis and fabrication of nanomaterials as shown in Fig. 12.1. The top-down approach involves the breaking down of the bulk material into nano-sized structures or particles and these techniques are an extension of those that have been used for producing micron-sized particles. An example of such a technique is high-energy wet ball milling. Bottom-up approach refers to the build-up of a material from the bottom: atom-by-atom, molecule-by-molecule, or cluster-by-cluster, e.g., through chemical synthesis. Many of the bottom-up approaches are still under development or are just beginning to be used for commercial production of nano-powders.

As they are mainly driven by the reduction of Gibbs free energy, bottom-up approaches promise a better chance to obtain nanostructures with fewer defects, more homogenous chemical composition, and better short- and long-range ordering than top-down approaches, which most likely introduce internal stress, in addition to surface defects and contaminations.

In spite of the progressively successful manufacturing of unique nanomaterials and their increasing demand of applications in various sectors (as shown in Table 12.1), their limitations cannot be neglected (Yao et al. 2013). The limitations of nanotechnology and its products have become an upcoming issue in research.

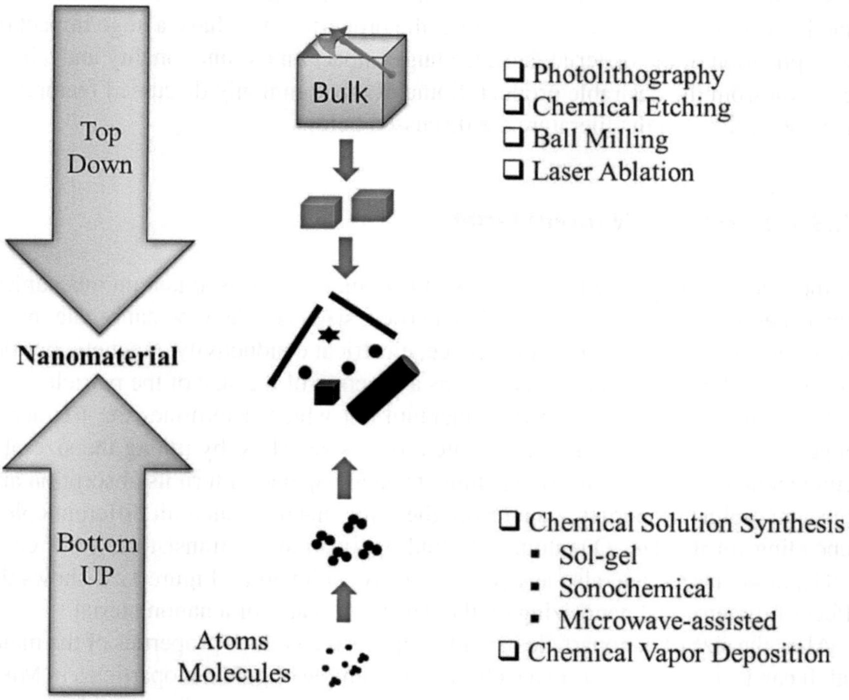

Fig. 12.1 A schematic illustration of the preparative methods for nanomaterials

In this chapter, a detailed overview of certain limitations and concerns related to nanotechnology shall be presented.

12.2 What Do You Mean by a Desirable Product?

A product is considered to be desirable when it possesses the anticipated functionality and properties that a scenario demands. In other words, it can be defined as a material that is supposed to behave in a certain manner when exposed to a particular environment. As far as nanotechnology is concerned, it aims at manufacturing of such an innovative material with explicit functionality and distinct applicability catering a pathway to secure a desired product.

12.3 How Do We Characterize a Desirable Product?

The material we intend to produce is mainly characterized on various features which include composition, size, shape, charge, surface area, porosity, etc. as the functionality of that material is attributed to its characteristic properties. The same material with different properties may behave differently instead of what was desired. It points out to the fact that desirable products must acquire those characteristic features which tend to make them work in a particular situation otherwise they may function differently. The slight change in the properties may have a huge impact on the application of the material, hence, a huge impact on its functionality and a huge deviation from the desirable product. Some of the commonly discussed features of the nanoparticles in the literature are discussed below:

12.3.1 Size of the Nanomaterial

As the size of matter is reduced to tens of nanometers or less, quantum mechanical effects begin to play a role. Thus, when particle size is made to be nanoscale, properties such as melting point, fluorescence, electrical conductivity, magnetic permeability, and chemical reactivity change as a function of the size of the particle.

For example, the band gap in a quantum dot which determines the frequency range of emitted light is inversely related to its size. Thus, by tuning the size of a semiconductor nanocrystal, we can tune its band gap and in turn its absorption and emission (color) properties. As a result, the same material can emit different colors depending on its size. Quantum dots find applications in transistors, solar cells, LEDs, diode lasers, and also as agents for medical imaging. Figure 12.2 shows the effect of dimensional nanosizing on the density of states of a nanomaterial.

Also, the size of nanoparticles greatly impacts the various properties of the material. It can be understood from a well-known example of gold nanoparticles in Mie's theory given in 1902. According to this theory, the correlation between the optical absorbance and size was made for the gold particles from the classical

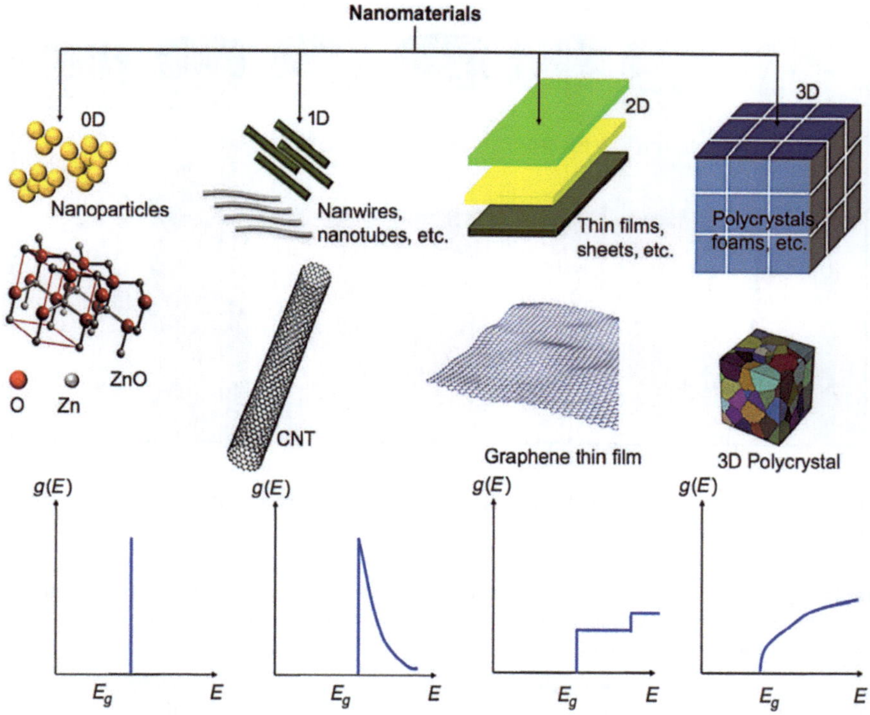

Fig. 12.2 Effect of dimensional nanosizing on the density of states of a nanomaterial. (Malhotra and Ali 2018)

electromagnetic theory and it was calculated that there is a significant change in the absorbance as the size of particles is altered (Turkevich 1985). In 1952, Turkevich and Garton using a monodisperse gold-platinum sols of varying particle size, experimentally confirmed the calculations made by Mie in 1902 (Njoki et al. 2007). Peter N. Njoki et al. also observed a clear decrease in absorbed wavelength (λ_{max}) as the particle size of gold nanoparticles increase from 30 to 90 nm and an evident transition of color from dark red to pink occurred as shown in Fig. 12.3 while as the same material has altogether a difference in optical properties (Njoki et al. 2007). Therefore, it becomes a clear evidence of size dependence of optical properties of materials which is dominant in the nano region.

12.3.2 Shape of the Nanomaterial

The shape of nanoparticles is considered as another factor on which the properties of the material depend largely which has been illustrated in the literature several times. Radha Narayanan and Mostafa A. El-Sayed et al. studied the dependence of catalytic activity of transition metals on their shape using different shaped platinum

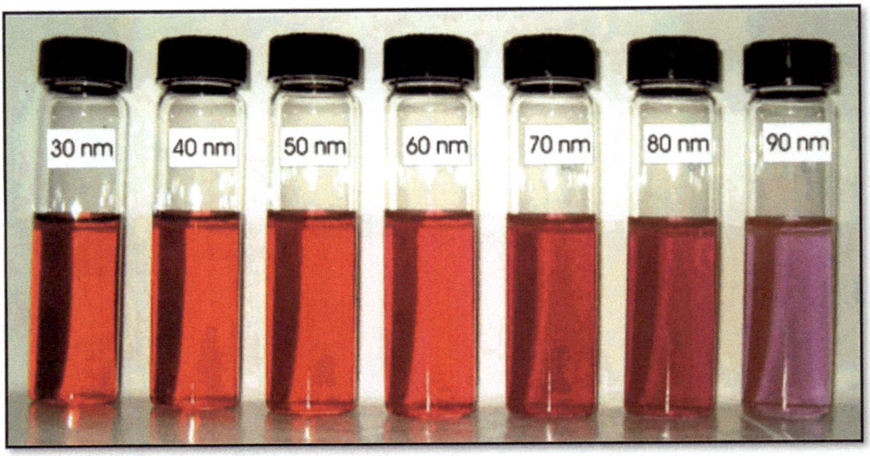

Fig. 12.3 Variation in the color of gold colloidal solutions with the size of dispersed gold nanoparticles (Njoki et al. 2007)

Table 12.2 Comparison of fraction of active surface sites (fraction of surface atoms located on defective corners and edges) and activation energy for the three different types of nanoparticles (Narayanan and El-Sayed 2005)

Nanoparticle shape	Nanoparticle size (nm)	Fraction of active surface sites (%)	Activation energy (KJ/mol)
Dominantly tetrahedral (76%)	48 ± 0.1	35	14.0 ± 0.6
Dominantly cubic (61%)	7.1 ± 0.2	4	26.4 ± 1.3
Dominantly near spherical (85%)	4.9 ± 0.1	13	22.6 ± 1.2

nanoparticles as catalyst in an electron transfer reaction (Narayanan and El-Sayed 2005). It was determined that the activation energy of the reaction was lowest when tetrahedral platinum nanoparticles were used as catalyst, while it was highest when cubic platinum nanoparticles were used. An intermediate activation energy was obtained when spherical nanoparticles were used as catalyst in the same reaction. Table 12.2 will help to understand the effect of shape on its catalytic activity (Fig. 12.4).

12.3.3 Surface Effects

Nanomaterials differ from bulk materials in many aspects primarily their surface area to volume ratio which exceeds substantially from the bulk materials leaving a huge impact on the properties of nanomaterials. Tremendous increase in surface area can be understood with an example: consider a cube with edge length 1 m

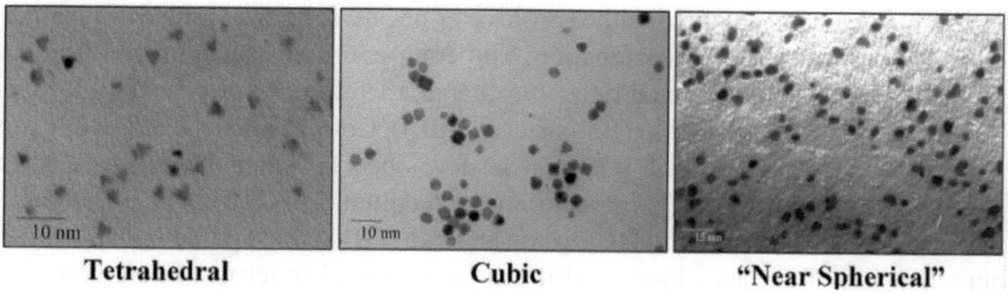

Fig. 12.4 Different shapes of Pt nanoparticles, tetrahedral, cubic and near spherical (Narayanan and El-Sayed 2005)

which is cut into similar cubes of dimension 10^{-9} m. The new total surface area of all such cubes will become a billion (109) times. Such an increase in surface area results in a tremendous increase in surface atoms. Those surface atoms possess dangling bonds or unsatisfied valances making the surface atoms more reactive as compared to those in the bulk leading them to interact with the environment more effectively (Malhotra and Ali 2018; Asha and Narain 2020). Hence, the physiochemical properties like melting-boiling point, catalytic activity, mechanical properties, etc. will alter dramatically with the change in the fraction of surface atoms (Roduner 2006).

12.3.4 Porosity

Porosity has a profound influence on the properties of nanomaterials, fundamentally altering their behavior and functionality. As nanomaterials possess a high surface area-to-volume ratio, even minute variations in porosity substantially impact their surface reactivity, adsorption capacity, cytotoxicity, and mechanical strength (Gatoo et al. 2014). Enhanced porosity often translates to improved gas or liquid adsorption, making these materials invaluable for applications like catalysis and gas storage. However, excessive porosity can compromise structural integrity and reduce mechanical robustness. By tailoring porosity, scientists can finely tune nanomaterial properties to achieve desired outcomes, whether it's boosting specific surface interactions or engineering structural sturdiness, opening vistas for innovative nanotechnology solutions.

12.3.5 Surface Charge

Surface charge plays an important role in defining the properties of nanomaterials. The magnitude of the surface charge defines the stability, agglomeration,and ability of the nanomaterials to interact with the surrounding environment. Organism response in the environment is majorly dependent on the surface charge of nanomaterials in the colloidal solution. When nanomaterials are immersed in an aqueous medium a charge is acquired, either positive or negative, by the surface of nanoparticles by protonation or deprotonation. The counterions are accumulated near the surface of nanoparticles to neutralize the charge, therefore, an electric double layer (EDL) is formed at the surface (Barisik et al. 2014). Co-ions are depleted within this EDL; hence, an electrostatic potential is developed between the surface charge (static layer) and diffused charge layer in the medium. This is known as electrokinetic potential or zeta potential (Clogston and Patri 2011). Zeta potential has a direct apprehension of surface charge and it can be measured in terms of zeta potential. The magnitude of zeta potential in milli volts directly indicates the measure of repulsive forces between the similarly charged nanoparticles at any point in the colloid. The more the zeta potential, the greater is the force of repulsion, greater would be the stability and less ability to agglomerate (Clogston and Patri 2011; 2018). Nanoparticles with a zeta potential of greater than +30 mV are considered to be strongly cationic and those with zeta potential less than −30 mV are said to be strongly anionic, whereas, the nanoparticles with the values of zeta potential between −10 and + 10 mV are considered as neutral (Clogston and Patri 2011).

12.3.6 Composition and Chemical Structure

It is quite obvious that the functionality of the material is a direct consequence of the nature of material, i.e., its composition. As the material composition varies properties will differ due to the intrinsic behavior of different types of substances. Various substances (elements/molecules/phases) of the nanoparticles are arranged in different ways due to the different conditions provided during synthesis which imparts a distinct character in the synthesized material. Depending upon the various conditions during synthesis process material with the same composition can arrange itself in different crystalline or non-crystalline structures resulting in the distinct physiochemical properties, hence, distinct functionality of the material.

For example, titania (TiO_2) nanoparticles are a type of white oxide ceramic that can exist in various crystalline forms, including rutile, anatase, and brookite. These different crystalline forms have distinct properties because of different crystal structures though composition being the same (Parvizi-Majidi 2000).

Rutile is one of the crystalline forms of titania and has a tetragonal structure. It exhibits slightly distorted octahedral coordination of titanium atoms with six oxygen atoms. The surrounding octahedra in rutile are in contact with 10 neighboring octahedra (Scarpelli et al. 2018). Rutile nanoparticles have specific characteristics and properties associated with their structure.

Anatase is another tetragonal crystalline form of titania. It consists of strongly distorted octahedral units, where titanium atoms are coordinated with six oxygen atoms. In anatase, each octahedron is in contact with eight neighboring octahedra. Compared to rutile, anatase nanoparticles have different interatomic distances and connectivity of the octahedral chains, resulting in varied properties.

Brookite is the third crystalline form of titania, characterized by an orthorhombic structure. It also exhibits distorted octahedral coordination of titanium atoms with

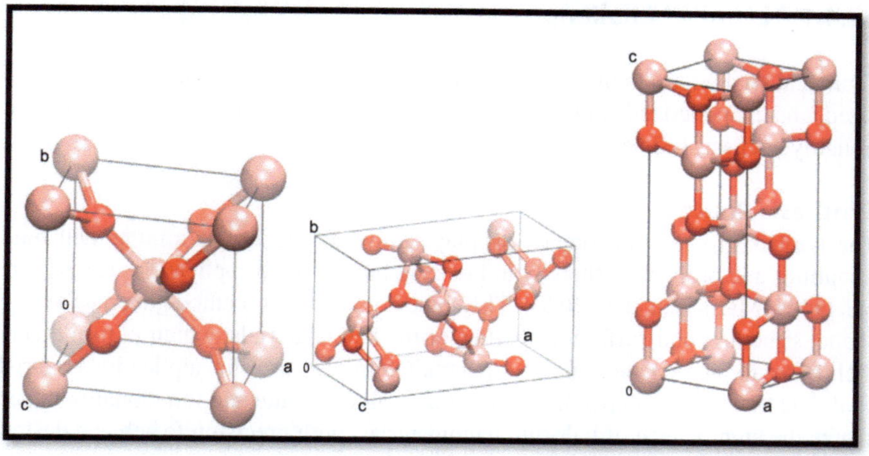

Fig. 12.5 Different crystalline forms of titania: a. Rutile, b. Brookite, and c. Anatase Jonas (Moellmann et al. 2012)

six oxygen atoms. The specific arrangement and connectivity of atoms in brookite nanoparticles give rise to distinct properties (Fig. 12.5).

The choice of crystalline form in titania nanoparticles can significantly impact their physical, chemical, and optical properties. Each form offers unique characteristics that make it suitable for different applications, such as photocatalysis, solar cells, sensors, and energy storage. The selection of a particular crystalline form depends on the desired functionality and performance requirements in specific applications.

Similarly, other properties of nanoparticles can be correlated with various factors such as shape, surface area, surface charge, etc. which will be discussed later in this chapter and consequently, the applications as a function of these characteristic features.

12.4 What Concerns Are Taken into Consideration to Obtain a Desirable Product?

The following concerns need to be taken into consideration to obtain a desirable product:

- *Technical challenges.*
- *Mass production.*
- *Cost.*
- *Environmental and health concerns.*
- *Characterization and standardization.*
- *Limited understanding of nanoscale phenomena.*
- *Integration with existing technologies.*
- *Ethical and societal implications.*

12.4.1 Technical Challenges

Technical challenges are mostly concerned with how the nanomaterials are synthesized, and characterized, and the effect of various reaction parameters on their functionality as discussed below.

Synthesis Route

The synthesis route of nanomaterials plays a pivotal role in determining their functionalities and properties, thereby influencing their diverse applications across various fields (Marinescu et al. 2020). The choice of synthesis method governs the size, shape, structure, and surface properties of the nanomaterials, which in turn dictate their behavior at the nanoscale and their suitability for specific applications (Harish et al. 2023). This interdependence between synthesis route and functionality underscores the significance of tailoring nanomaterials with precision to achieve desired characteristics (Jauhar et al. 2016).

The functionality of nanomaterials is intricately tied to their structural features, and different synthesis routes offer distinct ways to control these attributes. For instance, bottom-up methods like sol-gel synthesis, hydrothermal synthesis, and chemical vapor deposition allow for precise control over the nucleation and growth processes, enabling the production of nanoparticles with well-defined sizes and shapes (Paras et al. 2022). This control over particle morphology is particularly crucial in catalytic applications, where the exposed crystal facets of nanoparticles can significantly impact catalytic activity (Sun et al. 2016).

Similarly, top-down approaches such as mechanical milling and lithography are valuable for producing nanomaterials with controlled size reduction and patterning. These methods are crucial for applications in electronics, where precise feature sizes are essential for device performance. For example, the lithographic techniques used in semiconductor fabrication rely on precise manipulation of materials at the nanoscale to create transistors and other components (Shi and Boden 2016).

The synthesis route also plays a critical role in determining the crystallinity and phase of nanomaterials, which can profoundly influence their properties (Abedini et al. 2016). High-temperature methods like arc discharge and laser ablation enable the synthesis of nanoparticles with unique crystalline structures that exhibit enhanced mechanical and electronic properties. These nanomaterials find applications in strengthening materials and enhancing the performance of electronics.

Furthermore, surface functionalization is a key aspect of nanomaterial synthesis that directly impacts their interactions with the surrounding environment. Methods such as chemical functionalization and ligand exchange enable the modification of nanoparticle surfaces with specific molecules, making them suitable for applications in drug delivery, sensing, and biotechnology. The choice of ligands and functional groups can alter the surface charge, hydrophobicity, and reactivity of the nanomaterials, influencing their behavior in biological systems (Huang et al. 2023).

In the context of nanocomposites, the synthesis route determines the dispersion and alignment of nanomaterials within the matrix material, directly influencing mechanical, thermal, and electrical properties. Techniques like in-situ polymerization and solution blending are used to incorporate nanoparticles into polymers, ceramics, and metals, resulting in materials with enhanced properties for applications in aerospace, automotive, and structural engineering.

In conclusion, the synthesis route of nanomaterials intricately dictates their functionality by influencing their size, shape, structure, surface properties, and dispersion within composites. The choice of synthesis method must be carefully considered based on the desired functionality for a specific application. By tailoring the synthesis approach to match the intended use, researchers can harness the full potential of nanomaterials and advance various technological domains (Harish et al. 2023). As nanotechnology continues to evolve, understanding this profound dependence between synthesis route and functionality will remain essential for unlocking novel capabilities and innovative applications of nanomaterials.

Effect of Various Reaction Parameters (Temperature, Pressure, Time, Medium, Capping Agents, Precursor, and Reagents)

The functionality of nanomaterials is intricately linked to a range of synthesis parameters, including temperature, pressure, time, medium, capping agents, precursors, and reagents (Patra and Baek 2014). These parameters collectively shape the physical, chemical, and structural properties of the nanomaterials, ultimately determining their suitability for specific applications. The precise control of these parameters is essential for tailoring nanomaterials with desired characteristics and unlocking their potential in various fields.

Temperature is a critical parameter that profoundly influences nanomaterial synthesis. Higher temperatures often promote faster reaction kinetics, resulting in the formation of smaller nanoparticles. For instance, the synthesis of metal nanoparticles through reduction methods is highly temperature-dependent. By adjusting the temperature, researchers can manipulate the size and morphology of nanoparticles, impacting their optical and catalytic properties (Rai et al. 2006; Almatroudi 2020; Harish et al. 2022).

Pressure also plays a significant role in nanomaterial synthesis, particularly in processes involving gas-phase reactions. High-pressure conditions can enhance the nucleation and growth of nanoparticles, leading to unique structures and properties. Pressure-driven methods such as supercritical fluid deposition have been utilized to control the size and distribution of nanoparticles, enabling the creation of materials with tailored properties (Harish et al. 2022).

Time is another critical factor that governs nanomaterial synthesis. Reaction time determines the extent of particle nucleation, growth, and aggregation. Longer reaction times can lead to the formation of larger nanoparticles with modified properties. Conversely, shorter reaction times might yield nanoparticles with distinct crystal phases. Optimal control of reaction time is essential to achieve the desired particle characteristics (Rai et al. 2006).

The choice of medium, whether it is a solvent or a gas, significantly impacts nanomaterial synthesis. Solvents with different polarities and viscosities can influence nanoparticle formation and growth mechanisms. Additionally, gas-phase synthesis methods, such as chemical vapor deposition, allow for precise control over particle growth in controlled environments. The medium can affect particle size distribution, crystallinity, and surface properties, directly impacting functionality.

Capping agents and ligands play a crucial role in stabilizing nanoparticles and controlling their surface properties. These agents prevent nanoparticle aggregation and can also influence the interactions between nanoparticles and surrounding molecules or environments (Javed et al. 2020). Functionalized ligands can enhance dispersibility in solvents, tailor surface chemistry, and enable targeted applications in areas such as drug delivery and catalysis.

The selection of precursors and reagents directly determines the composition and crystal structure of nanomaterials. Different precursors can lead to the formation of nanoparticles with distinct properties. For example, the use of various metal salts as precursors in solution-based methods can result in nanoparticles with different electronic and optical behaviors. By carefully choosing precursors and reagents, researchers can engineer nanomaterials with specific functionalities.

In conclusion, the parameters involved in nanomaterial synthesis, including temperature, pressure, time, medium, capping agents, precursors, and reagents, collectively exert a profound influence on the functionality of the resulting materials. Understanding and controlling these parameters are essential for tailoring nanomaterials to suit various applications. Through systematic investigation and optimization of these synthesis parameters, researchers can achieve the desired characteristics and unlock the potential of nanomaterials in fields ranging from electronics and energy to medicine and environmental remediation.

12.4.2 Mass Production

Indeed, a primary limitation of nanotechnology lies in the manufacturing process. Despite significant progress in developing nanoscale structures and materials, mass production of nanoscale products remains a formidable challenge. The fabrication techniques commonly employed in nanotechnology, such as lithography and self-assembly, pose inherent limitations in terms of scalability, efficiency, and cost-effectiveness (Saini and Ledwani, 2022).

One of the major hurdles in scaling up nanomanufacturing processes is the inherent complexity and intricacy involved. Techniques such as lithography, while effective at producing precise nanostructures, are often time-consuming and require specialized equipment and expertise. Moreover, the high cost associated with lithography and similar techniques can be prohibitive, particularly when aiming for large-scale production.

Another challenge in nanomanufacturing is achieving uniformity and reproducibility across large batches of nanoscale products. Variations in fabrication conditions and material properties can result in inconsistencies in product quality, leading to reliability issues and increased manufacturing costs.

Furthermore, the integration of nanomanufacturing processes with existing manufacturing infrastructure poses additional challenges. Adapting conventional manufacturing techniques to accommodate nanoscale materials and processes requires significant investment in equipment, training, and process optimization.

Addressing these challenges requires concerted efforts from researchers, industry partners, and policymakers to develop innovative manufacturing techniques and overcome scalability barriers. This includes exploring alternative fabrication methods that offer improved scalability, efficiency, and cost-effectiveness, as well as developing standardized protocols and quality control measures to ensure consistency and reliability in nanoscale production.

Moreover, investments in infrastructure, workforce development, and collaborative research initiatives are essential for accelerating progress in nanomanufacturing and unlocking the full potential of nanotechnology across various industries and applications.

In summary, while nanotechnology holds immense promise, its widespread adoption is hindered by limitations in the manufacturing process. Overcoming these challenges requires innovative approaches, strategic investments, and collaborative efforts to develop scalable, efficient, and cost-effective nanomanufacturing techniques that can meet the demands of commercial production. By addressing these hurdles, we can realize the transformative impact of nanotechnology and unlock new opportunities for innovation and advancement.

12.4.3 Cost

The expense associated with integrating nanotechnology across diverse industries stands out as a substantial barrier. The extensive research and development efforts essential for crafting nanoscale materials, tools, and methodologies demand considerable financial investment. Furthermore, the procurement and maintenance of specialized equipment and facilities tailored for nanofabrication significantly escalate the overall costs involved, as noted by Qu et al. in their 2013 study.

These elevated costs pose a formidable challenge to the widespread incorporation of nanotechnology into various applications, particularly within developing nations or sectors constrained by limited financial resources. The financial burden associated with nanotechnology implementation not only impedes access to cutting-edge advancements but also exacerbates disparities in technological capabilities and economic opportunities across different regions and industries.

In light of these challenges, efforts to mitigate the cost barrier to nanotechnology adoption are essential for fostering equitable access and promoting innovation on a global scale. Strategies such as collaborative research initiatives, public-private partnerships, and technology transfer programs can facilitate resource-sharing and knowledge dissemination, thereby lowering the financial barriers to entry and fostering broader participation in the nanotechnology revolution.

Furthermore, targeted investments in infrastructure development and capacity-building initiatives can enhance the readiness of developing countries to embrace nanotechnology, enabling them to leverage its transformative potential for economic growth, sustainable development, and societal benefit.

In essence, while the cost of implementing nanotechnology remains a significant limitation, proactive measures aimed at reducing financial barriers and promoting inclusive innovation can pave the way for a more equitable and sustainable adoption of nanotechnology across diverse industries and regions.

12.4.4 Environmental and Health Concerns

Nanomaterials, characterized by their distinctive properties at the atomic and molecular level, often exhibit behaviors that diverge from those of their bulk counterparts. This discrepancy in behavior underscores the need for thorough examination and consideration of potential environmental and health ramifications associated with their usage. The heightened reactivity and expanded surface area inherent in nanomaterials present notable concerns regarding their potential toxicity, persistence in the environment, and propensity for bioaccumulation, as highlighted by studies conducted by Handy and Shaw (2007) and Ray et al. (2009).

Given these concerns, it becomes imperative to adopt a comprehensive approach to understanding and mitigating the risks associated with nanotechnology. This entails conducting extensive research to elucidate the mechanisms underlying the interaction between nanomaterials and biological systems, as well as their fate in various environmental settings. Furthermore, the development and implementation of robust regulatory frameworks are essential to ensure the safe and responsible utilization of nanotechnology.

As emphasized by O'Brien and Cummins (2008), effective risk management strategies necessitate a multidisciplinary effort involving scientists, policymakers, industry stakeholders, and regulatory agencies. Collaborative endeavors can facilitate the establishment of guidelines and standards governing the production, handling, and disposal of nanomaterials, thereby promoting transparency, accountability, and the protection of human health and the environment.

In essence, addressing the potential risks associated with nanomaterials requires a proactive and multifaceted approach that integrates scientific research, regulatory oversight, and stakeholder engagement. By fostering a culture of responsible innovation and governance, society can harness the transformative potential of nanotechnology while minimizing adverse impacts on both human health and the environment.

12.4.5 Characterization and Standardization

Facilitating effective communication between the macroscopic realm and the nanoscale presents a significant hurdle in achieving practical manufacturing processes. Heisenberg's uncertainty principle underscores this challenge, as it reveals that the act of measuring a system inherently alters or perturbs it. Consequently, when attempting to measure nanosystems, any connection to a microscopic detector inevitably disrupts the nanosystem's pristine state. This disruption raises concerns because it suggests that the very act of measurement could inadvertently stabilize the nanocomponents, leading to energy loss. Consequently, this interference could potentially diminish or altogether remove the unique properties or phenomena associated with nanomaterials.

Expanding on this idea, it's crucial to recognize that the delicate balance of forces and interactions at the nanoscale can be easily disrupted by external influences. Even the most subtle perturbations introduced during measurement procedures can have profound consequences on the behavior and characteristics of nanomaterials. This phenomenon underscores the intricate relationship between observation and manipulation at the nanoscale and highlights the need for innovative approaches to minimize interference while gathering accurate data.

Moreover, as we strive to bridge the gap between the macroscopic and nanoscopic worlds, it becomes imperative to develop measurement techniques that not only provide insights into nanosystems but also preserve their intrinsic properties. This necessitates the exploration of non-invasive or minimally invasive measurement methods that can capture the nuances of nanoscale phenomena without unduly influencing the systems under study.

In essence, addressing the communication challenges between the macroscopic and nanoscale domains requires a nuanced understanding of quantum principles and their implications for measurement and manipulation. By developing strategies to mitigate the disruptive effects of measurement on nanosystems, we can pave the way for more accurate characterization and manipulation of nanomaterials, ultimately advancing the field of realistic manufacturing.

Hence accurate characterization and standardization of nanomaterials and devices are essential for their reliable and reproducible use. However, characterizing nanoparticles and nanoscale structures is challenging due to their size and complexity. Traditional analytical techniques often struggle to provide accurate measurements at the nanoscale. Developing reliable characterization techniques and establishing standardized protocols are ongoing areas of research to address this limitation.

12.4.6 Limited Understanding of Nanoscale Phenomena

Even with notable advancements in nanotechnology, numerous facets of nanoscale phenomena continue to elude comprehensive understanding. The behavior of materials and devices at the nanoscale often manifests novel and unpredictable properties, presenting a persistent challenge for researchers and engineers alike. This inherent complexity and variability hinder our ability to fully leverage the potential of nanotechnology and constrain our capacity to design and engineer nanoscale systems with precision.

At the nanoscale, materials exhibit unique characteristics that deviate from their bulk counterparts, owing to quantum effects and surface interactions. These emergent properties can give rise to unexpected behaviors, such as enhanced reactivity, altered mechanical properties, and novel electronic phenomena, which defy conventional understanding and prediction.

Moreover, the intricate interplay between nanomaterials and their environment further complicates the picture, as factors such as surface chemistry, nanoparticle aggregation, and environmental conditions can profoundly influence their behavior and performance. This complexity poses formidable challenges in terms of characterizing and controlling nanoscale systems, impeding progress in various fields ranging from nanoelectronics and nanomedicine to environmental remediation and energy storage.

Addressing these knowledge gaps requires interdisciplinary collaboration and innovative approaches that combine experimental techniques, theoretical modeling, and advanced simulation methods. By unraveling the underlying principles governing nanoscale phenomena, researchers can gain deeper insights into the behavior of nanomaterials and devise strategies to harness their unique properties for practical applications.

Furthermore, investments in fundamental research and education are crucial for cultivating the next generation of scientists and engineers equipped with the skills and knowledge needed to tackle the complexities of nanotechnology. Through continuous exploration and discovery, we can gradually unravel the mysteries of the nanoscale world and unlock its full potential for technological innovation and societal advancement.

12.4.7 Integration with Existing Technologies

Integrating nanotechnology with established technologies and infrastructure presents a significant challenge, primarily due to compatibility and interface issues between nanoscale components and macroscopic systems. This challenge spans various applications, from incorporating nanoscale sensors into existing electronics to integrating nanoscale drug delivery systems into the human body.

One key consideration in this integration process is ensuring compatibility between nanoscale and macroscopic systems. Nanomaterials often exhibit unique properties that differ from those of conventional materials, necessitating careful evaluation of how they interact with existing components and systems. For instance, when incorporating nanoscale sensors into electronic devices, it's crucial to assess their compatibility with different substrates, wiring materials, and fabrication processes to ensure seamless integration and optimal performance.

Another critical factor to address is the interface between nanoscale and macroscopic systems. This interface encompasses various aspects, including power requirements, communication protocols, and biocompatibility. For instance, in the context of nanoscale drug delivery systems, ensuring compatibility with biological tissues and minimizing adverse reactions requires meticulous attention to design considerations such as surface functionalization, targeting specificity, and controlled release mechanisms.

Furthermore, addressing interface issues involves developing robust communication protocols to facilitate seamless interaction between nanoscale and macroscopic components. This entails establishing standardized protocols for data transmission, signal processing, and device control, ensuring interoperability and compatibility across different systems and platforms.

Additionally, integrating nanotechnology with existing infrastructure often entails overcoming practical challenges such as manufacturing scalability, reliability, and cost-effectiveness. Addressing these challenges requires collaborative efforts involving researchers, engineers, policymakers, and industry stakeholders to develop innovative solutions and streamline the integration process.

Overall, integrating nanotechnology with existing technologies and infrastructure requires careful consideration of compatibility, interface issues, and practical challenges. By addressing these concerns through interdisciplinary collaboration and innovation, we can unlock the transformative potential of nanotechnology across various applications, paving the way for advancements in electronics, healthcare, energy, and beyond.

12.4.8 Ethical and Societal Implications

As nanotechnology continues to advance, it brings with it a host of ethical and societal considerations that cannot be overlooked. The remarkable capabilities and potential of nanotechnology also raise significant concerns about its responsible use and potential consequences.

One prominent concern revolves around the possibility of misuse or unintended consequences stemming from nanotechnology. The ability to manipulate matter at the nanoscale opens up new avenues for innovation and progress, but it also raises questions about privacy, security, and equity. For instance, the development of nanoscale surveillance devices could infringe upon individual privacy rights, while the use of nanomaterials in weaponry could pose significant security risks. Moreover, disparities in access to nanotechnology and its benefits could exacerbate existing inequalities, further widening the gap between the haves and have-nots.

To address these concerns, it is imperative to adopt a proactive approach to the development and governance of nanotechnology. Responsible development practices that prioritize safety, transparency, and ethical considerations are essential to ensure that nanotechnology benefits society as a whole. This entails robust risk assessment procedures, stringent regulatory frameworks, and ongoing monitoring of nanotechnology applications to mitigate potential risks and prevent adverse outcomes.

Furthermore, public engagement and dialogue are crucial for fostering awareness and understanding of the ethical and societal implications of nanotechnology. By involving stakeholders from diverse backgrounds, including scientists, policymakers, industry representatives, and members of the public, we can ensure that the development and deployment of nanotechnology align with societal values and priorities.

Ultimately, responsible development and governance are paramount to harnessing the full potential of nanotechnology while safeguarding against its potential risks and ethical dilemmas. By working collaboratively to address these challenges, we can strive toward a future where nanotechnology contributes to societal well-being and sustainable development for all (Bainbridge 2004).

12.5 Conclusion and Outlook

Indeed, while the potential of nanotechnology is vast, it is essential to acknowledge and tackle its limitations and challenges head-on. Progress in this field demands persistent efforts, ongoing research, substantial investment, collaborative endeavors, and a multidisciplinary approach. One of the primary challenges in nanotechnology is the need for precise control and manipulation at the nanoscale. Achieving this level of precision requires sophisticated tools, techniques, and methodologies, which necessitate continual advancements and innovation. Moreover, there are concerns surrounding the environmental and health impacts of nanomaterials. Understanding the potential risks associated with nanotechnology and developing strategies to mitigate them is paramount for ensuring its safe and responsible implementation. Additionally, the scalability of nanotechnology presents a significant hurdle. While nanoscale phenomena exhibit remarkable properties, translating these capabilities to larger scales remains a formidable task. Overcoming this challenge requires interdisciplinary collaboration and innovative solutions. Furthermore, ethical and societal implications must be carefully considered. Issues such as privacy, security, equity, and accessibility need to be addressed to ensure that nanotechnology benefits society as a whole and does not exacerbate existing inequalities. By actively acknowledging and addressing these limitations and challenges, we can unlock the full potential of nanotechnology. Continued research, investment, collaboration, and a multidisciplinary approach will be instrumental in overcoming these hurdles and realizing the transformative impact of nanotechnology across various fields. Ultimately, by working together to understand and mitigate the limitations of nanotechnology, we can pave the way for its safe, responsible, and beneficial implementation. This concerted effort will contribute to a brighter future for science, technology, and society as a whole, fostering innovation, sustainability, and societal well-being.

References

Abedini A, Bakar AAA, Larki F, Menon PS, Islam MS, Shaari S (2016) Recent advances in shape-controlled synthesis of noble metal nanoparticles by radiolysis route. Nanoscale Res Lett 11(1):1–13

Almatroudi A (2020) Silver nanoparticles: synthesis, characterization and biomedical applications. Open Life Sci 15(1):819

Asha AB, Narain R (2020) Nanomaterials properties. In: Polymer science and nanotechnology: fundamentals and applications. Elsevier, pp 343–359
Bainbridge WS (2004) Social and ethical implications of nanotechnology, Springer handbook of nanotechnology, pp 1135–1151
Barisik M, Atalay S, Beskok A, Qian S (2014) Size dependent surface charge properties of silica nanoparticles. J Phys Chem C 118(4):1836–1842
Clogston JD, Patri AK (2011) Zeta potential measurement. Methods Mol Biol 697:63–70
Gatoo MA, Naseem S, Arfat MY, Mahmood Dar A, Qasim K, Zubair S (2014) Physicochemical properties of nanomaterials: implication in associated toxic manifestations. Biomed Res Int (2014):1–8
Handy RD, Shaw BJ (2007) Toxic effects of nanoparticles and nanomaterials: implications for public health, risk assessment and the public perception of nanotechnology. Health Risk Soc 9(2):125–144
Harish V, Ansari MM, Tewari D, Gaur M, Yadav AB, García-Betancourt ML, Abdel-Haleem FM, Bechelany M, Barhoum A (2022) Nanoparticle and nanostructure synthesis and controlled growth methods. Nano 12(18):3226
Harish V, Ansari MM, Tewari D, Yadav AB, Sharma N, Bawarig S, García-Betancourt M-L, Karatutlu A, Bechelany M, Barhoum A (2023) Cutting-edge advances in tailoring size, shape, and functionality of nanoparticles and nanostructures: a review. J Taiwan Inst Chem Eng 149:105010
Huang H, Qiao Y, Yuan Y, Zhang J (2023) Surface functionalization for heterogeneous catalysis, Encyclopedia of nanomaterials. Elsevier, pp 407–419
Jauhar S, Kaur J, Goyal A, Singhal S (2016) Tuning the properties of cobalt ferrite: a road towards diverse applications. RSC Adv 6(100):97694–97719
Javed R, Zia M, Naz S, Aisida SO, ul Ain N, Ao Q (2020) Role of capping agents in the application of nanoparticles in biomedicine and environmental remediation: recent trends and future prospects. J Nanobiotechnol 18(1):1–15
Khan AH, Masood S, Majeed A, et al (2022) Green synthesis of Gd (OH) 3 and Eu: Gd (OH) 3 Nanorods: effect of reaction parameters on morphology, crystallinity and thermal conversion to Photoluminescent oxide Nanorods. Nano 17:2250070
Malhotra BD, Ali MA (2018) Nanomaterials in biosensors: fundamentals and applications. In: Nanomaterials for biosensors, pp 1–74
Marinescu L, Ficai D, Oprea O, Marin A, Ficai A, Andronescu E, Holban AM (2020) Optimized synthesis approaches of metal nanoparticles with antimicrobial applications. J Nanomater 2020:1–14
Majeed S, Shivashankar SA (2013) Pr6O11 micro-spherical nano-assemblies: microwave-assisted synthesis, characterization and optical properties. Mater Chem Phys 143:155–160
Majeed S, Shivashankar SA (2014a) Novel spherical hierarchical structures of GdOOH and Eu:GdOOH: rapid microwaveassisted synthesis through self-assembly, thermal conversion to oxides, and optical studies. J Mater Chem C Mater 2:2965–2974. https://doi.org/10.1039/c3tc32492c
Majeed S, Shivashankar SA (2014b) Rapid, microwave-assisted synthesis of Gd2O3 and Eu:Gd2O3 nanocrystals: characterization, magnetic, optical and biological studies. J Mater Chem B 2:5585–5593
Majeed S, Shivashankar SA (2014c) Microspherical, hierarchical structures of blue–green-emitting Dy:GdOOH and Dy:Gd2O3. Mater Lett 125:136–139. https://doi.org/10.1016/j.matlet.2014.03.167
Majeed S, Bashir M, Shivashankar SA (2015) Dispersible crystalline nanobundles of YPO4 and ln (Eu, Tb)-doped YPO4: rapid synthesis, optical properties and bio-probe applications. J Nanopart Res 17:1–15
Minakshi GA, Singh BSR, Venkateswarlu B (2012) Nanotechnology: scope and limitations in Agriculture. Int J Nanotechnol Appl 2(1):2277–4777
Moellmann J, Ehrlich S, Tonner R, Grimme S (2012) A DFT-D study of structural and energetic properties of TiO2 modifications. J Phys Condens Matter 24(42):424206

Narayanan R, El-Sayed MA (2005) Catalysis with transition metal nanoparticles in colloidal solution: nanoparticle shape dependence and stability. J Phys Chem B 109(26):12663–12676

Nazir S, Masood S, Shivashankar SA et al (2024) A facile route to synthesize β-Ga2O3 nanoparticles from γ-polymorph through a rapid microwave route and their optical properties. Vietnam J Chem 62:500–505

Njoki PN, Lim IIS, Mott D, Park HY, Khan B, Mishra S, Sujakumar R, Luo J, Zhong CJ (2007) Size correlation of optical and spectroscopic properties for gold nanoparticles. J Phys Chem C 111(40):14664–14669

O'Brien N, Cummins E (2008) Recent developments in nanotechnology and risk assessment strategies for addressing public and environmental health concerns. Hum Ecol Risk Assess 14(3):568–592

Paras YK, Kumar P, Teja DR, Chakraborty S, Chakraborty M, Mohapatra SS, Sahoo A, Chou MMC, Liang CT, Hang DR (2022) A review on low-dimensional nanomaterials: nanofabrication, characterization and applications. Nanomaterials 13(1):160

Parvizi-Majidi A (2000) Whiskers and particulates. In: Comprehensive composite materials, pp 175–198

Patra JK, Baek KH (2014) Green Nanobiotechnology: factors affecting synthesis and characterization techniques. J Nanomater 2014:219

Qu X, Brame J, Li Q, Alvarez PJJ (2013) Nanotechnology for a safe and sustainable water supply: enabling integrated water treatment and reuse. Acc Chem Res 46(3):834–843

Rafiq M, Rather AH, Khan RS et al (2024) Enhancing nanofiber characteristics through solvothermal approaches for environmental cleaning and energy-saving applications. Mater Today Commun 110120

Rai A, Singh A, Ahmad A, Sastry M (2006) Role of halide ions and temperature on the morphology of biologically synthesized gold nanotriangles. Langmuir 22(2):736–741

Ray PC, Yu H, Fu PP (2009) Toxicity and environmental risks of nanomaterials: challenges and future needs. J Environ Sci Health C 27(1):1–35

Roduner E (2006) Size matters: why nanomaterials are different. Chem Soc Rev 35(7):583–592

Saini N, Ledwani L (2022) Potential applications of nanotechnology in agriculture: conceptions, characteristics, prospects, and limitations-a review. NanoWorld J 8(S1):S147–S161

Scarpelli F, Mastropietro TF, Poerio T, Godbert N (2018) Mesoporous TiO_2 thin films: state of the art. Titanium Dioxide – Mater Sustain Environ 508(1):3–6

Shi X, Boden SA (2016) Scanning helium ion beam lithography. Front Nanosci 11:563–594

Sun Y, Chen L, Bao Y, Zhang Y, Wang J, Fu M, Wu J, Ye D (2016) The applications of morphology controlled ZnO in catalysis. Catalysts 6(12):188

Turkevich J (1985) Colloidal gold. Part II color, coagulation, adhesion, alloying and catalytic properties. Gold Bull 18(4):125–131

Wani S, Sofi HS, Sheikh FA et al (2017) ZnGa2O4 nanophosphors: rapid synthesis, characterization and luminescence properties. Mat Sci Res India 14:116–122

Wani S, Shivashankar SA, Majeed S (2020) Rapid synthesis of Tb 3+-doped gadolinium oxyhydroxide and oxide green phosphors and their biological behaviour. Indian J of Chemistry-Section A (IJCA) 56:1285–1292

Yao D, Chen Z, Zhao K, Yang Q, Zhang W (2013) Limitation and challenge faced to the researches on environmental risk of nanotechnology. Procedia Environ Sci 18:149–156

Correction to: Application of Nanoparticles as Innate Immune Stimulator

Ifeanyi Elibe Mba, Paul Ekene Chidebelu, and Emeka Innocent Nweze

Correction to:
Chapter 4 in: S. Rehman, E. Al-Suhaimi (eds.), ***Nanotechnology Based Microbicides and Immune Stimulators,*** **https://doi.org/10.1007/978-981-96-0397-8_4**

The original version of this book was inadvertently published with the below error:

Chapter 4 was originally published with author's first, middle and last name appearing in reversed sequence. This has now been corrected and updated as follows "Ifeanyi Elibe Mba, Paul Ekene Chidebelu, and Emeka Innocent Nweze."

The updated version of this chapter can be found at https://doi.org/10.1007/978-981-96-0397-8_4

S. Rehman, E. Al-Suhaimi (eds.), *Nanotechnology Based Microbicides and Immune Stimulators*, https://doi.org/10.1007/978-981-96-0397-8_13

MIX
Papier aus verantwortungsvollen Quellen
Paper from responsible sources
FSC® C105338

If you have any concerns about our products,
you can contact us on
ProductSafety@springernature.com

In case Publisher is established outside the EU,
the EU authorized representative is:
Springer Nature Customer Service Center GmbH
Europaplatz 3, 69115 Heidelberg, Germany

Printed by Libri Plureos GmbH
in Hamburg, Germany